Stratosphere Troposphere Interactions

T0184919

K. Mohanakumar

Stratosphere Troposphere Interactions

An Introduction

 Springer

Prof. K. Mohanakumar
Dean, Faculty of Marine Sciences
Department of Atmospheric Sciences
Cochin University of Science and Technology
Lakeside Campus, Fine Arts Avenue
Cochin 682016, India
kmk@cusat.ac.in
kmkcusat@gmail.com

ISBN 978-90-481-7809-4   e-ISBN 978-1-4020-8217-7

© 2010 Springer Science + Business Media B.V.
No part of this work may be reproduced, stored in a retrieval system, or transmitted in any form or by any means, electronic, mechanical, photocopying, microfilming, recording or otherwise, without written permission from the Publisher, with the exception of any material supplied specifically for the purpose of being entered and executed on a computer system, for exclusive use by the purchaser of the work.

Printed on acid-free paper

9 8 7 6 5 4 3 2 1

springer.com

Preface

Stratospheric processes play a significant role in regulating the weather and climate of the Earth system. Solar radiation, which is the primary source of energy for the tropospheric weather systems, is absorbed by ozone when it passes through the stratosphere, thereby modulating the solar-forcing energy reaching into the troposphere. The concentrations of the radiatively sensitive greenhouse gases present in the lower atmosphere, such as water vapor, carbon dioxide, and ozone, control the radiation balance of the atmosphere by the two-way interaction between the stratosphere and troposphere.

The stratosphere is the transition region which interacts with the weather systems in the lower atmosphere and the richly ionized upper atmosphere. Therefore, this part of the atmosphere provides a long list of challenging scientific problems of basic nature involving its thermal structure, energetics, composition, dynamics, chemistry, and modeling. The lower stratosphere is very much linked dynamically, radiatively, and chemically with the upper troposphere, even though the temperature characteristics of these regions are different.

The stratosphere is a region of high stability, rich in ozone and poor in water vapor and temperature increases with altitude. The lower stratospheric ozone absorbs the harmful ultraviolet (UV) radiation from the sun and protects life on the Earth. On the other hand, the troposphere has high concentrations of water vapor, is low in ozone, and temperature decreases with altitude. The convective activity is more in the troposphere than in the stratosphere.

Stratospheric changes can affect the climate in a complex way through radiative and dynamical interactions with the troposphere. The climate could be changed as a result of alterations in the incoming and outgoing radiative fluxes. There is also the possibility that changes in ozone could lead to changes in the stratospheric distribution of wind and temperature and thus affect the dynamical interactions between the troposphere and stratosphere.

Even though stratosphere and troposphere are distinctly opposite in characteristics, the changes which occur in these layers contribute to the climate variability. The decrease in stratospheric ozone and increase in tropospheric ozone are a major concern in studies on climate change and variability. The stratospheric ozone is transported to the upper troposphere through stratosphere–troposphere exchange processes (STE). The ozone absorbs UV radiation in the lower stratosphere and keeps radiative balance (thermal structure) of the earth atmosphere system. Hence the study of stratosphere–troposphere interactions is significantly important to understand the climate variability.

Recognizing the importance of these processes for the climate system, the World Climate Research Programme (WCRP) set up, in 1992, a research project to study Stratospheric Processes and their Role in Climate (SPARC). Activities of SPARC include the construction of a stratospheric reference climatology and the improvement of understanding of trends in temperature, ozone, and water vapor. The principal objective of this project is to help the stratospheric research community focus on issues of particular interest to climate. The present book covers most of the SPARC themes and should be useful to the researchers of SPARC community for reference.

This book describes various physical, radiative, dynamical, and chemical processes involved in the coupling between these two layers of different characteristics. The book is intended for undergraduate and graduate students in atmospheric science/middle atmosphere/stratospheric physics/atmospheric dynamics courses, and will be useful to all research workers in meteorology, aeronomy, atmospheric physics, atmospheric chemistry, and environmental sciences. Current areas of interest, such as Antarctic ozone hole, global warming effects on stratospheric cooling, stratospheric influence on climate change are also addressed. Stratosphere–troposphere exchange, transport processes in the upper troposphere and lower stratosphere, and the role of stratosphere on tropospheric weather systems are presented. The information provided in the book has broad applicability in other branches of atmospheric science, and will be of interest to those studying such areas as climate change, extreme weather events and has potential for the prediction of tropospheric weather systems.

Cochin, India   *K. Mohanakumar*

Acknowledgments

While teaching graduate classes, I looked for books on middle atmosphere which contain necessary background information in physics, dynamics, meteorology, and chemistry to advise my students to refer. Not many books are available in this specialized area, though there are some dealing with higher-level atmospheric dynamics or chemistry. A comprehensive basic book dealing with middle and lower atmosphere, especially linked with weather and climate, is not available.

In the last two decades, with the observational evidences of Antarctic ozone hole, stratospheric cooling, global warming, and the linkage between stratospheric phenomena, such as the equatorial quasi biennial oscillation and high-latitude sudden stratospheric warming and Arctic oscillation's influence on tropospheric weather and climate, it is important to provide necessary information to the atmospheric scientific community about the importance of the effect of the stratosphere on tropospheric changes. This state of affairs prompted me to write an introductory book on stratosphere–troposphere interactions with basic information about physics, dynamics, chemistry, and meteorology.

I am really indebted to my university, Cochin University of Science and Technology, for granting me sabbatical leave for a period of one year to take up this challenging project. During this one-year period, I could make a basic outline of the book. After joining duty, I was involved in my regular teaching, research supervision, science projects, and other administrative assignments, which slowed down the progress of the book writing project.

After completing a draft form, I approached international publishing agencies enquiring about the possibility of publishing the book. The quick response from Springer Publishing Company was very welcome. I am extremely thankful to Dr. Robert Doe, the Publishing Editor, Springer, who has given me constant encouragement and full support, and responded quickly to my queries and doubts. My sincere thanks are also due to Dr. Christian Witschel, Executive Editor, Springer, and their associate Ms. Nina Bennink, and the graphic designer for the beautiful design of the cover page of this book.

I extend my sincere thanks to Professor B. V. Krishnamurthy, Former Director, Space Physics Laboratory, Vikram Sarabhai Space Centre, Trivandrum; Professor

P. N. Sen, Former Director, India Meteorological Department, and Visiting Professor, Pukyong National University, South Korea; Professor B. H. Subbarayya, Former Professor, Physical Research Laboratory, Ahmedabad, and ISRO Scientist, Bangalore; and Professor P. V. Joseph, Former Director, India Meteorological Department, and Emeritus Professor, Cochin University of Science and Technology, for spending their precious time in patiently going through various chapters of the book and providing me their valuable suggestions and comments for refining the text. Their constant encouragement and wholehearted support made it possible for me to complete this book in time.

With the help of Dr. Robert Doe, Springer arranged a reviewer to go through the entire book for a final review before submitting it to the Editorial Board for publication. I would like to express my sincere gratitude to this anonymous reviewer for the valuable comments and timely advice for finalizing the manuscript.

I extend my sincere thanks to Dr. Bill Randel, NCAR, for providing me the high-resolution figures on climatology for reproduction in the book. Thanks are also due to Ms. Victoria De Luca, SPARC IPO, University of Toronto, for providing high-resolution figures published in the SPARC Newsletter. Necessary permission granted to reproduce the figures and tables published by various publishers and authors is also acknowledged.

Sincere thanks are due to my colleagues, especially Professor H. S. Ram Mohan, Dr. K. R. Santosh, Mr. Baby Chakrapani, Dr. V. Madhu, and Mr. K. S. Appu, for their support and encouragement. I am really indebted to my student, Mr. S. Abhilash, who took a lot of pain in the preparation of the entire manuscript in LaTex format as well as the figures. Because of his dedicated and sincere effort I could submit the manuscript in time. I also extend my sincere thanks to Shuaib, Liju, Prasanth, Nithin, Sabeer, Sabeerali, Resmi, Dr. G. Mrudula and Dr. Venu G. Nair for their support in the preparation of diagrams, materials, and data used in this book.

Finally, my sincere thanks are due to my wife, Ajitha, and my daughters, Meera and Meenu, for their forbearance during many evenings, holidays, and weekends during which I was fully occupied with the writing project. Many personal and domestic programs were postponed to give priority to this assignment. Full cooperation and encouragement from my family members really helped me to complete this task in the stipulated time.

During the course of the preparation of the book, I felt that scientific book-writing is really a tough experience and requires a lot of concentration, as well as physical and mental strain. I could successfully complete this challenging assignment on time by God's grace. Even now I consider that more information needs to be included in the book, although, of course, it will be an unending task. I hope that at the time of a second edition, I will be able to update the book.

Contents

Chapter 1
Structure and Composition of the Lower and Middle Atmosphere

1.1 The Evolution of the Earth's Atmosphere

The history of the Earth's atmosphere prior to one billion years ago is not clearly known. Scientists have studied fossils and made chemical analysis of rocks to find out how life on Earth evolved to its present form. Several theories have been suggested. It is hypothesized that life developed in two phases over billions of years.

In the first phase, explosions of dying stars scattered through the galaxy and created swirling clouds of dust particles and hot gases. These extended trillions of kilometers across space. As the cloud cooled, fragmented small particles adhered to each other. Over 4 billion years ago, the cloud had formed into a flattened and slowly rotating disk. The Sun was born in the center of this disk. Away from the disk, Earth and the other planets formed as tiny pieces of matter which were drawn together. Earth started out as a molten mass that did not cool for millions of years. As it cooled it formed a thin, hard crust with no atmosphere or oceans.

Molten rock frequently exploded through the crust. Water vapor was released from the breakdown of rocks during volcanic eruptions. Eventually, the crust cooled enough for this vapor to condense and come down as rain to form the oceans that covered most part of the Earth.

In the second phase, scientists have hypothesized that bubbles floating on the ancient ocean trapped carbon-containing molecules and other chemicals essential for life. These bubbles may have burst and released these chemicals into the atmosphere. Organic compounds formed and dissolved in the early atmosphere, collecting in the shallow waters of the Earth. However, no one is aware how the first living cells developed between 3.6 and 3.8 billion years ago. Eventually, these protocells developed into cells having the properties presently known as life.

These unicellular bacteria multiplied in the warm shallow waters, where they mutated and developed into a variety of protests and fungi. About 600 million years ago plants and animals were formed on the Earth. Life could not develop then on land since there was no ozone layer to shield early life from damaging ultraviolet (UV) radiation. The photosynthetic bacteria which emerged about 2.3–2.5 billion

1

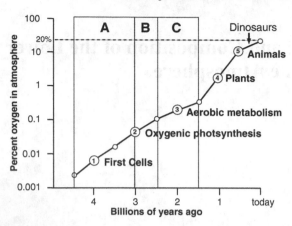

Fig. 1.1 Evolution of the concentration of oxygen in the Earth's atmosphere: (**A**) No oxygen produced by the biosphere; (**B**) Oxygen produced but absorbed by Oceans and sea bed rock; and (**C**) Oxygen absorbed by land surfaces and formation of ozone layer (Courtesy: Tameera, Wikipedia)

years ago, could remove carbon dioxide (CO_2) from the atmosphere and, using sunlight, combine it with water to make carbohydrates. In the process they created oxygen (O_2) and released it into the ocean. Some of the oxygen escaped into the atmosphere.

The evolution of geological and biological events leading to changes of oxygen contents in the Earth's atmosphere since the formation of the planet are shown in Fig. 1.1. Evidence of oxygen in the ocean is noted between 2.5 and 3 billion years ago, by identifying oxidized iron bands in the seabed rock. Ozone layer formation was substantiated from the oxidized iron bands in the land 1.5–2.5 billion years ago. Biological events show that the photosynthetic bacteria started producing oxygen 3 billion years ago and the aerobic metabolism evolved about 2 billion years ago. Evolution of multicellular plants and animals started in the later parts as illustrated in Fig. 1.1.

Earth's atmosphere was formed over a period of 2 billion years. Some of the oxygen was converted into ozone (O_3), which was produced in the lower stratosphere and protected life from harmful UV radiation. This allowed green plants to live closer to the surface of the ocean, making it easier for oxygen to escape into the atmosphere. About 400–500 million years ago the first plants began to exist on land. Over the following millions of years a variety of land plants and animals evolved. As more plants appeared, the levels of oxygen increased significantly, whereas the carbon dioxide levels dropped. At first it combined with various elements, such as iron, but eventually oxygen accumulated in the atmosphere resulting in mass extinctions and further evolution. With the appearance of an ozone layer life-forms were better protected from UV radiation. The present nitrogen-oxygen enriched atmosphere is sometimes referred to as Earth's third atmosphere, in order to distinguish the current chemical composition from two notably different previous compositions.

1.1.1 Living Earth

Earth is really a wonderful planet. It is the only planet in our solar system that has all the necessary components to support life. Our Earth is only a tiny part of the universe, but it is the home of human beings and many other organisms. These organisms can live on Earth because it has an atmosphere. Animals and plants live almost everywhere on the surface of Earth.

The atmosphere moderates daytime and nighttime temperature swings. The atmosphere filters radiant energy during the day, preventing the surface from over-heating. At night the atmosphere prevents a large part of the radiant heat from es-caping back into space, keeping the surface warm. Most organisms, both plants and animals, essentially need water to live. Seventy-one percent of the Earth's surface is covered by water. Living things also need nitrogen, oxygen, and carbon dioxide. Earth's thin layer of atmosphere provides all of these elements. The atmosphere also screens out lethal levels of the Sun's UV radiation. The atmosphere, however, could not exist if Earth were not at the existing distance from the Sun.

1.2 Earth's Atmosphere and Its Composition

The gaseous envelope surrounding the Earth is known as its atmosphere. It is a relatively stable mixture of several types of gases from different origins. It has a mass of about 5.15×10^{15} tons held to the planet by gravitational attraction. The mean molecular mass of air is $28.966 \, \text{g mol}^{-1}$.

The atmosphere is a mixture of gases, some with fairly constant concentrations, others that are variable in space and time. In addition, there are suspended particles (e.g. aerosol, smoke, ash, etc.) and hydrometeors (e.g. cloud droplets, raindrops, snow, ice crystals, etc.). Table 1.1 shows the composition of dry air in the Earth's atmosphere below 100 km.

Nitrogen, oxygen, and argon account for about 99.96% of the permanent gases. Of the variable constituents, carbon dioxide can be somewhat variable in concentration on a localized basis at low levels. Water vapor is a highly variable constituent, with concentrations ranging from nearly zero in the coldest and dry regions of the atmosphere up to as much as 4% by volume in hot and humid air masses. Ozone, the other major greenhouse gas, also varies distinctly. In addition to these variable constituents there are also aerosols and hydrometeors which can vary widely in space and time.

The proportions of gases, excluding water vapor and ozone, are nearly uniform up to a height of about 100 km above the Earth's surface. Eddies are effective at mixing gases in the lowest 100 km of the atmosphere. This part of the atmosphere is called *homosphere*. More than 99.9% of the total atmospheric mass, however, is concentrated in the first 50 km from Earth's surface.

Table 1.1 Major constituents of the Earth's atmosphere upto 100 km (dry air)

Constituents	Percentage by volume	Molecular weight (g mol^{-1})
(A) *Constant concentrations*		
Nitrogen (N_2)	78.08	28.01
Oxygen (O_2)	20.95	32.00
Argon (Ar)	0.933	39.95
Carbon dioxide (CO_2)	0.033	44.01
Neon (Ne)	18.2×10^{-4}	20.18
Helium (He)	5.2×10^{-4}	04.02
Krypton (Kr)	1.1×10^{-4}	83.80
Xenon (Xe)	0.089×10^{-4}	131.29
Hydrogen (H)	0.5×10^{-4}	02.02
Methane (CH_4)	1.5×10^{-4}	16.04
Nitrous oxide (N_2O)	0.27×10^{-4}	44.01
Carbon monoxide (CO)	0.19×10^{-4}	28.01
(B) *Variable concentrations*		
Water vapor (H_2O)	0–4	18.02
Ozone (O_3)	$0–4 \times 10^{-4}$	48.02
Ammonia (NH_4)	0.004×10^{-4}	17.02
Sulphur dioxide (SO_2)	0.001×10^{-4}	64.06
Nitrogen dioxide (NO_2)	0.001×10^{-4}	46.05
Other gases	Trace amounts	—
Aerosol, dust, gases	Highly variable	—

Nitrogen and oxygen make up to 99% of the atmosphere at sea level, with the remainder comprising CO_2, noble gases, and traces of many gaseous substances. Commonly, the unit of concentration used when referring to trace substances is parts per million (ppm). The volume fraction is equal to the mole fraction. Hence, a mole fraction of 3.55×10^{-4} for CO_2 is equal to 355 ppm. There is increasing evidence that the percentages of trace gases are changing because of both natural and human factors. Carbon dioxide, nitrous oxide, and methane (CH_4) are produced by the burning of fossil fuels, expelled from living and dead biomass, and released by the metabolic processes of microorganisms in the soil, wetlands, and oceans. Atmospheric temperature and chemistry are generally influenced by the trace gases.

Some of the gases do not have uniform mixing ratios (e.g. ozone, water vapour, etc.) in the lowest 100 km. They can have a source at the surface or in the atmosphere, or a sink at the surface or in the atmosphere. If the gas's lifetime is shorter than the time it takes to get transported from one place to another, then the gas may not be uniformly distributed throughout the atmosphere.

Above 100 km, the mixing of air parcels is dominated by molecular diffusion. This part of the atmosphere is subjected to bombardment by radiation and high-energy particles from the Sun and outer space. This barrage has profound chemical effects on the composition of the atmosphere, especially the outer layers. In addition, gaseous molecules are influenced by gravity, leading to lighter molecules being found in the outer layers and heavier molecules closer to the Earth. Consequently, the composition of the atmosphere beyond the middle atmosphere is not uniform and is known as *heterosphere*. The upper boundary at which gases disperse into space lies at an altitude of approximately 1,000 km above sea level.

1.2.1 Formation of Homosphere and Heterosphere

Formation of the homosphere and heterosphere in the atmosphere can be explained as follows. Based on the concept of diffusion, the vertical molecular flux is given as

$$F = -D(\nabla N) \tag{1.1}$$

where D is the diffusion coefficient and N is the molecular number density. The negative sign indicates that the fluxes by diffusion are down-gradient.

The rate of change of N with respect to time is given by

$$\frac{\partial N}{\partial t} = -D(\nabla N) = D\nabla^2 N \tag{1.2}$$

The vertical molecular flux in terms of the rate of change of N with height is

$$F = -D\frac{\partial N}{\partial z} \tag{1.3}$$

The diffusion caused by eddies can be estimated by:

$$F_{\text{eddy}} = -K_{\text{eddy}}\frac{\partial N}{\partial z} \tag{1.4}$$

The solution to this diffusion equation for atmospheric conditions results in a time scale for diffusive travel, as time equals $<X^2>/2D$.

Now, let us use kinetic theory to understand the molecular diffusion coefficient D. The resulting expression for D is:

$$D = k_B \frac{(T)^{3/2}}{(p)(1/m)^{1/2}} \tag{1.5}$$

where k_B is the Boltzmann's constant (1.38×10^{-23} J K^{-1}), T is the temperature, p is the pressure, and m is the molecular mass. At the surface, $D = 2 \times 10^5$ m^2 s^{-1}. When $p = 5 \times 10^{-7}$ atm, T is constant, $D \sim 40$ m s^{-1}.

When eddy diffusion dominates over molecular diffusion, the gases are well mixed and form the homosphere. When molecular diffusion dominates over eddy diffusion, the gases separate according to mass, as we see in the heterosphere.

1.3 Atmospheric Pressure

Atmospheric pressure is the pressure above any area of the Earth's atmosphere caused by the the weight of the air. The air pressure varies spatially and temporally, because the amount and weight of air above the Earth vary with location and time. Atmospheric pressure shows a semidiurnal variation caused by global atmospheric tides. The tidal effect is stronger in tropical zones and almost absent in polar regions. The average atmospheric pressure at sea level is about 1,013.25 hectapascals (hPa).

1.3.1 Vertical Structure of Pressure and Density

Meteorological parameters, such as pressure and air density, vary dramatically with height in the atmosphere. The variation can be over many orders of magnitude and is very much larger than horizontal or temporal variations. It is therefore necessary to define a standard atmosphere in which geophysical quantities have been averaged horizontally and in time, and which vary as a function of height only. The quasi-exponential height dependence of pressure and density can be inferred from the fact that the observed vertical profiles of pressure and density on the semilog plots closely resemble straight lines.

The standard atmospheric values specified by the International Civil Aviation Organisation (ICAO) are: (i) sea level pressure (p) is 1,013.2 hPa; (ii) sea level density (ρ) is 1.225 kg m^{-3}; (iii) sea level temperature (T) is 288.15 K; and fixed lapse rates for p and T.

The vertical variation of pressure (p) with height (z) may be derived as approximately (see Wallace and Hobbs 2006) as:

$$p(z) = p(0)\exp\left(\frac{-z}{H}\right) \tag{1.6}$$

where $p(z)$ is the pressure at height z above sea level, $p(0)$ is the sea level pressure, and H is a constant called the scale height of the atmosphere. Pressure decreases by a factor of e in passing upward through a layer of depth H. For the Earth's atmosphere, H is about 8.4 km. This equation is valid only for an isothermal atmosphere, in which the temperature remains constant with height.

A similar approximate expression may be derived for density ρ as follows:

$$\rho(z) = \rho(0)\exp\left(-\frac{z}{H}\right) \tag{1.7}$$

Note that density also decreases rapidly with height. It can be shown that half of the mass of the Earth's atmosphere is below the 500 hPa level or an altitude of about 5.5 km.

Figure 1.2 illustrates the vertical profiles of pressure in the troposphere and stratosphere. As elevation increases, fewer air particles are above. Therefore the

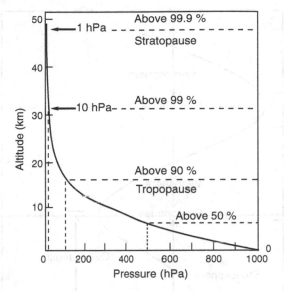

Fig. 1.2 Vertical distribution of atmospheric pressure (US Standard Atmosphere)

atmospheric pressure decreases quasi-exponentially with increasing altitude. The atmospheric pressure drops by ~50% at a height of about 5 km above the Earth's surface (see Fig. 1.2). At an altitude of 50 km the pressure (i.e., mass of particles above unit area at that level) is about 1 hPa so that only about 0.1% of the mass of the atmosphere lies above that level. Similarly, because the pressure at 90 km is about 0.001 hPa, only about one millionth of the mass of the atmosphere lies above that level.

1.4 Atmospheric Thermal Structure

The atmospheric layers of the Earth are characterized by variations in temperature produced by differences in the radiative and chemical composition of the atmosphere at different altitudes. With increasing distance from Earth's surface the chemical composition of air becomes progressively more dependent on altitude and the atmosphere becomes enriched with lighter gases. Based on the temperature changes with height, the Earth's atmosphere is divided into mainly four concentric spherical strata by narrow transition zones. Each layer is a region where the change in temperature with respect to height has a constant trend. The layers are called *spheres* and the transition zones between concentric layers are called *pauses*. The four concentric layers of the atmosphere are termed as troposphere, stratosphere, mesosphere, and thermosphere. The vertical distribution of temperature in the Earth's atmosphere is shown in Fig. 1.3.

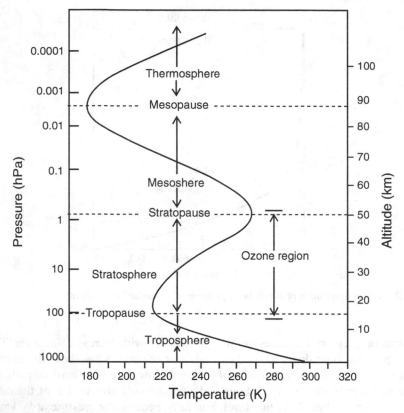

Fig. 1.3 Vertical thermal structure of Earth's atmosphere upto 120 km (Adapted from G. Brasseur and S. Solomon 1984)

1.4.1 Troposphere

Troposphere is the lowest part of the atmosphere and is closer to the Earth, and extends about 8 km above the poles and 18 km over the equator. It is the densest part of the atmosphere which contains almost all the water vapor, clouds, and precipitation. Temperature generally decreases with height in the troposphere at about 6–7°C km^{-1} in the lower half and 7–8°C km^{-1} in the upper half. Because of the general decrease of temperature with height and the presence of weather systems, the troposphere is often characterized by fairly significant localized vertical motions, although these are generally much smaller than horizontal motions. Sometimes shallow layers may be present in the troposphere in which temperature increases with height. These *inversions* inhibit vertical motion. The presence of water vapor, clouds, storms, and weather, contributes to the significance of the troposphere.

Water vapor plays a major role in regulating air temperature because it absorbs solar energy and thermal radiation from the planet's surface. The troposphere contains 99% of the water vapor in the atmosphere. The water vapor content, however, decreases rapidly with altitude, thus reflecting the change in temperature. Water vapor concentrations also vary with latitudinal position, being greatest above the tropics and decreasing toward the polar regions. All weather phenomena occur within the troposphere, although turbulence may extend into the lower portion of the stratosphere. Troposphere means region of *turning* or *mixing*, and is so named because of vigorous convective air currents within the layer.

The troposphere is bounded at the top by the *tropopause*, whose altitude varies considerably depending on the location and type of weather systems, latitude, etc. The temperature and altitude of the tropopause at a given location can vary rapidly depending on prevalent weather systems.

The tropopause may be considered as the base of a large inversion layer, i.e., the stratosphere, which inhibits vertical mixing. Consequently there are often significant concentration gradients across the tropopause. For example, the concentration of water vapor, which results largely from evaporation from the Earth's surface, decreases distinctly above the tropopause while ozone concentration increases noticeably. The moist ozone-poor tropospheric air does not mix much with the dry, ozone-rich stratospheric air.

1.4.2 Stratosphere

The stratosphere is the second major stratum in the atmosphere. It resides above the tropopause upto 50 km as shown in Fig. 1.4. The air temperature in the stratosphere increases gradually to around 273 K at the stratopause ($\sim$50 km), which is marked by a reversal in the temperature trend. Because the air temperature in the stratosphere increases with altitude, it does not cause convection and has a stabilizing effect on atmospheric conditions in the region and confines turbulence to the

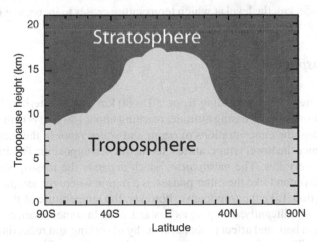

Fig. 1.4 Latitudinal variation of the vertical extension of troposphere and lower stratosphere (Adapted from B. Geerts and E. Linacre)

troposphere. As water vapor content within the stratosphere is very low, ozone plays the major role in regulating the thermal regime of this layer. Temperature increases with ozone concentration. Solar energy is converted to kinetic energy when ozone molecules absorb ultraviolet radiation, resulting in heating of the stratosphere.

The vertical temperature gradient in the stratosphere strongly inhibits vertical mixing, in contrast to the situation in the troposphere. From about 20 to 32 km there is usually a near-isothermal layer, whereas the temperature above rises with height. The stability of the stratosphere results in a strongly layered structure in which thin layers of aerosol can persist for a long time. The small concentrations of water vapor mean that latent heat release or condensation becomes unimportant, so weather and clouds are rare. However, *mother-of-pearl clouds* are sometimes seen at altitudes of 20–30 km.

The stratosphere is a region of intense interactions among radiative, dynamical, and chemical processes, in which horizontal mixing of gaseous components proceeds much more rapidly than vertical mixing. The stratosphere is warmer than the upper troposphere, primarily because of a stratospheric ozone layer that absorbs solar ultraviolet radiation.

The chemical composition of the stratosphere is generally similar to that of the troposphere with some exceptions, the most notable of which are ozone and water vapor. The stratosphere is relatively dry. However, it is rich in ozone as it is the main region of ozone production. Ozone absorbs ultraviolet radiation from the Sun and with the low densities present at stratospheric altitudes, this absorption is an efficient mechanism of transferring kinetic energy to a relatively small number of molecules due to which the air temperature becomes high. Ozone in the upper stratosphere therefore acts as a heat source. Some of the heat is transferred down by subsidence and radiation, although the stratosphere as a whole remains warm at the top, where the temperatures are close to those at the Earth's surface, and cold at the bottom and therefore very stable.

The upper limit of the stratosphere is called the *stratopause*, which occurs at an altitude of 50–55 km, the level at which temperature ceases to increase with altitude.

1.4.3 Mesosphere

The mesosphere, a layer extending from 50 to 80 km, is characterized by decreasing temperatures with increasing altitude, reaching about 180 K at 80 km. Compared to lower regions, the concentrations of ozone and water vapor in the mesophere are negligible, hence the lower temperatures. Its chemical composition is fairly uniform. Pressures are very low. The *mesopause*, which separate the mesosphere from the next highest layer and like the other pauses, is a region where the temperature trend changes direction. Like the tropopause, however, the temperature of the mesopause can vary quite significantly, dropping as low as 150 K. In some instances *noctilucent clouds* can form here and affect global climate by absorbing and reflecting incoming solar radiation. The mesopause is the level at which the lowest atmospheric temperatures are usually found.

Middle atmosphere is the region lying between the troposphere and thermosphere which extends from approximately 10 to 100 km, comprising the stratosphere and mesosphere.

1.4.4 Thermosphere

The thermosphere is a region of high temperatures above the mesosphere. It includes the ionosphere and extends out to several hundred kilometers. The temperatures in this region are of the order of 500–2,000 K and the densities are very low. The thermosphere is that part of the heterosphere which does not have a constant chemical composition with increasing altitude. Rather, atoms tend to congregate in layers with the heavier species at lower altitudes.

The increase in temperature in the thermosphere is due to the absorption of intense solar radiation by the limited amount of molecular oxygen present. At an altitude of 100–200 km, the major atmospheric components are still nitrogen and oxygen, although at this high altitude gas molecules are widely separated. *Auroras* normally occur in this region between 80 and 160 km.

The *thermopause* is the level at which the temperature stops rising with height. Its height depends on the solar activity and is located between 250 and 500 km.

1.4.5 Exosphere

The exosphere is the most distant atmospheric region from Earth's surface. The upper boundary of the layer extends to heights of perhaps 960–1,000 km. The exosphere is a transitional zone between Earth's atmosphere and interplanetary space.

1.5 Structure of the Upper Atmosphere

The upper atmosphere is divided into regions based on the behavior and number of free electrons and other charged particles. The importance of the upper atmosphere is that instead of absorbing or reflecting radiation it deflects ionized particles.

1.5.1 Ionosphere

The ionosphere is the region of the Earth's atmosphere in which the number of ions, or electrically charged particles, is large enough to affect the propagation of radio waves. The ionosphere begins at an altitude of about 50 km but is most distinct

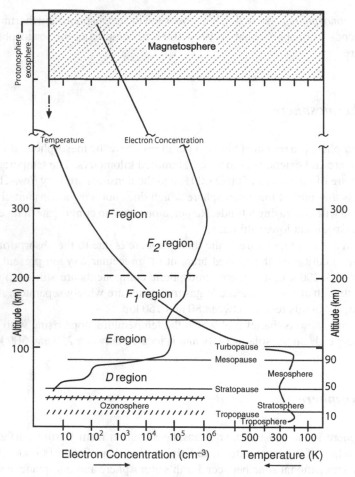

Fig. 1.5 A typical daytime profile of electron concentration of the ionosphere and the neutral atmosphere. A typical temperature profile in the reverse direction is also shown (Adapted from University of Leicester)

above about 80 km. The ionization is caused mainly by solar radiation at X-ray and ultraviolet wavelengths. The ionosphere is responsible for the long-distance propagation, by reflection, of radio signals in the shortwave and broadcast bands.

Vertical variation of the ionospheric layers in the Earth's atmosphere is depicted in Fig. 1.5. It can be seen that the ionosphere is highly structured in the vertical direction. It was first thought that discrete layers were involved, referred to as the D, E, F1, and F2 layers. However, the layers actually merge with one another to such an extent that they are now referred to as regions rather than layers. The very high temperatures in the Earth's upper atmosphere are colocated with the upper ionosphere since both are related to the effect of X-rays from the Sun. That is, the X-rays both ionize and heat the uppermost portion of the Earth's atmosphere. Tremendous variations occur in the ionosphere at high latitudes because of the dynamical effects of electrical forces and because of the additional sources of plasma production. The most notable is the visual aurora, one of the most spectacular natural sights.

The aurora has a poleward and equatorward limit during times of magnetic storms. Residents of the arctic regions of the northern hemisphere see the *Northern* lights in their southern sky. The aurora forms two rings around the poles of the Earth. The size of the rings waxes and wanes while wavelike disturbances propagate along its extent.

1.5.2 Plasmasphere

The plasmasphere is not really spherical but a doughnut-shaped region (a torus) with the hole aligned with Earth's magnetic axis. In this case the use of the suffix sphere is more in the figurative sense, like a *sphere of influence*. It is composed mostly of hydrogen ions (protons) and electrons, and is essentially an extension of the ionosphere. The torus has a very sharp outer edge called the *plasmapause*, which is usually some 4–6 Earth radii (19,000–32,000 km) above the equator. The inner edge of the plasmasphere is taken as the altitude at which protons replace oxygen as the dominant species in the ionospheric plasma, which usually occurs at about 1,000 km altitude.

1.5.3 Magnetosphere

The magnetic field of the Earth to a large extent shields it from the continual supersonic outflow of the Sun's ionized upper atmosphere, known as the solar wind. Inside of the plasmapause, geomagnetic field lines rotate with the Earth. Outside the plasmapause, however, magnetic field lines are unable to corotate because they are influenced strongly by electric fields produced by the solar wind. The magnetosphere is thus a nonspherical cavity in which the Earth's magnetic field is constrained by the solar wind and interplanetary magnetic field and shaped by the passage of the Earth. Consequently, the magnetosphere is shaped like an elongated teardrop with the tail pointing away from the Sun, as shown in the Fig. 1.6.

The outer boundary of the magnetosphere is called the *magnetopause* and marks the outer limit of Earth's gaseous envelope. The magnetopause is typically located at about 10 Earth radii (about 56,000 km) above the Earth's surface on the day side and stretches into a long tail, the *magnetotail*, a few million kilometers long (about 1,000 Earth radii) on the night side of the Earth. This is well past the orbit of the Moon (which is at 60 Earth radii), but the Moon itself is usually not within the magnetosphere except for a couple of days around the Full Moon.

Physical processes occurring within the magnetosphere modulate the energy flow carried by the solar wind to the Earth. But it is very responsive to changing solar wind conditions. At times the magnetosphere acts as a shield deflecting the incident

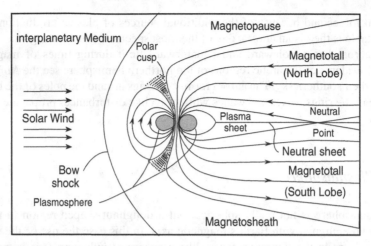

Fig. 1.6 Earth's magnetosphere with principal particle regions (C.T. Russel 1987, Courtesy: Terra Scientific Publishing Company, Tokyo)

energy; at other times it acts as an accelerator, driven by the solar wind, creating charged particle beams that hit the neutral upper atmosphere, causing it to light up with the brilliant forms of the polar aurora.

Since the scope of the book is on stratosphere–troposphere interactions, further discussion will be confined to the characteristics of Earth's lower and middle atmosphere.

1.6 The Tropopause

The tropopause is an important meteorological concept. It separates the troposphere from the stratosphere, i.e., two volumes of air with significantly different properties (Holton et al., 1995). In this region the air ceases to cool, and the air becomes almost completely dry. Basically, it is the boundary between the upper troposphere and the lower stratosphere that varies in altitude between the poles and the equator.

Tropopause is the transition layer between the troposphere and the stratosphere, where an abrupt change in temperature lapse rate usually occurs. It is defined by the World Meteorological Organization (WMO) as the lowest level at which the lapse rate decreases to $2\,\mathrm{K\,km^{-1}}$ or less, provided that the average lapse rate between this level and all higher levels within 2 km does not exceed $2\,\mathrm{K\,km^{-1}}$. Occasionally, a second tropopause may be found if the lapse rate above the first tropopause exceeds $3\,\mathrm{K\,km^{-1}}$ fall below this value. This so-defined thermal tropopause can be obtained from single temperature profiles and can be applied in both the tropics and the extratropics.

The tropopause is not a fixed boundary. Severe thunderstorms in the intertropical convergence zone (ITCZ) and over midlatitude continents in summer continuously push the tropopause upwards and as such deepen the troposphere. A pushing up of tropopause by 1 km reduces the tropopause temperature by about 10 K. Thus in areas and also in times when the tropopause is exceptionally high, the tropopause temperature becomes very low, sometimes below 190 K. Intense convective clouds in the tropics often overshoot the tropopause and penetrate into the lower stratosphere and undergo low-frequency vertical oscillations.

Tropopause height shows large variations with latitude, season, and even day-to-day. Latitudinal variation of the tropopause from the poles to the equator is schematically illustrated in Fig. 1.7. The tropopause height varies from 7–10 km in polar regions to 16–18 km in the tropics. Tropical tropopause is higher and colder, whereas polar tropopause is lower and warmer. The characteristic features of the tropopause over the tropics, midlatitude, and polar regions are illustrated in Table 1.2. The tropopause height also varies from troughs to ridges, with low tropopause height in cold troughs and high in warm ridges. Since these troughs and ridges propagate, the tropopause height exhibits frequent fluctuations at a particular location during midlatitude winters.

The highest tropopause is seen over south Asia during the summer monsoon season, where the tropopause occasionally peaks above 18 km. The oceanic warm pool of the western equatorial Pacific also exhibits higher tropopause height of 17.5 km. On the other hand, cold conditions lead to lower tropopause, evidently due to weak convection.

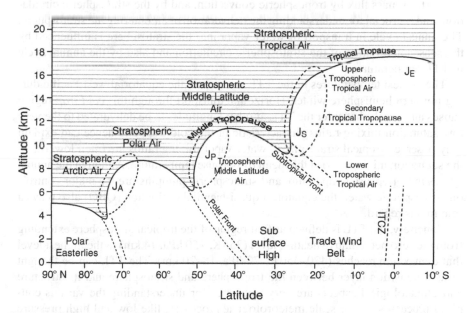

Fig. 1.7 Diagram showing the latitudinal variation of tropopause levels (Shapiro et al. 1987, Courtesy: American Meteorological Society)

Table 1.2 Characteristic features of tropopause at various latitude zones

Features	Tropical tropopause	Midlatitude tropopause	Polar tropopause
Location	Over tropics, between the two subtropical jet streams	Between polar and subtropical jet streams	North of polar jet
Height	~18 km	~12 km	6–9 km
Altitude	~80–100 hPa	~200 hPa	~300–400 hPa
Temperature	~ −80°C	~ −60°C	~ −45°C
Potential temperature	~375–400 K	~325–340 K	~300–310 K
Character	Sharply defined, highest and coldest	Higher in summer and lower in winter	Often difficult to identify

1.6.1 Tropical Tropopause

The tropical tropopause layer (TTL) is the region of the tropical atmosphere that lies between the top of the main cumulus outflow layer (~12 km) and the thermal tropopause (~16 km). This layer is a transition layer between dynamical control of the vertical mass flux by tropospheric convection, and by the stratospheric circulation, and is crucial for understanding the dehydration of air entering the stratosphere. The annual cycle in transport of water vapor into the stratosphere is influenced by the seasonal variation of the stratospheric circulation and also by the annual cycle in TTL temperatures.

The coldest temperatures in the TTL occur over the equatorial west Pacific during northern hemisphere winter. Horizontal transport through this *cold trap* region causes air parcels that enter the TTL at other longitudes to be dehydrated to the very low saturation mixing ratios characteristic of the cold trap, and hence can explain why observed tropical stratospheric water vapor mixing ratios are often lower than the saturation mixing ratio at the mean tropopause temperature. Although the annual cycles in tropopause temperature and stratospheric pumping are the major controls on stratospheric water, the equatorial quasi-biennial oscillation (QBO) also plays a role in this regard.

Concisely, the TTL is defined as that region of the tropical atmosphere extending from the zero net radiative heating level (355 K, 150 hPa, 14 km) to the highest level that convection reaches (420–450 K, 70 hPa, 18–20 km). The TTL can be thought of as a transition layer between the troposphere and stratosphere and its structure and climatological aspects are very important for understanding the various coupling processes. Large-scale meteorological processes, like low and high pressure systems, can cause day-to-day variations in the tropopause height.

1.6.2 Tropopause Acronyms

The tropopause region exhibits a complex interplay between dynamics, transport, radiation, chemistry, and microphysics. This is particularly highlighted in the case of ozone and water vapor, which provide much of the climate sensitivity in this region. The tropopause region is therefore considered as a critical region for climate. Various acronyms used for the definition of tropopause are listed below (Haynes and Shepherd 2001).

Lapse-rate tropopause (LRT): LRT is the conventional meteorological definition of the tropopause, in both tropics and extratropics, as the base of a layer at least 2 km thick, in which the rate of decrease of temperature with height is less than $2\,\mathrm{K\,km^{-1}}$.

Cold-point tropopause (CPT): CPT is the level of minimum temperature. This is useful and significant in the tropics.

Tropical thermal tropopause (TTT): Since in the tropics the LRT and the CPT are usually less than 0.5 km apart (LRT being the lower) we ignore the distinction between them and refer to the TTT. The TTT is typically at 16–17 km.

Secondary tropical tropopause (STT): STT is the level of maximum convective outflow, above which the lapse rate departs from the moist adiabat. The STT is typically at 11–12 km.

Clear-sky radiative tropopause (CSRT): CSRT is the level at which the clear-sky heating is zero. Below the CSRT there is descent on average (outside convective clouds). Above the CSRT there is ascent on average. The CSRT is typically at 14–16 km.

It is believed that the tropical tropopause is a source region of tropospheric moist air entering the stratosphere. Observational studies (Gettelman and Forster 2002) indicate that most of the tropical deep convection does not reach the cold point tropopause but ceases a few hundred meters lower, sandwiching the TTL, i.e., a relatively undisturbed body of air, subject to prolonged chemical processing of air parcels which may slowly ascend into the stratosphere.

1.6.3 Dynamic Tropopause

More recently the so-called *dynamical tropopause* has become popular. Dynamic tropopause is used with potential vorticity instead of vertical temperature gradient as the defining variable. There is no universally used threshold, but the most common ones are that the tropopause lies at the 2 PVU (potential vorticity unit) or 1.5 PVU surface. This threshold will be taken as a positive or negative value (e.g., 2 and -2 PVU), giving surfaces located in the northern and southern hemisphere respectively. To define a global tropopause in this way, the two surfaces arising from the positive and negative thresholds need to be joined near the equator using another type of surface such as a constant potential temperature surface as illustrated in Fig. 1.8.

A PV anomaly is produced by the intrusion of stratospheric air into the upper troposphere. An upper level PV anomaly advected down to middle troposphere is

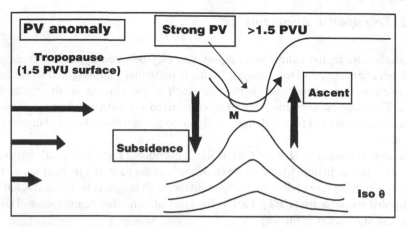

Fig. 1.8 A typical dynamical tropopause (Adapted from EUMeTrain)

called tropopause dynamic anomaly or folding of the dynamical tropopause. Due to PV conservation, the anomaly leads to deformations in vertical distribution of potential temperature and vorticity. In a baroclinic flow increasing with height, the intrusion of PV anomaly in the troposphere produces a vertical motion. The deformation of the isentropes imposes ascending motion ahead of the anomaly and subsiding motion behind it.

For conservative flow, the dynamical tropopause (unlike the thermal tropopause) is a material surface; this is an advantage for instance when considering the exchange of mass across the tropopause (Wirth 2003; Wirth and Szabo 2007). Despite overall similarities between the thermal and dynamical tropopause, they are certainly not identical and in specific situations there may be significant differences. For an atmosphere at rest *potential vorticity (PV)* is essentially a measure of static stability, and one can basically enforce both tropopauses to be at the same altitude through a suitable choice of the PV value for the dynamical tropopause.

1.6.4 Ozone Tropopause

Apart from *thermal* and *dynamical* tropopauses, there is another category for the definition of the tropopause based on the ozone content (Bethan et al. 1996) called *ozone tropopause*. In most seasons, ozone-mixing ratio similar to PV features a sharp positive vertical gradient at a particular altitude somewhere in the tropopause region. It can be used for defining an ozone tropopause from a single ozone sounding. In addition, ozone-mixing ratio like PV is approximately materially conserved on synoptic timescales. Therefore one may expect that the ozone tropopause would behave like the dynamical tropopause.

1.6.5 Tropopause Folds

Tropopause fold is the extrusion of stratospheric air within an upper-tropospheric baroclinic zone which slopes downward from a normal tropopause level ($\sim$200–300 hPa) to the middle troposphere ($\sim$500–700 hPa). The tropopause fold is a mesoscale feature which forms in response to strong descent at the tropopause level. It constitutes an intense phase of upper tropospheric frontal development in which the tropopause undulation collapses. In regions of confluent flow the tropopause may be deformed such that it will form a fold (as shown in Fig. 1.9) which will decay after 1 or 2 days. During the buildup phase of a fold the flow is generally conservative, whereas the decay phase is characterized through nonconservative flow, e.g., diabatic heating and turbulent mixing. It is these nonconservative processes which achieve the stratosphere–troposphere exchange.

The most vigorous tropopause folds occur during the winter and spring and are less frequent than cyclone development. They are usually observed downstream from a ridge, where there is large-scale descent in the entrance region of an upper-level jet streak. Ozone-rich air originates in the lower stratosphere, west of the trough axis on the cyclonic side of the upper-level jet streak. This airstream then descends anticyclonically to the lower troposphere east of the surface high-pressure system or crosses the trough axis and ascends.

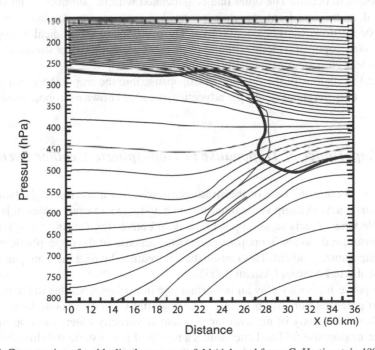

Fig. 1.9 Cross section of an idealized tropopause fold (Adapted from: G. Hartjenstein 1999)

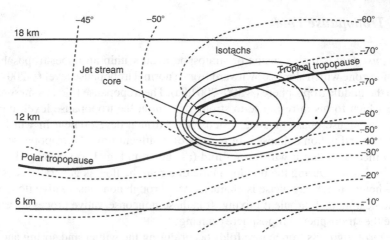

Fig. 1.10 Discontinuity in the tropical and polar tropopauses and the formation of jetstreams (Adapted from The Green Lane, Environmental Canada)

Two major types of tropopause folds are noted. One is associated with the polar front jet (PFJ), which may extend deep into the troposphere along the polar front. In some PFJ tropopause folds, significant intrusion of stratospheric air deep into the troposphere occurs. The other one is associated with the subtropical jet stream (STJ) and subtropical front, which is confined in the upper troposphere only and rarely extends downward below 500 hPa. The positions of the tropical tropopause and the polar tropopause along with the formation of subtropical jet stream over the midlatitude region are schematically shown in Fig. 1.10.

Vertical intrusions of the dynamical tropopause into the troposphere, which are folded due to differential isentropic advection, are also known as tropopause folds.

1.6.6 Importance of Tropopause to Tropospheric Weather Events

Tropopause locates the base of the stratosphere which is a layer of high static stability. Thus it acts to damp vertical motion on both large and small scales. It is most noticeable when it acts as an upper lid to deep convection causing the spreading out of cirrus anvil clouds. Tropopause marks the reversal of the tropospheric north–south temperature gradient. This causes the jet stream to be near the tropopause, via the thermal wind concept (Asnani 2005).

Tropopause folds may play an important role in cyclogenesis due to the conservation of potential vorticity principle. The potential vorticity is quite large in the stratosphere. If a body of air with large potential vorticity enters the troposphere through a tropopause fold and encounters a region of lower static stability, then relative vorticity must increase to conserve the potential vorticity, which affects the circulation pattern of the troposphere.

1.7 Climatology of the Lower and Middle Atmosphere

The vertical distribution of temperature and wind structure of the lower and middle atmosphere has been studied extensively for several decades using a variety of techniques. A large number of measurements have been made by using balloons, aircrafts, radiosonde, rocket and satellite observations, both spatially and temporally, on a global scale. Based on the global observations, and modeling information, Stratospheric Process and Its Influence on Climate (SPARC) has prepared a reference climatology of the middle atmosphere (SPARC 2002). In this section, the temperature and wind climatology of the troposphere and stratosphere are mainly discussed.

1.7.1 Temperature

Figure 1.11 illustrates the zonal mean temperature extent from surface to 90 km and from south pole to north pole for January derived from METEO analyses (1,000–1.5 hPa), and a combination of HALOE plus MLS data above 1.5 hPa (Randel et al. 2004). The thick dashed lines in the figure denote the mean tropopause (taken from NCEP reanalyses) and mean stratopause, obtained by the local temperature maximum near 50 km.

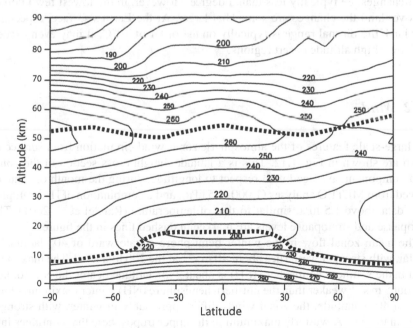

Fig. 1.11 Climatological zonal mean temperature for January Courtesy (Randel et al. 2004; Courtesy SPARC)

The temporal and latitudinal variability of mean temperature profile shows that there is considerable latitudinal and seasonal variability. Temperature decreases with latitude in the troposphere. The latitudinal gradient is about twice as steep in the winter hemisphere compared to that in the summer hemisphere. The tropopause is much higher and colder over the tropics than over the polar regions.

The latitudinal distribution of temperature in the lower stratosphere is rather complicated. The summer hemisphere has a cold equator and a warm pole. The winter hemisphere is cold at both equator and pole with a warmer region in middle latitudes. The cold pool of stratospheric air over the winter pole is highly variable. On occasions, it disappears for a period of a few weeks during midwinter. During these so-called *sudden stratospheric warmings*, the stratospheric temperatures over individual stations have been observed to rise by as much as 70°C in 1 week (Labitzke and van Loon 1999).

At the stratopause, there is a monotonic temperature gradient between the warm summer pole and the cold winter pole. At the mesopause, the situation is exactly opposite: the summer pole is cold, the winter pole is warm. Temperature has a pronounced diurnal variability in certain regions of the atmosphere. The strongest variability is observed in the upper thermosphere. In this region, day to night temperature differences are on the order of several hundred degrees.

There are also significant, but much smaller, diurnal variations around the stratopause level. These give rise to strong tidal motions in the Earth's upper atmosphere. The tidal motions manifest as regular oscillations in surface pressure which are prominent in the tropics. In the middle and upper troposphere, the day to night changes are typically less than a degree. However, in the lowest few kilometers over land the changes are somewhat larger. At the Earth's surface, especially over land, the diurnal range is typically on the order of 10°C. It may even exceed 20°C over high altitude desert regions.

1.7.2 Wind

The large-scale features of the atmospheric zonal wind circulation from surface to 90 km are shown in Fig. 1.12. This is a latitude–height cross section of the zonal wind component, averaged with respect to longitude, during the month of January derived from METEO analyses (1,000–1.5 hPa), and a combination of HALOE plus MLS data above 1.5 hPa, similar to that of temperature (Randel et al. 2004). The tropopause and stratopause levels are marked in dashed lines in the figure.

The mean zonal flow in the winter hemisphere equatorward of 40° latitude is similar, with stronger westerlies about 40 m s^{-1} at the 200 hPa level. The maximum wind in the southern hemisphere (SH) is about 2–3° latitude nearer the equator and is about 5 m s^{-1} weaker than the northern hemisphere (NH) winter maximum. Poleward of 40°S latitude, the zonal winds differ appreciably in winter, with stronger winds in the SH. A westerly maximum in the upper troposphere that continues into

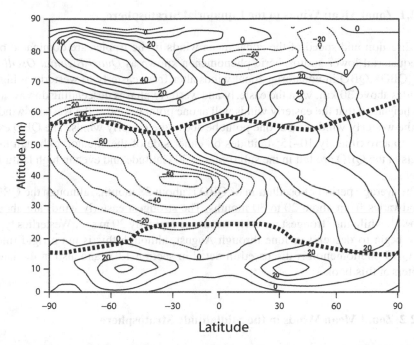

Fig. 1.12 Climatological zonal mean temperature for January (Randel et al. 2004, Courtesy: SPARC)

the stratosphere is evident between 50° and 60°S in accordance with the upward-increasing meridional temperature contrast poleward of 45°S. The distribution of wind differs considerably between the summer hemispheres. The upper troposphere westerly maximum is nearly twice as strong in the SH and is located farther pole-ward than the peak in the NH. In the middle and upper troposphere the tropical easterlies are much stronger in the NH than in the SH, and in the subtropics the westerlies are much stronger in the SH.

Prominent features are cores of strong westerly winds in middle latitudes at an altitude of 10 km. However, the strongest zonal winds occur in the mesosphere at an elevation of 60 km. Again there are two jet cores in middle latitudes, the stronger in the winter hemispheres westerly; the other in the summer hemisphere is easterly. During the equinoxes, these jets undergo dramatic reversals as the latitudinal tem-perature gradient reverses. Certain important features of the longitudinally averaged zonal wind field do not show up explicitly in Fig. 1.12. For example, the *sudden stratospheric warming* phenomenon is accompanied by large changes in the longi-tudinally averaged zonal wind at high latitudes in the winter stratosphere. The mid-winter warmings are accompanied by a pronounced weakening of the westerlies at stratospheric levels. Sometimes the westerlies disappear altogether. These changes in the stratosphere have little effect on the wind structure in the troposphere.

1.7.2.1 Zonal Mean Winds in the Equatorial Stratosphere

The direction and speed of the zonal mean winds in the tropics are dominated by the equatorial lower stratospheric phenomenon known as *Quasi Biennial Oscillation* (QBO). QBO in equatorial zonal winds alter from easterly to westerly at high altitudes above 30 km, with the easterly and westerly phases descending downward with height, so that the easterly winds will at one point be above the westerly winds and the westerly winds will at one point be above the easterly winds. The QBO extends to aproximately 10–15° latitudes on either side of the equator, although the effects of the QBO are felt in the subtropics to 30° latitude and even in high latitude regions.

The average period of oscillation of QBO is about 28 months, although the QBO period varies from about 20 to 30 months. The strongest easterly winds are about $30 \, \mathrm{m \, s^{-1}}$, while the strongest westerly winds are typically $20 \, \mathrm{m \, s^{-1}}$. Westerlies typically begin to descend in June through August, although this is not a rigid rule. The QBO phenomenon is discussed in detail in section 1.7.8 and also in the later chapters of this book.

1.7.2.2 Zonal Mean Winds in the Midlatitude Stratosphere

The zonal mean winds in the midlatitude winter hemisphere are westerly, with a maximum velocity of $80 \, \mathrm{m \, s^{-1}}$ at 65 km altitude and 40° latitude. In summer hemisphere it becomes easterly, with a maximum velocity of about $50 \, \mathrm{m \, s^{-1}}$ at 65 km altitude and 40° latitude.

In the northern hemisphere, zonal mean winds change from westerly to easterly in May, starting at the highest latitudes and altitudes and moving downward towards the tropics. Winds change from easterly to westerly in September, once again starting from high latitudes and altitudes.

As in the troposphere, the actual winds in the stratosphere have significant meridional components. Air parcels in midlatitudes typically traverse the globe in 1–2 weeks, depending on the location and the circumstances.

1.7.3 Diurnal Cycle

Time variations of the atmospheric state have two periodic components known as a *diurnal cycle* and an *annual cycle*, which are responses to the periodic variations of the external forcings due to the Earth's rotation and revolution about the Sun, respectively. Examples of the former are land and sea breeze, mountain and valley winds, and thermal tides, which are periodic responses of the atmosphere to the diurnal differential heating by the Sun on local or global scales. A monsoon climate with dry and rainy seasons is a clear example of the annual cycle.

1.7.4 Annual Oscillation

The *Annual Oscillation (AO)* is defined as the tendency of the lower stratospheric winds to become easterlies in the summer hemisphere and westerlies in the winter hemisphere. The AO is primarily an extratropical phenomenon and does not interact strongly with tropical circulation systems.

The seasonal march and the year-to-year variation of the stratospheric circulation are significantly different between the northern and southern hemispheres. In the northern hemisphere, monthly mean (steady state) planetary waves show the maximum amplitudes in midwinter associated with the occurrence of sudden warmings, but in general the phase of east–west wave number 1 is almost fixed due to the topographic effect of the surface.

On the other hand, planetary waves in the southern hemisphere stratosphere show large amplitudes in late winter or early spring (September and October), and the off-pole pattern of the polar vortex, which means the phase and amplitude of wave number 1 varies from year to year, as can be seen in the Antarctic ozone hole. The generation and maintenance of SH stratospheric waves are quite sensitive to the transient wave activity in the troposphere (Hirota and Yasuko 2000).

1.7.5 Semiannual Oscillation

One of the significant features of the low latitude upper stratosphere is the semiannual oscillation (SAO) of zonal winds. It is reported that the SAO is not generated by the solar declination changes in the low latitudes during the course of the year. It is however known that changes in the eddy momentum deposited to the zonal winds in the upper stratosphere and lower mesosphere is responsible for the equatorial westward current phase of the SAO which occurs shortly after the equinoxes.

Vertical variation of the amplitudes of QBO and SAO in temperature in the troposphere and stratosphere over the equatorial region derived from the ERA40 reanalysis data are shown in Fig. 1.13. In temperature, the QBO and SAO are generally weak in their amplitude but have the same phase in the troposphere. The amplitude of QBO increases steeply in the lower stratosphere and it peaks around 25 km. Above this level, the amplitude of QBO decreases with height. In the case of SAO, the amplitude increases steadily from the lower stratosphere and reaches the maximum around 40 km. At this level the amplitude of SAO in temperature is higher than the peak amplitude of QBO at 25 km. SAO in temperature decreases at higher levels.

Figure 1.14 gives the zonal wind amplitudes of QBO and SAO with height in the troposphere and stratosphere over the equator. In zonal wind also, amplitudes of QBO and SAO are weak and in the same phase in the entire troposphere. The QBO increases in amplitude rapidly with height in the lower stratosphere and reaches the maximum around 30 km altitude, and decreases in the upper stratosphere. On the other hand, the amplitude of SAO remains weak in the lower and middle

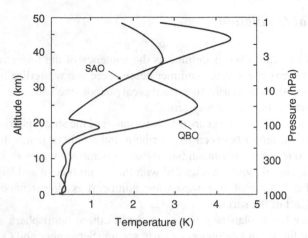

Fig. 1.13 Vertical variations in the amplitudes of SAO and QBO in temperature (Courtesy: W.J. Randel)

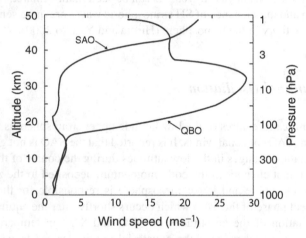

Fig. 1.14 Vertical distribution of the amplitudes of SAO and QBO in zonal wind over the equator (Courtesy: W.J. Randel)

stratosphere, but increases rapidly with height in the upper stratosphere. In this region, the amplitudes of the QBO and SAO are out of phase.

The semiannual oscillation is confined to tropics, with its phase decreasing with altitude in a manner similar to QBO. These features strongly suggest that the westerly phase of SAO should be due to the westerly momentum deposition at the relevant altitudes by Kelvin waves. The very rapid dissipation with heights of the long-period Kelvin waves producing the westerly phase of the QBO implies that the longer-period Kelvin modes would not be available to transport energy and momentum to the mesospheric levels. However, the shorter-period Kelvin waves are not absorbed significantly in the lower stratosphere and these waves with very small amplitudes carry large enough westerly momentum for the westerly phase of the SAO. So far there does not exist a satisfactory mechanistic model for the SAO in the upper stratosphere and lower mesosphere.

1.7.6 Interannual and Intraseasonal Oscillations

Interannual variation is a year-to-year variation which is defined as a deviation from the climatological annual cycle of a meteorological quantity. It can be caused by a variation of an external forcing of the atmospheric circulation system, or can be generated internally within the system. On the other hand, *intraseasonal variation* is a low-frequency variation within a season, and it is considered to be the result of internal processes which may exist even under constant external conditions.

Intraseasonal and interannual variations are defined as deviations from the periodic annual response. In general, intraseasonal variation means low-frequency variation with week-to-week or month-to-month timescales, while interannual variation means year-to-year variation. Some part of these variations is a response to the time variations of the external forcings or boundary conditions of the atmospheric circulation system, while the rest is generated internally within the system.

The time variations of the troposphere and the stratosphere are to be considered independently, partly because the heat capacity is entirely different between the layers and the adjacent ocean and land. In recent years, however, interactions between the troposphere and the stratosphere in the intraseasonal and interannual timescales have drawn attention to a possible stratospheric role in climate.

1.7.7 Jet Streams

Among the fascinating features of upper-air circulations are discontinuous bands of relatively strong winds of more than 30 m s^{-1}, called jet streams. It is a narrow band of air that moves around the globe at relatively very high speed near the tropopause height. Strong vertical shear is experienced in the vicinity of the jet streams. As with other wind fields that increase with increasing height, jet streams can be explained as an application of the thermal wind equation. They are located above areas of particularly strong temperature gradients (e.g., frontal zones). In such areas, the pressure gradients and the resulting wind speeds increase with increasing height so long as the temperature gradients persist in the same direction. In general, this will extend to the tropopause, after which the temperature gradient reverses direction and the wind speeds diminish. Thus, jet streams are usually found in the upper troposphere (i.e., at levels of 9–18 km). The positions of jet streams in the atmosphere are illustrated in Fig. 1.15.

Because regions of strong temperature gradients can be created in different ways, there are several classes of jet streams. Perhaps the most familiar is the polar-front jet

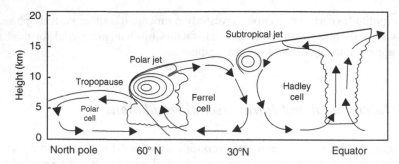

Fig. 1.15 Positions of jet streams in the meridional circulation (adapted from National Weather Service, NOAA)

stream. As noted earlier, the polar front is the boundary between polar and midlatitude air. In winter this boundary may extend equatorward to 30°, while in summer it retreats to 50–60°. Winter fronts are also distinguished by stronger temperature contrasts than summer fronts. Thus, jet streams are located more equatorward in winter and are more intense during that time with jet core wind speed exceeding 75 m s^{-1}.

A second jet stream is located at the poleward limit of the equatorial tropical air above the transition zone between tropical and midlatitude air. This *subtropical jet stream (STJ)* is usually found at latitudes of 30–40° in general westerly flow. This jet may not be marked by pronounced surface temperature contrasts but rather by relatively strong temperature gradients in the mid-troposphere. Moreover, when the *polar-front jet (PFJ)* penetrates to subtropical latitudes, it may merge with the subtropical jet to form a single band.

Also peculiar to the tropical latitudes of the northern hemisphere is a high-level jet called the *tropical easterly jet stream (TEJ)*. Such jets are located about 15° N over continental regions due to the latitudinal heating contrasts over tropical landmasses that are not found over the tropical oceans.

The location of all three jet streams in relation to other mean meridional circulation features is shown in the above figure. Jet streams occur in both hemispheres. Those in the southern hemisphere resemble the northern hemispheric systems, though they exhibit less day-to-day variability due to the presence of smaller landmasses.

1.7.8 Quasi-Biennial Oscillation (QBO)

The QBO was discovered in the 1950s, but its origin remained unclear for some time. Rawinsonde soundings showed that its phase was not related to the annual cycle, as is the case for all other stratospheric circulation patterns. In the 1970s it was recognized that the periodic wind change was driven by atmospheric waves emanating from the tropical troposphere that travel upwards and are dissipated in the stratosphere by radiative cooling.

As mentioned in section 1.7.2.1, QBO is an east–west oscillation in stratospheric winds characterized by an irregular period averaging 28 months. In the equatorial region where the QBO is dominant, easterlies are typically 30 m s^{-1} and westerlies near 20 m s^{-1}. The QBO is seen between 100 hPa and 2 hPa, with maximum amplitude near 10 hPa (Hamilton et al. 2004). Figure 1.16 is the update of the time–height cross section based on radiosonde observations from stations near the equator since 1953 (Marquardt and Naujokat 1997). The equatorial winds are dominated by alternating easterly and westerly wind regimes, with a period varying from 22 to 36 months. These wind regimes propagate irregularly downward, with easterly shear zones tending to propagate more slowly and less irregularly. The QBO may also be seen in temperature, and it dominates the interannual variability of total ozone in the tropics.

The amplitude of the QBO decreases rapidly away from the equator. However, observations and theory show that the QBO affects a much larger region of the atmosphere. Through wave coupling, the QBO affects the extratropical stratosphere during the winter season, especially in the northern hemisphere where planetary wave amplitudes are large. These effects also appear in constituents such as ozone. In the high-latitude Northern winter, the QBO's modulation of the polar vortex may affect the troposphere through downward penetration. Tropical tropospheric observations show intriguing quasi-biennial signals which may be related to the stratospheric QBO (Baldwin et al. 2001). The QBO has been even linked to variability in the upper stratosphere, mesosphere, and ionospheric F layer.

1.7.9 Mean Meridional Winds

Schematic representation of meridional circulation is illustrated in Fig. 1.17. The arrows indicate the direction of air movement in the Hadley cell, Ferrel cell, and the polar cell in the meridional plane. The position and the direction of flow of the tropical easterly jet stream, subtropical jet stream, and the polar jet stream are also depicted.

The Hadley cell ends at about 30° north and south of the equator because it becomes dynamically unstable, creating eddies that are the reason for the weather disturbances of the midlatitude belts. These eddies force a downward motion just south of the jet axis and an upward motion between 40° and 60° north and south of the equator, forming the Ferrel cell. The eddies are also responsible for spreading the westerlies down to the surface.

1.7.10 Zonally Averaged Mass Circulation

The annually averaged atmospheric mass circulation in the meridional plane is depicted in Fig. 1.18. The arrows indicate the direction of air movement in the longitudinal plane. The total amount of mass circulating around each cell is given by the largest value in that cell, based on the NCEP-NCAR reanalysis data for the period 1958–2007.

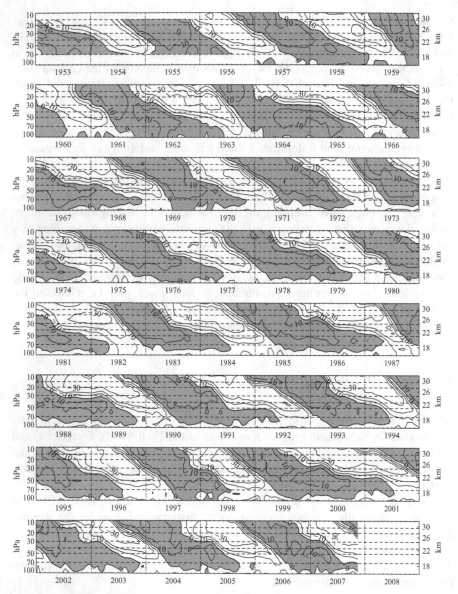

Fig. 1.16 Time–height section of zonal wind at equatorial stations showing the QBO. Isopleths are in meters per second. The data are from Canton Island (January 1953 to August 1967), Canton Island (September 1967 to December 1975), Gan/Maldive Islands (January 1976 to April 1985), and Singapore (May 1985 to August 1997) and updated upto 2007 November (Courtesy: Marquadt and Naujokat, Stratospheric Research Group, FSU Berlin)

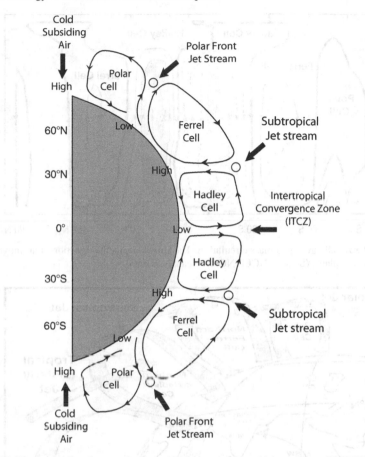

Fig. 1.17 Mean meridional mass circulation in the atmosphere (Adapted from RMIT University)

The rising motion in the tropics is capped from above by the stratosphere, where the air warms with height, thus suppressing upward motion. The law of mass continuity requires the air to move away from the tropics, northward and southward as in the diagram. This motion amounts to an upper-level mass divergence, forced by the rising motion. Again, for reasons of mass continuity, the diverging upper-level tropical air must return to the surface poleward of the equator. At the same time, mass continuity at the surface requires low-level convergence and the movement of air toward the equator (Andrews et al. 1987; McPhaden et al. 1998). A typical pattern of mean meridional circulation and surface winds during the boreal summer season is illustrated in Fig. 1.19.

The meridional winds are much weaker than the zonal winds. The typical transit time for air parcels to travel from the tropics to the pole is many months in the lower stratosphere, and longer at higher altitudes because the tropospheric air enters the stratosphere through the tropical tropopause and leaves the stratosphere at high latitudes near and above 60° latitude.

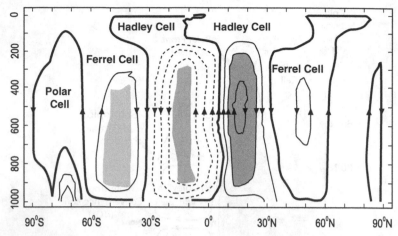

Fig. 1.18 The zonally averaged mass circulation. The arrows depict the direction of air movement in the meridional plane. (Source: NCEP/NCAR Reanalysis data)

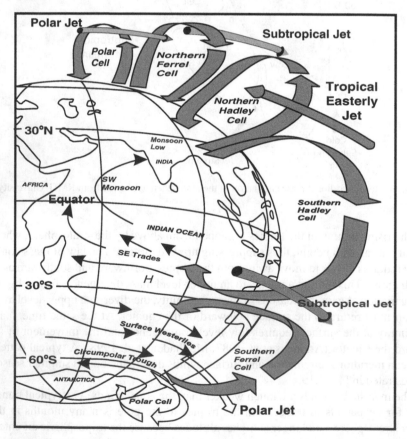

Fig. 1.19 Illustration of mean meridional mass circulation (broader shaded arrows), jetstreams (thick shaded arrows) and surface winds (thin arrows) during the northern hemispheric summer (Adapted from RMIT University)

The mean zonal stratospheric wind is roughly 20 m s^{-1}. That means the zonal wind in the stratosphere travels at a speed of $1,700 \text{ km day}^{-1}$ and it takes about 10 days to circle the globe. On the other hand, the mean meridional stratospheric wind is relatively very weak. The mean meridional stratospheric wind is approximately 0.1 m s^{-1} and it takes nearly 3 years to travel 6,000 km. The mean vertical wind in the lower stratosphere is $2 \times 10^{-4} \text{ m s}^{-1}$, which shows that the vertical wind in the stratosphere moves at a speed of 5 km year^{-1}.

1.7.11 The Polar Vortex

During the winter polar night, sunlight does not reach the south pole. A strong circumpolar wind develops in the middle to lower stratosphere. These strong winds constitute the *polar vortex* (see Fig. 1.20). Wind speeds around the vortex may reach 100 m s^{-1}. The vortex establishes itself in the middle to lower stratosphere. It is important because it isolates the very cold air within it. In the absence of sunlight the air within the polar vortex becomes very cold. Special clouds can therefore form once the air temperature gets to below $-80°C$.

Near the Arctic circle the situation is considered to be less severe because its polar vortex is not as well defined as that of the Antarctic; thus, the Arctic stratosphere is warmer than its Antarctic counterpart.

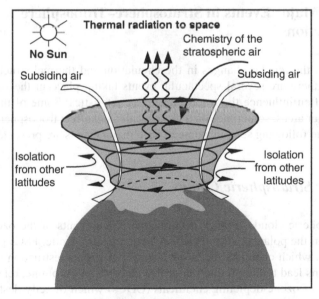

Fig. 1.20 Schematic representation of Polar Vortex (Adapted from J. Hays and P. deMenocal, University of Columbia)

The stratosphere is cooled by the greenhouse gases. The extent of the cooling is more pronounced in the higher levels of the stratosphere. The differential cooling is suspected to have contributed to the tightening of the vortex. Contraction of the polar vortex will affect the weather changes throughout the northern hemisphere (Kodera and Kuroda 2000). Prior to 1970 the polar vortex was volatile in nature. Strengthening and weakening of the polar vortex occurred from week to week or month to month, especially during winter. After the 1970s, the vortex has shown a considerable preference toward strengthening.

The eastward circulation of the Antarctic polar vortex is strongest in the upper stratosphere and strengthens over the course of the winter. The polar night jet is important because it blocks the transport between the southern polar region and the southern midlatitudes. It acts as a barrier and effectively blocks any mixing of air between inside and outside the vortex during the winter (Baldwin et al. 2003). Thus the ozone-rich air in the midlatitudes cannot be transported into the polar region.

Isolation of polar air allows the ozone loss processes to proceed without hindrance and replenishment by intrusions of ozone-rich air from midlatitudes. This isolation of the polar vortex is a key ingredient to polar ozone loss, since the vortex region can evolve without being disturbed by the more conventional chemistry of the midlatitudes. The polar night jet over the Arctic is not as effective in keeping out intrusions of warmer, ozone-rich midlatitude air. This is because there is more wave activity and hence more north–south mixing of air in the northern hemisphere than in the southern hemisphere.

1.8 Other Major Events in Stratosphere–Troposphere Interactions

In addition to the regular changes in the circulation and thermal structure of the stratosphere, there are several spectacular events taking place in the stratosphere which may often influence the tropospheric characteristics. Some of these events extend to larger areas, sometimes on a global scale, and affect tropospheric weather systems. In the following section, some of such major events are presented.

1.8.1 Polar Stratospheric Clouds

Polar stratospheric clouds (PSCs) are important components of the ozone depletion process in the polar regions of the Antarctic and the Arctic. PSCs provide the surfaces upon which chemical reactions involved in ozone destruction take place. These reactions lead to the production of free chlorine and bromine, released from CFCs and other ozone-depleting chemicals (ODCs) which directly destroy ozone molecules.

Since the stratosphere is very dry, the clouds formed in this region entirely differ in character to those formed in the moist troposphere. In the extreme cold condition

Fig. 1.21 Polar stratospheric clouds over Kiruna, Sweden (photo by H. Berg, Forschungszebntrum, Karlsrube)

of the polar winter, two types of PSCs may form. Type I clouds, consisting of mainly nitric acid and sulphuric acid, are more frequent. Whereas, Type II clouds are rarer and contain water or ice which form only below $-90°C$. Photograph of a polar stratospheric cloud is shown in Fig. 1.21.

Since these clouds are located at high altitudes and due to the curvature of the surface of the Earth, these clouds may receive sunlight below the horizon and reflect them to the ground. They appear shining brightly well before dawn or after dusk.

1.8.2 Sudden Stratospheric Warming

Sudden stratospheric warming (SSW) is an event where the polar vortex of westerly winds in the northern winter hemisphere abruptly (i.e., in a few days' time) slows down or even reverses direction, accompanied by a rise of stratospheric tempera-ture by several tens of degrees Celsius. The first observation of sudden stratospheric warming (SSW) was reported by Scherhag (1952) and the first theoretical explana-tion proposed by Matsuno (1971).

During northern winter, occasionally the circulation becomes highly disturbed, accompanied by a marked amplification of planetary waves. The disturbed motion is characterized by marked deceleration of zonal mean westerlies or even a rever-sal into zonal mean easterlies. At the same time, temperature over the polar cap

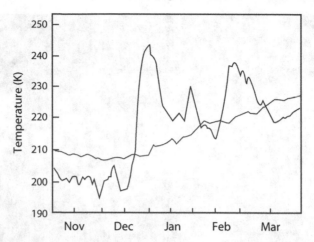

Fig. 1.22 An abrupt change in stratospheric temperature observed 90–50°N area weighted polar cap temperature at 10 hPa during a sudden stratospheric event. Bold line indicates daily changes in temperature and thin line is the mean (1958–2002) temperature from NCEP/NCAR reanalysis data sets (Adapted from A.J. Charlton and L. Polvani)

increases sharply by as much as 50 K, so the dark winter pole actually becomes warmer than the sunlit tropics. This dramatic sequence of events takes place in just a few days and is hence known as a sudden stratospheric warming.

Figure 1.22 shows the 90–50° N area averaged polar cap temperature at 10°N during November 2001 to March 2002. It can be seen that in the event of a major startospheric warming observed in December 2001, the temperature increased by about 50°C and returned back to normal after the event. In northern hemisphere winter, a few of these so-called major SSWs take place along with several minor events. Major SSWs normally occur in the northern hemisphere because orography and land–sea temperature contrasts are responsible for the generation of long (wave number 1 or 2) Rossby waves in the troposphere. These waves travel upward to the stratosphere and are dissipated there, producing the warming by decelerating the mean flow (Matsuno 1971). Since the sudden warming is observed only in the northern hemisphere, it is logical to conclude that topographically forced waves are responsible for the vertical energy propagation. The southern hemisphere with its relatively small landmasses at middle latitudes has much smaller-amplitude stationary planetary waves.

During major SSW, the north pole warms dramatically with reversal of meridional temperature gradient, and breakdown of polar vortex occurs. The polar vortex is replaced by *blocking high* over this region. The westerlies in the Arctic at 10 hPa are replaced by easterlies so that the center of the vortex moves south of 60–65°N during the breakdown of polar vortex. The vortex is either displaced entirely or split into two. This type of warming has not been observed in the Antarctic, except in 2002. Minor warming can indeed be intense and sometimes reverse the temperature gradient, but it does not result in a reversal of the circulation at 10 hPa level. This is found in the Antarctic as well as the Arctic regions.

One of the dynamic aspects of the stratospheric winter warming is the coupling effect of the warming with the tropical stratosphere. Stratospheric warming is accompanied by a slight cooling in the tropical stratosphere of both the hemispheres. This was first observed from the radiance data obtained from the Nimbus satellite (Fritz and Soules 1972). Other studies based on the satellite data also showed the presence of slight cooling in the tropical stratosphere during the warming periods (Houghton 1978). Analysis on this aspect based on rocket data from an equatorial station (Thumba, 8°N, 76°E, India) revealed occurrence of strong cooling in the equatorial stratosphere and found that the cooling penetrates to tropospheric layers especially during the peak intensity of major warming (Appu 1984). The strato-tropospheric temperatures during such occasions attain the lowest temperature of the year.

There exists a link between sudden stratospheric warmings and the QBO (Labitzke and van Loon 1999). If the QBO is in its easterly phase, the atmospheric wave-guide is modified in such a way that upward-propagating Rossby waves are focused on the polar vortex, intensifying their interaction with the mean flow. The number of warming episodes is such that it agrees reasonably well with the number of QBO cycles in the equatorial lowermost stratosphere. A possible relationship with the equatorial QBO has been noted by several authors (Holton et al. 1995; Baldwin et al. 2001). A statistically significant relationship exists between the frequeny of the occurrence of sudden stratospheric warmings in the high latitude region and the change in phases of the equatorial QBO.

1.8.3 Arctic Oscillation

Arctic oscillation (AO) is an atmospheric circulation pattern in which the atmospheric pressure over the polar regions varies out of phase with that over middle latitudes (about 45°N) on timescales ranging from weeks to decades. The oscillation extends through the entire depth of the troposphere. During late winter and early spring (January–March) it extends upward into the stratosphere where it modulates in the strength of the westerly vortex that encircles the Arctic polar cap region.

The Arctic oscillation exhibits a *negative phase* with relatively high pressure over the polar region and low pressure at midlatitudes, and a *positive phase* in which the pattern is reversed. In the positive phase, higher pressure at midlatitudes drives ocean storms farther north, and changes in the circulation pattern bring wetter weather to Alaska, Scotland, and Scandinavia, as well as drier conditions to the western United States and the Mediterranean. In the positive phase, frigid winter air does not extend as far into the middle of North America as it would during the negative phase of the oscillation. This keeps much of the United States east of the Rocky Mountains warmer than normal, but leaves Greenland and Newfoundland colder than usual. Weather patterns in the negative phase are in general opposite to those of the positive phase, as illustrated in Fig. 1.23.

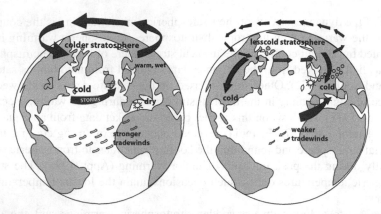

Fig. 1.23 Effects of the positive and negative phases of the Arctic Oscillation (Adapted from R. R. Stuwart 2005, Courtesy: J.M. Wallace)

In the earlier part of the last century, the Arctic oscillation alternated between its positive and negative phases. Beginning from the 1970s, however, the oscillation has tended to stay mainly in the positive phase, causing lower than normal Arctic air pressure and higher than normal temperatures in much of the United States and Northern Eurasia (Kodera and Kuroda 2000).

1.8.4 North Atlantic Oscillation

North Atlantic Oscillation (NAO) is the large-scale fluctuation in atmospheric pressure between the subtropical high pressure system located near the Azores in the Atlantic Ocean and the subpolar low pressure system near Iceland. The surface pressure drives surface winds and winter storms from west to east across the north Atlantic affecting climate from England to Western Europe as far eastward as central Siberia and the eastern Mediterranean and southward to West Africa.

In midlatitudes, NAO is the leading mode of variability over the Atlantic region. NAO is profoundly linked to the leading mode of variability of the whole northern hemisphere circulation, the Arctic oscillation. The North Atlantic oscillation and Arctic oscillation are different ways of describing the same phenomenon.

1.9 Atmospheric Tides

Atmospheric pressure, temperature, density, and winds are subject to variations with 24-hour (diurnal) and 12-hour (semidiurnal) periods. The minute but measurable variations of atmospheric parameters with lunar semidiurnal period are also caused by the gravitational attraction between the Moon and the Earth. But the variation

of the atmospheric parameters with the solar diurnal and solar semidiurnal periods
are caused predominantly by the heating of the atmosphere due to the absorption of
solar radiation by water vapor in the troposphere and by ozone in the stratosphere
and mesosphere. The heating generates pressure changes with peculiar patterns of
variation with latitude, longitude, and altitude. In particular, the maximum heating
rate and the associated pressure change at any given altitude travel with the subsolar
point in the atmosphere, and for this reason the tides generated by solar heating are
known as *migrating tides*.

The migrating meridional and zonal pressure gradients generate accelerations of
air parcels which are subject to Coriolis forcing as they progress. In addition to the
hydrostatic balance, mass continuity, and thermodynamic energy conservation, the
ensuing equilibrium between global distribution of pressure and velocity fields associated with the tides is subject to the principal boundary conditions of the Earth's
poles. The natural boundaries provided by the poles result in specific modes of oscillations with specific latitudinal structures only being possible for each tidal period.

Basically tides are generated by the daily variation of the atmospheric heating by
solar radiation. The heating is generated by the absorption of solar UV radiation by
ozone in the stratosphere and mesosphere, and the absorption of near-infrared bands
by water vapor in the troposphere.

1.10 Major Greenhouse Gases in the Troposphere
    and Stratosphere

Atmospheric ozone, water vapor, and carbon dioxide are the major greenhouse gases
which are abundant in the atmosphere and control the radiation balance of the Earth
atmosphere system. These greenhouse gases are very good absorbers in the infrared
part of the spectrum and regulate the temperature of the atmosphere. In this section
the characteristics and distribution of these major greenhouse gases are discussed.

1.10.1 Stratospheric Ozone

Stratospheric ozone is the most important minor constituent present in the Earth's
atmosphere. The more or less continuous increase in temperature with height in
the stratosphere is mainly due to the absorption of solar ultraviolet radiation by a
layer of ozone molecules with peak abundance near 25 km. Although ozone is a
minor constituent in the atmosphere, it absorbs ultraviolet radiation very effectively
at wavelengths between 200 nm and 300 nm. This property of the ozone protects
the life on Earth by preventing the harmful radiation reaching the Earth's surface.

The word *ozone* is derived from the Greek word *ozein*, meaning to smell. Ozone
has a pungent odour that allows ozone to be detected even in very low amounts.
Ozone will rapidly react with many chemical compounds and is explosive in concentrated amounts.

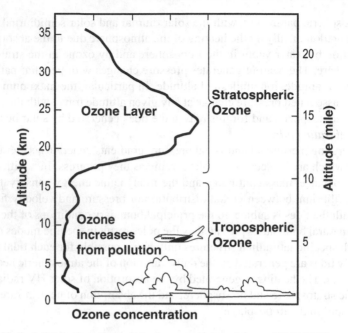

Fig. 1.24 Vertical distribution of Atmospheric ozone (Adapted from : WMO 2007)

Vertical distribution of ozone in the atmosphere is shown in Fig. 1.24. More than 90% of ozone resides in the stratosphere in what is commonly known as the *ozone layer* or *ozonosphere*. The remaining ozone is found in the troposphere.

Ozone molecules have a relatively low abundance in the atmosphere. In the stratosphere near the peak of the ozone layer, there are up to 12,000 ozone molecules for every billion air molecules. Most air molecules are in the form of molecular oxygen (O_2) or molecular nitrogen (N_2). Near to the Earth's surface, ozone is even less abundant, with a typical range of 20–100 ozone molecules formed in each billion air molecules. The highest surface ozone values are due to the formation of ozone in polluted air by anthropogenic activities.

The absorption of UV_b radiation by ozone is a source of heat in the stratosphere. This helps to maintain the stratosphere as a stable region of the atmosphere with temperatures increasing with altitude. As a result, ozone plays a key role in controlling the temperature structure of Earth's atmosphere.

Stratospheric ozone is considered good for humans and other lifeforms because it absorbs UV_b radiation from the Sun. Otherwise, UV_b would reach Earth's surface in amounts that are harmful to a variety of lifeforms. In humans, as their exposure to UV_b increases, so does their risk of skin cancer, cataracts, and a suppressed immune system. Excessive UV_b exposure also can damage terrestrial plant life, single-cell organisms, and aquatic ecosystems. Other UV radiation, UV_a, which is not absorbed significantly by ozone, causes premature aging of the skin (WMO 2007).

Ozone is also formed near Earth's surface in natural chemical reactions and in reactions caused by the presence of man-made pollutant gases. Ozone produced by pollutants is *bad* because more ozone comes in direct contact with humans, plants, and animals. Increased levels of ozone are generally harmful to living systems because ozone reacts strongly to destroy or alter many other molecules. Excessive ozone exposure reduces crop yields and forest growth. In humans, ozone exposure can reduce lung capacity, and cause chest pains, throat irritation, and coughing, thereby worsening preexisting health conditions related to the heart and lungs. In addition, increases in tropospheric ozone lead to a warming of Earth's surface. The negative effects of increasing tropospheric ozone contrast sharply with the positive effects of stratospheric ozone as an absorber of harmful UV_b radiation from the Sun.

Ozone is a natural component of the clean atmosphere. In the absence of human activities on Earth's surface, ozone would still be present near the surface and throughout the troposphere and stratosphere. Ozone's chemical role in the atmosphere includes helping to remove other gases, both those occurring naturally and those emitted by human activities. If all the ozone were to be removed from the lower atmosphere, other gases such as methane, carbon monoxide, and nitrogen oxide would increase in abundance.

Ozone is constantly being produced and destroyed in a natural cycle. However, the overall amount of ozone is essentially stable. Large increases in stratospheric chlorine and bromine have upset that balance. Therefore, ozone levels are beginning to fall. The ozone depletion process begins when chloroflurocarbons (CFCs) and other ozone-depleting substances (ODS) leak or are released from equipments. Winds efficiently mix the troposphere and evenly distribute the gases. CFCs are extremely stable, and they do not dissolve in rain. After a period of a few years, ODS molecules reach the stratosphere. Strong UV light breaks apart the ODS molecule. CFCs release chlorine atoms and halons release bromine atoms. It is these atoms that actually destroy ozone, not the intact ODS molecule. It is estimated that one chlorine atom can destroy over 100,000 ozone molecules before finally being removed from the stratosphere.

Ozone photochemistry, transportation, ozone hole, and their influence on tropospheric weather systems are discussed in detail in later chapters of this book.

1.10.2 Carbon Dioxide

Carbon dioxide has a relatively constant mixing ratio with height in the atmosphere, and is more or less evenly distributed. The main sources of carbon dioxide are burning of fossil fuels, human and animal respiration, the oceans, and volcanic activity. The main sinks are photosynthesis and the production of carbonates (limestones) in the ocean/land system. The rate of removal of carbon dioxide is observed to be less than the generation (from fossil fuel burning) because the concentration of carbon dioxide in the atmosphere has been rising steadily during the 20th century as illustrated in Fig. 1.25.

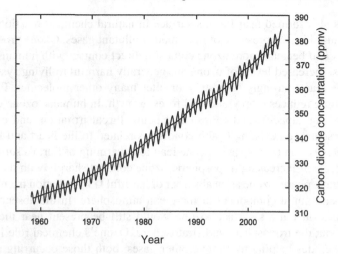

Fig. 1.25 Carbon dioxide concentration in the Earth's atmosphere, observed at Mauna Loa, Hawaii (Courtesy: Robert A. Rhode, Global Warming Art)

About 99% of the Earth's carbon dioxide is dissolved in the oceans. Because solubility is temperature-dependent the gas therefore enters or leaves the oceans. It is estimated that the annual amount of carbon dioxide entering or leaving the air by all mechanisms is about one tenth of the total carbon dioxide content of the atmosphere.

1.10.3 Water Vapor

Water vapor is unique among atmospheric trace constituents in that conditions for saturation are common in the atmosphere. This property is the most important factor governing the distribution of water vapor in the atmosphere, both in the troposphere, where it varies by as much as four orders of magnitude in a vertical profile, and in the stratosphere, where variations are much smaller but still significant (see Fig. 1.26).

Water vapor is extremely important in radiative absorption and emission processes in the atmosphere. Its concentration is highly variable. Although always present, in some localities it is difficult to measure, but in the tropics its concentration can be as high as 3% or 4% by volume. Water vapor content of air is a strong function of air temperature. For example, air at 40°C can hold up to 49.8 g of water per kilogram of dry air, while at 5°C this reduces to 5.5 g per kilogram of dry air.

The release of latent heat from condensation of water in the atmosphere is significant in the global energy budget and climate. Relatively small amounts of water vapor can produce great variations in weather. This is largely due to changes in its concentration and in latent heat release, particularly below 6 km where a high proportion of moisture lies.

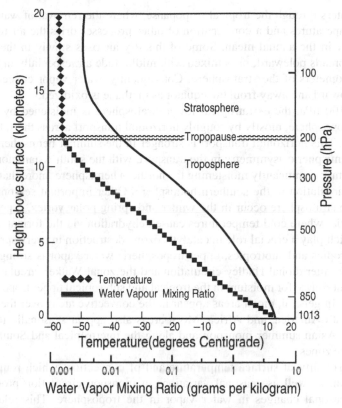

Fig. 1.26 Vertical distribution of water vapor (Courtesy: D. Siedel)

The major sources of water vapor are evaporation from water surfaces and transpiration from plant life. The main sink is condensation in clouds with resulting precipitation over oceans and land. On an average the concentration of atmospheric water vapor decreases with altitude, although this distribution may vary with space and time.

1.10.4 Water Vapor in the Stratosphere

In the stratosphere above 100 hPa, the distribution of water vapor can be explained as a balance between dry air entering through the tropical tropopause and a source of water vapor from methane oxidation in the upper stratosphere. The stratospheric circulation helps to determine the distribution, along with wave-induced mixing and upward extension of tropospheric circulation. Variations in the zonal direction are rapidly mixed so that water vapor is nearly constant following a fluid element. Nearly all air passing from the troposphere to the part of the stratosphere above

100 hPa enters through the tropical tropopause, where the removal of water vapor by low temperatures and a combination of other processes dries the air to around 3.5–4 ppmv in the annual mean. Some of this dry air rises slowly in the tropics, but most spreads poleward, or is mixed with midlatitude air, especially in the lowest few kilometers of the stratosphere. Consequently, water vapor concentrations increase upward and away from the equator as methane is oxidized.

Below 100 hPa, the extratropical lower stratosphere is moistened by seepage from the troposphere, mostly by roughly horizontal transport across the subtropical tropopause. This horizontal transport is stronger in the summer hemisphere. There is also a hemispheric asymmetry in the transport, with the south Asian monsoon in boreal summer significantly moistening the northern hemisphere, more than similar monsoon circulation in the southern hemisphere. Other important seasonal variations in the stratosphere occur in the winter and spring polar vortex, especially in the Antarctic, where cold temperatures cause dehydration via the formation of ice clouds, which play a crucial role in catalytic ozone destruction in every spring.

In the tropics and subtropics, upper-tropospheric water vapor is strongly influenced by the meridional Hadley circulation and the zonal Walker circulation. The predominant source for moisture in the tropical and subtropical upper troposphere is convection. In general, moist areas appear in the convective areas over the western Pacific, South America, and Africa. Moist areas also appear seasonally in the region of the Asian summer monsoon and along the intertropical and South Pacific convergence zones.

The seasonality of surface temperature and of convection, which roughly follow the Sun, as well as seasonal variations in monsoon circulation produce associated seasonal changes in water vapor in the troposphere. This relationship between convection and upper tropospheric moisture changes sign near the tropical tropopause, somewhere between 150 hPa and 100 hPa, so that convection dries the tropopause region. Water vapor is also influenced by fluctuations at both shorter and longer timescales, including the quasi-biennial oscillation in the stratosphere, and El Niño and the southern oscillation and the tropical intraseasonal oscillation in the troposphere.

Water vapor is highly variable in middle to high latitudes in the upper troposphere, and can be supplied by transport from the tropics, by mesoscale convection, or by extratropical cyclones. Dry air can be transported from the subtropics or from the extratropical lower stratosphere. These transport phenomena tend to be periodic rather than continual.

A complex mix of processes takes place at the tropical tropopause to remove water vapor from air, as it enters the stratosphere. Within the background of large-scale mean ascent, the dehydration processes include smaller-scale ascent, radiative and microphysical processes within clouds, and wave-driven fluctuations in temperature. The location, strength, and relative importance of these processes vary seasonally. But the seasonal variation in tropopause level water vapor is influenced by the seasonal variation in tropical tropopause temperatures. Air rising through the tropopause is marked with seasonally varying mixing ratio, and retains these markings as it spreads rapidly poleward and more slowly upward into the stratosphere.

1.11 Upper Troposphere and Lower Stratosphere

The upper troposphere and lower stratosphere (UT/LS) is a complex region and the exchange process take places through the interface between the troposphere and stratosphere. Since the radiative and chemical timescales are relatively long, the transport is highly significant. The radiative properties and phase changes of atmospheric moisture link the hydrological and energy cycles of the Earth system. As the average residence time of water vapor in the atmosphere is around 10 days, the atmospheric branch is a relatively fast component of the global hydrological cycle. Tropical rainfall is particularly important as a forcing mechanism for the atmospheric large-scale circulation and climate.

The UT/LS has some distinct characteristics that influence Earth's climate (see Fig. 1.27). The pressures are still high enough to influence the course of reactions and photochemical processes. This region is below the ozone layer such that the radiation is limited to wavelengths greater than ~290 nm. Upper troposphere and lower stratosphere contain the coldest parts of the lower atmosphere, to the extent that highly reactive particles can be produced. These particles facilitate heterogeneous reactions taking place on a solid substrate and multiphase reactions happening in a liquid droplet, thereby altering the composition of this region. The particles,

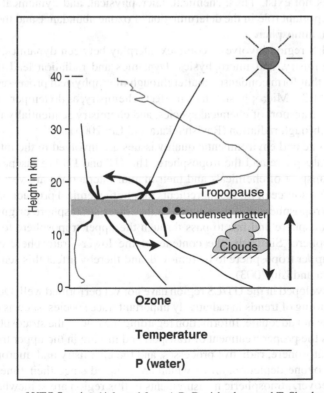

Fig. 1.27 Issues of UT/LS region (Adapted from A.R. Ravishankara and T. Shepherd 2001)

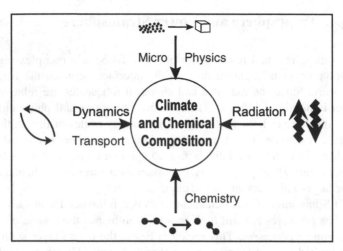

Fig. 1.28 UT/LS Region-complex interplay between dynamics, transport, radiation, chemistry, and microphysics (Adapted from A.R. Ravishankara and T. Shepherd 2001)

especially cirrus clouds, also directly interact with radiation. However, these are the regions where a clear separation of the timescales for chemistry and dynamical transport does not exist. Thus, chemical, microphysical, and dynamical processes all play an important role in the determination of ozone abundance and the radiative balance of the atmosphere.

Thus UT/LS region involves a complex interplay between dynamics, transport, radiation, chemistry, and microphysics. Dynamics and radiation lead to the low temperatures that form condensed matter through microphysical processes as represented in Fig. 1.28. Microphysics in turn affects chemistry, as do temperatures, solar radiation, and transport of chemical species, and chemistry sequentially feeds back onto climate through radiation (Ravishankara and Liu 2003).

Many climate and environmental quality issues are involved in the interface between the stratosphere and the troposphere. The UT and LS are inseparably connected via transport of chemicals and their mutual interactions are very large and significant. The source of all the ingredients for photochemical production of ozone in the upper troposphere and destruction in the lower stratosphere originates from the lower atmosphere and has to pass through the upper troposphere to reach the lower stratosphere. Similarly, the contents of the lower stratosphere are passed through the upper troposphere to be removed, and thereby affect this sensitive area (Ravishankara and Liu 2003).

Models developed in the UT/LS region have not yet performed well. Our present-day understanding of trends in radiatively important trace species, such as ozone and water vapor, is not adequate. Information regarding the role of the stratosphere in climate requires the proper treatment of transport and mixing in the upper troposphere and lower stratosphere, radiative processes, and the chemistry and microphysics of stratospheric ozone depletion, all of which are coupled since their timescales are similar. Moreover, atmospheric measurements in this region are somewhat difficult and incomplete.

1.12 Atmospheric Aerosols

Atmospheric aerosols are small airborne particles of widely differing chemical composition. They can either be of natural or of anthropogenic origin and it is estimated that anthropogenic aerosols constitute around 50% of the global mean aerosol optical thickness. The major sources of anthropogenic aerosols are from fossil fuel burning and biomass burning.

Atmospheric aerosols are important in many ways. Aerosol content affects the Earth's albedo and plays a major role in the global radiation balance and climate. Various aerosols act as cloud condensation nuclei and are important in the formation of clouds and precipitation. In addition, the scattering property of the aerosols can be used by a number of next-generation active remote-sensing instruments in derivation of geophysical parameters.

Most aerosol particles originate from blowing soil, smoke, volcanoes, and the oceans. Particles made of sodium chloride or magnesium chloride are hygroscopic and therefore act as good sites for the condensation of water to form cloud droplets. The concentration of the aerosols varies considerably but is typically on the order of 103 cm^{-3} over oceans, 104 cm^{-3} over rural land, and 105 cm^{-3} over cities. The concentrations generally decrease with altitude.

The size of aerosol particles is usually given as the diameter of the particle assuming a spherical shape. The sizes of different aerosol particles in the atmosphere are illustrated in Fig. 1.29. Aerosols are usually assigned into three size categories: (i) aitken particles, or *nucleation mode* (0.001–0.1 μm diameter); (ii) large particles, or *accumulation mode* (0.1–1 μm diameter); and (iii) giant particles, or *coarse particle mode* (>1 μm diameter).

The terms *nucleation mode* and *accumulation mode* refer to the mechanical and chemical processes by which aerosol particles in those size ranges are usually produced. The smallest aerosols, in the nucleation mode, are principally produced by gas-to-particle conversion, which occurs in the atmosphere. Aerosols in

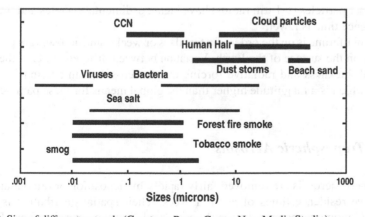

Fig. 1.29 Size of different aerosols (Courtesy: Bruce Caron, New Media Studio)

the accumulation mode are generally produced by the *coagulation* of smaller particles and by the *heterogeneous condensation* of gas vapor onto existing aerosol particles.

1.12.1 Water-soluble Aerosols

Aerosols consist of water-soluble compounds, such as sulphate, nitrate, and sea salt, and are efficient cloud condensation nuclei (CCN). In unpolluted continental conditions, smaller particles are more likely to be water-soluble. Nearly 80% of the particles in the 0.1–0.3 μm size range are comprised of water-soluble particles. Over oceans, however, much of the coarse particle mode comprised of sea salt aerosols is water-soluble. Water-soluble aerosols are *hygroscopic* and they are capable of attracting water vapor from the air. The size of hygroscopic particles varies with relative humidity, leading to changes in optical properties as well. The presence of polar functional groups on organic aerosols, particularly carboxylic and dicarboxylic acids, makes many of the organic compounds in aerosols water-soluble and allows them to participate in cloud droplet nucleation. Aerosols, such as metal oxides, silicates, and clay minerals, originate from soil dust or volcanoes and are insoluble.

1.12.2 Residence Time of Aerosols

The residence time of aerosols depends on their size, chemistry, and height in the atmosphere. Particle residence times range from minutes to hundreds of days. Aerosols between 0.1 and 1.0 μm, known as the accumulation mode, remain in the atmosphere for a longer time. Smaller aerosols (the nucleation mode) are subject to Brownian motion. As a result, higher rates of particle collision and coagulation increase the size of individual particles and remove them from the nucleation mode. The coarser particles (>1 μm radius) have higher sedimentation rates and therefore the residence time is lower.

Aerosol lifetime is on the order of a few days or weeks, and aerosols are produced unevenly on the surface of the Earth. Variation between different parts of the globe in optical thickness and radiative forcing due to aerosols, can amount to tens of $W\,m^{-2}$, which is a magnitude higher than the global means for these parameters.

1.12.3 Tropospheric Aerosols

Tropospheric aerosols are removed fairly rapidly by deposition or rainout and typically have residence times of about 1 week. Their spatial distribution is therefore very inhomogeneous and strongly correlated with the source regions. Present knowledge on the optical properties of tropospheric aerosols, as well as their spatial and temporal evolution, is inadequate. Routine satellite monitoring of tropospheric aerosols, when not obstructed by clouds, should provide insights into not only these aspects but also the way aerosols alter cloud optical properties.

1.12.4 Stratospheric Aerosols

Stratospheric aerosols have much longer residence times, on the order of about 1 year, and therefore have a more uniform distribution. As a result, volcanic explosions in which the debris reaches into the stratosphere can perturb global climate for several years. Satellite and sonde measurements indicate a steady increase in the background concentration of stratospheric sulphate of about 40–60% over the last decade. It seems that this record is not affected by volcanic aerosols but could reflect an influx of increasing tropospheric aerosols. Emissions from subsonic aircraft flying in the stratosphere are also contributing aerosols in the stratosphere. Considering the expected growth in future air traffic, it is indeed necessary to regularly monitor the stratospheric aerosol trends on a global basis.

Problems and Questions

1.1. Find the global mean surface pressure in hPa, if the mass of the atmosphere is 5.10×10^{18} kg. Assume that the mean value of the acceleration due to gravity is 9.807 m s^{-2}. How much is the global mean surface pressure different from the standard mean sea level pressure?

1.2. Determine the mean molecular weight of a sample of air consisting of nitrogen, oxygen, and argon, given that the molecular weight of nitrogen, oxygen, and argon is 28.01, 32.00, and 39.85, respectively.

1.3. An inflated balloon at the ground is taken in an aircraft which is flying at 8 km height. If the cabin is not pressurized, what would happen to the balloon? Explain.

1.4. What is the pressure at the top of the Eiffel Tower roof, if the pressure and density at the surface are 1,013.25 hPa and 1.225 kg m^{-3}, respectively. (Hint: Eiffel Tower roof height is 330 m)

1.5. Estimate the height at which the density falls to 0.5 kg m^{-3}, and the pressure decreases to 5 hPa. Assume that the pressure and density decrease exponentially with a scale height of 7.5 km, given that surface level pressure is 1,013 hPa and density is 1.225 kg m^{-3}.

1.6. Let the mean lapse rate of the troposphere be 6.5°C km^{-1}. If the surface temperature is 30°C, what is the temperature at the top of Mt. Everest, which is 8,848 m above the surface?

1.7. A hot air balloon is traveling eastward along $30°N$ at a mean speed of 12 m s^{-1} in half part of the globe, and 16 m s^{-1} in the remaining portion. If the prevailing wind is calm, estimate how much time it takes to circumnavigate the entire globe at this latitude belt.

1.8. Consider an aircraft flying at a height of 12 km from Paris to Tokyo at a speed of 800 km per hour. If the aircraft enters into the centre of the subtropical jet stream, which has an average core speed of 50 m s^{-1}, estimate the time gained by the aircraft on reaching the destination. The distance between Paris and Tokyo is approximately $10,000$ km.

1.9. The temperature of the equatorial tropopause at 18 km is $-87°C$ and that at $30°N$ is $-66°C$ at 14 km altitude. Calculate the temperature gradient between the equator and $30°N$. Neglect the curvature of the Earth and assume mean lapse rate of $6.5°C$.

1.10. In the equatorial region, on January 2007, the zonal wind at 30 km is westerly at a speed of 15 m s^{-1}. In January 2008, what would be phase of the zonal wind at 30 km and at 18 km? Explain.

1.11. It is assumed that during the early stages of its formation the Earth's atmosphere contained a large amount of hydrogen, but the present atmosphere does not have very much hydrogen gas in it. Where did that hydrogen go?

1.12. The tropopause temperature in the tropics is much colder than the tropopause temperature in middle and polar latitudes. What structural aspect of the atmospheric temperature profile makes this possible even though the surface temperature in the tropics is much warmer than the surface temperature in middle and polar latitudes?

1.13. What are the general latitudinal trends in temperatures as you move poleward from the equator in troposphere and stratosphere? Explain whether the tropospheric changes are more severe in the summer or winter hemisphere?

1.14. Why is the height of the tropopause different at different latitudinal zones? Suppose the height of the tropopause is same throughout the globe, what would be the wind structure and temperature pattern of the lower atmosphere?

1.15. If the tropical tropopause is raised to 25 km, discuss the changes that would occur in the tropospheric weather systems. Describe the role of tropopause in maintaining the general circulation of the atmosphere.

1.16. What would be the vertical thermal structure of the atmosphere in the absence of ozone? Why is stratospheric ozone called *good ozone* and tropospheric ozone called *bad ozone*?

1.17. In the Earth's atmosphere, the maximum ozone concentration is noted around 25 km altitude region. Why then is stratopause found at 50 km, and not at 25 km?

1.18. How does the stratospheric northern hemisphere winter temperature structure compare to that of the southern hemisphere winter? What are regions of coldest temperatures in the stratosphere termed?

1.19. What latitude regions experience least and greatest changes in zonal winds? Explain. What is the main driving mechanism responsible for the Earth's large-scale atmospheric circulations?

1.20. What causes semiannual oscillations in the tropical atmosphere? Why are they absent at higher latitudes?

1.21. Describe the charactertics of stratospheric warming occurrence in the northern and southern hemispheres. Does this high latitude phenomenon have influence on other altitude as well as latitude regions? Explain.

1.22. What are the various sources of aerosols in the atmosphere? How do aerosols reach the stratosphere? Suppose you remove the entire aerosols from the atmosphere, what would happen?

References

Andrews DG, Holton JR, Leovy CB (1987) Middle Atmosphere Dynamics, Academic, New York
Asnani GC (2005) Tropical Meteorology, Second Edition, Vol. 2, Pune, India
Appu KS (1984) On perturbations in the thermal structure of tropical stratosphere and mesosphere in winter, Indian J Radio Space Phys, 13: 35–41
Baldwin MP, Gray LJ, Dunkerton TJ, Hamilton K, Haynes PH, Randel WJ, Holton JR, Alexander MJ, Hirota I, Horinouchi T, Jones DBA, Kinnersley JS, Marquardt C, Sato K, Takahashi M (2001) The quasibiennial oscillation, Rev of Geophys, 39: 179–229
Baldwin MP, Stephenson DB, Thompson DWJ, Dunkerton J, Charlton AJ, ONeil A (2003) Stratospheric memory and skill of extended range weather forecasts, Science, 301: 636–640
Bethan S, Vaughan G, Reid SJ (1996) A comparison of ozone and thermal tropopause heights and the impact of tropopause definition on quantifying the ozone content of the troposphere, Quart J Royal Met Soc, 122: 929–944
Brasseur G, Solomon S (1984) Aeronomy of the Middle Atmosphere, D Riedel Publishing Company, Dordrecht
Caron B, What are aerosols, New Media Studio, The National Science Digital Library (http://www.newmediastudio.org/DataDiscovery/Aero_Ed_Center/Charact/size_of_aerosols.gif)
Charlton A, Polvani L (2006) Algorithm for identifying sudden stratospheric warming, Dept. of Applied Physics and Applied Mathematics, Columbia University
EUMeTrain, Vertical cross sections, PV anomaly (http://www.zamg.ac.at/eumetrain/CAL_Modules/VCS/Content/images/pv2.jpg)
Frits S, Soules S (1972) Planetary variations of stratospheric temperatures, Mon Wea Rev, 100: 582–589
Geerts B, Linacre E (1997) The height of the tropopause, University of Wyoming (http://www-das/~geerts/cwx/notes/chap01/trop-height01.gif)
Gettelman A, Forster PM. de F, (2002) A climatology of the tropical tropopause layer, J Met Soc Japan, 80 (4B), 911–924

Hamilton K, Hertzog A, Vial F, Stenchikov G (2004) Longitudinal variation of the stratospheric quasi-biennial oscillation, J Atmos Sci, 61: 383–402

Hartjenstein G (1999) Tropopause folds and the related stratosphere troposphere mass exchange, Project Report of University of Munich (*http://www.lrz-muenchen.de/projekte/ hlr-projects/1997-1999/cd/daten/pdf/uh22102.pdf*)

Haynes P, Shepherd T (2001) Report on the SPARC Tropopause Workshop, Bad Tolz, Germany, 17–21 April 2001

Hays J, deMenocal P, Lecture Notes on the Seasonality of ozone, Section 1.5, Columbia University (*http://www.ideo.columbia.edu/edu/dees/V1003/lectures/ozone/index.html*)

Hirota I, Yasuko H (2000) Interannual variations of planetary waves in the Southern Hemisphere stratosphere, 2nd SPARC General Assembly 2000, 26–30 October, Mardel Plata, Argentina

Holton JR, Haynes PH, McIntyre ME, Douglass AR, Rood RB, Pfister L (1995) Stratosphere-troposphere exchange, Rev Geophys, 33: 403–440

Houghton JT (1978) The stratosphere and mesosphere, Quart J Roy Met Soc, 104: 1–29

Kodera K, Kuroda Y (2000) Tropospheric and stratospheric aspects of the Artctic Oscillation, Geophys Res Lett, 27: 3349–3352

Labitzke K, van Loon H (1999) The Stratosphere, Phenomena, History and Relevance, Springer, Berlin

Marquardt C, Naujokat B (1997) An update of the equatorial QBO and its variability. 1st SPARC Gen. Assembly, Melbourne Australia, WMO/TD-No. 814: 87–90

Matsuno T (1971) A dynamical model of stratospheric warmings, J Atmos Sci, 28: 1479–1494

McPhaden MJ, Busalacchi AJ, Cheney R, Donguy J, Gage KS, Halpern D, Julian M Ji, Meyers PG, Mitchum GT, Niiler PP, Picaut J, Reynolds RW, Smith N, Takeuchi K (1998) The tropical ocean-global atmosphere observing system: A decade of progress, J Geophys Res, 103: 14169–14240

National Weather Service, Northern hemisphere cross section showing jetstreams and tropopause elevations, NOAA (*http://www.srh.noaa.gov/jetstream/global/image/jetstream3.jpg*)

Randel WJ, Udelhofen P, Fleming E, Geller MA, Gelman M, Hamilton K, Karoly D, Ortland D, Pawson S, Swinbank R, Wu F, Baldwin M, Chanin ML, Keckhut P, Labitzke K, Remsberg E, Simmons A, Wu D (2004) The SPARC Intercomparison of Middle Atmosphere Climatologies. J Climate, 17: 986–1003

Ravishankara AR, Liu S (2003) Highlights from the Joint SPARC-IGAC Workshop on Climate-Chemistry Interactions, held at Giens, France, during April 02–06, 2003

Ravishankara AR, Shepherd T (2001) Upper tropospheric and lower stratospheric processes, SPARC/IGAC Activities, SPARC Brochure – The SPARC Initiative

RMIT University, Jetstreams, Current Global Climates (*http://users.gs.mit.edu.caa/global/ graphics/jetstreams.jpg*)

RMIT University, Vertical circulation cells, Current Global Climates (*http://users.gs.rmit.edu.au/ caa/global/graphics/hadley.jpg*)

Russel CT (1987) The magnetosphere, in The Solar Wind and the Earth, edited by S.-I Akasofu and Y. Kamide, Terra Scientific Publishing Co., Tokyo

Shapiro MA, Hampel T, Kruger AJ (1987) The tropopause fold, Mon Wea Rev, 115: 444–454

Sherhag R (1952) Die explosion sartigen stratospharener warmungen des spatwinters 1951–1952, Ber Deut Wetterd, 6: 51–53

Siedel D, Annual cycles of tropospheric water vapor, Air Resource Laboratory, NOAA

Stuwart RR (2005) The oceanic influence on North American drought, in Oceanography in the 21st Century – An Online Textbook, Department of Oceanography, Texas A&M University

Tameeria (2007) Oxygen buid up in Earth's atmosphere, Wikipedia (*http://en. wikipedia.org/wiki/User.tameeria/images*)

The Green Lane, Environmental Canada (*http://www.qc.ec.gc.ca/Meteo/images/Fig_5-9_a.jpg*)

University of Leicester, Medium frequency radio frequency theory notes (*http://www.k1ttt. net/technote/kn41f/kn41f8_files/ionosphereprofile.gif*)

US Standard Atmosphere (1976) US Government Printing Office, Washington DC

Wallace JM, Hobbs PV (2006) Atmospheric Science – An Introductory Survey, Second edition, Elsevier, New York

Wirth V (2003) Static stability in the extratropical tropopause region. J Atmos Sci, 60:1395–1409

Wirth V, Szabo T (2007) Sharpness of the extratropical tropopause in baroclinic life cycle experiments, Geophys Res Letts, 34: L02809, doi:10.1029/2006GL028369

WMO (2007) Scientific Assessment of Ozone Depletion : 2006, Global Ozone Research and Monitoring Project, WMO Report No. 50, Geneva

Chapter 2
Radiative Processes in the Lower and Middle Atmosphere

2.1 Introduction

The circulation and dynamics of the Earth's atmosphere depend mainly on the magnitude and distribution of the net radiative heating of the Earth atmosphere system. In the troposphere, the net diabatic heating rate is dominated by the imbalance between the transfer of heat from the surface and the thermal emission of radiation to space. Latent heat is the major component of the flux from the surface to the atmosphere, and clouds play a major role in the emission of radiation to space.

In the stratosphere, the net heating depends solely on the imbalance between local absorption of solar UV radiation and infrared radiative loss. In this region, ozone is the principal absorber and carbon dioxide is the dominant emitter. Infrared emission by ozone and water vapor, and solar radiation absorption by water vapor, molecular oxygen, carbon dioxide, and nitrogen dioxide play secondary roles. The distribution of the radiative sources and sinks due to the above gases exerts a zero-order control on the large-scale seasonally varying mean temperature and zonal wind fields of the stratosphere (Andrews et al. 1987). These radiative processes are therefore significant to understand the stratosphere–troposphere interactions.

In this chapter, we will discuss the basic principles of radiation, solar radiation, and their processes in the stratosphere and troposphere. Detailed information on the radiative processes in the atmosphere is available elsewhere (Chandrasekhar 1950, Goody 1964, Kondratyev 1972, Liou 1980, Paltridge and Platt 1976, Smith 1985, etc.).

2.2 Basic Principles of Radiation

Radiation is the process through which energy moves through space from the source, without any material medium. Radiation sources are generally collections of matter or devices that convert other forms of energy into radiative energy. In some cases the energy to be converted is stored within the object like the Sun and radioactive materials. In other cases the radiation source is only an energy converter, and other forms of energy must be applied in order to produce radiation.

Most forms of radiation can penetrate through a certain amount of matter. But in most situations, radiation energy is eventually absorbed by the material and converted into another form of energy.

2.2.1 Electromagnetic Energy

Solar radiation is composed of electromagnetic waves that travel through space. Electromagnetic waves are formed when an electric field couples with a magnetic field. The electric and magnetic fields of an electromagnetic wave are perpendicular to each other and to the direction of wave. Electromagnetic energy is a form of energy that is transferred by radiation from all things in nature.

Electromagnetic energy spectrum comprises a broad band of wavelengths that travels from its source through space in the form of harmonic waves at the uniform speed of light in vacuum (3×10^8 m s^{-1}). Radiation is the term that relates to the emission and propagation of electromagnetic energy in the form of waves. In other words, radiation may be defined as a process in which energy is transmitted across space whether or not a material medium (viz., as for conduction, convection, or advection) is present.

Approximate wavelengths, frequencies and energy levels of the various regions of the electromagnetic spectrum are listed in Table 2.1. The various types of electromagnetic radiation in order of increasing energy level are also illustrated in Fig. 2.1 and are briefly described as follows.

Radio waves are the region of the electromagnetic spectrum with very long wavelengths (>0.1 m). Radio, TV, and radar communications occur using these types of waves.

Microwaves are the spectrum lying between ultra high frequency (UHF) radio waves and (heat) infrared waves. Microwaves are used to generate heat and for communications. They may cause heat damage to tissues.

Table 2.1 Wavelength, frequency, energy, and blackbody temperatures of various regions in the electromagnetic spectrum (Adapted from American Museum of National History)

Electromagnetic region	Wavelength (nm)	Frequency (Hz)	Energy (eV)	Blackbody temperature (K)
Radio wave	$>1 \times 10^8$	$<3 \times 10^9$	$<10^{-5}$	<0.03
Microwave	$1 \times 10^6 - 1 \times 10^8$	$3 \times 10^9 - 3 \times 10^{11}$	$10^{-5} - 10^{-2}$	$0.03 - 3$
Infrared	$7 \times 10^2 - 1 \times 10^6$	$3 \times 10^{11} - 4 \times 10^{14}$	$10^{-2} - 2$	$3 - 4100$
Visible	$4 \times 10^2 - 7 \times 10^2$	$4 \times 10^{14} - 7.5 \times 10^{14}$	$2 - 3$	$4100 - 7300$
Ultraviolet	$10 - 4 \times 10^2$	$7.5 \times 10^{14} - 3 \times 10^{16}$	$3 - 10^3$	$7300 - 3 \times 10^5$
X-ray	$1 \times 10^{-2} - 10$	$3 \times 10^{16} - 3 \times 10^{19}$	$10^3 - 10^5$	$3 \times 10^5 - 3 \times 10^8$
Gamma ray	$<1 \times 10^{-2}$	$>3 \times 10^{19}$	$>10^5$	$>3 \times 10^8$

The Electromagnetic Spectrum

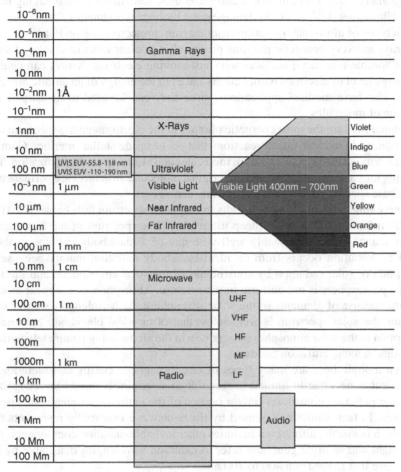

nm= nanometer, A=Angstrom, μm=micrometer, mm=millimeter
cm =centimeter, m=meter km=kilometer, Mm=Megameter

Fig. 2.1 Chart illustrating the spectrum of electromagnetic radiation (Adapted from University of Colorado)

Infrared radiation is the region with wavelengths long enough to cause molecules to vibrate around, increasing the temperature of the molecules. Infrared radiation is heat energy.

Visible light is the region where photons (see section 2.2.2) have enough energy to interact with certain pigment molecules in the retina of the eye to give us sight. This corresponds to the region of most intense solar output. All the colors of the rainbow fall into this small region, ranging from violet (400 nm) through indigo, blue, green, yellow, orange to red (700 nm).

Ultraviolet radiation is the region where photons are sufficiently energetic to change energy states within atoms and molecules, sometimes even breaking them apart. Ultraviolet (UV) light can damage some biological organisms. Ozone absorbs certain types of ultraviolet radiation from the Sun, protecting life on Earth.

X-rays are very energetic photons produced in nuclear reactions, solar storms, and in bombarding metal surfaces with fast-moving electrons. X-rays can change the energy level of electrons within atoms and even the energy of an atomic nucleus. X-rays also have medical applications, since they can be used to investigate the structure of molecules.

Gamma rays are the most energetic photons in the electromagnetic spectrum that are produced in nuclear fusion reactions that occur inside stellar interiors. Gamma rays can remove electrons away from molecules and atoms, inducing ionization. The resulting ions are very reactive and the original molecular structures are permanently lost. These spectral regions are biologically damaging.

Ninety-nine percent of the Sun's electromagnetic spectrum falls between 10,000 nm (far infrared) and 100 nm (deep ultraviolet). The spectrum of the Sun's spectral irradiance agrees reasonably well with that of a blackbody radiator at about 5,700 K. Variation occurs from an ideal blackbody radiation mainly because of absorption of solar radiation by constituents of the solar atmosphere, and the fact that the photosphere is not uniform in temperature. Absorption lines appear around 700 nm because of elements in the solar atmosphere. At wavelengths shorter than 280 nm, the solar spectrum is well below that of an ideal blackbody because of absorption in the solar atmosphere that results in the atoms being completely ionized with some notable emission bands.

Even though there are many types of electromagnetic energy consideration of the subject is necessarily limited to those of solar origin. It should be recognized that solar radiation constitutes only a portion of the entire electromagnetic energy spectrum. In fact, sunlight as sensed by the human eye essentially represents that part which is visible, although it includes other invisible radiation components. The terms light and sunlight generally refer to radiation wavelengths detectable by the human eye that range from 400 to 700 nm.

2.2.2 Energy of Radiation

Electromagnetic waves consist of radiant energy. The quantity of radiant energy may be calculated by integrating radiant flux with respect to time. It is generally thought of as radiation emitted by a source into the surrounding environment. Specific forms of radiant energy include electron space discharge, visible light, vacuum energy, and other wave types. Radiant energy is exhibited in the spontaneous nuclear disintegration with emission of particulate or electromagnetic radiations.

A photon is an elementary particle which is the carrier of electromagnetic radiation of all wavelengths. Its most important characteristic is the quantity of energy it contains. The photon energy is the energy of the individual photons that determines the type of electromagnetic radiation, such as light, X-ray, radio signals, etc.

One of the significant characteristics of photon energy is that it generally determines the penetrating ability of the radiation. The lower-energy X-ray photons are often referred to as soft radiation, whereas those at the higher-energy end of the spectrum would be termed as hard radiation. Generally, high-energy (hard) X-ray photons are more penetrating than the softer portion of the spectrum. The energy associated with a photon is given by

$$W = h\nu = \frac{h}{\lambda} \tag{2.1}$$

where h is the Planck's constant $(6.626 \times 10^{-26}\ \mathrm{J\ s^{-1}})$, ν is the frequency of light, and λ is the wavelength. So the energy of a photon varies inversely with wavelength of the radiation.

2.2.3 Photometry and Radiometry

Photometry is the science of measurement of light, in terms of its perceived brightness to the human eye. It is different from radiometry, which is the science of measurement of light in terms of absolute power. In photometry, the radiant power at each wavelength is weighted by the luminosity function that models human brightness sensitivity. Radiometry is important in astronomy, especially radio astronomy, and is important for remote-sensing applications.

2.2.3.1 Radiometric Quantities

Radiometry deals with quantities associated with the radiation field, and the interaction coefficients deal with quantities associated with the interaction of radiation and matter. Commonly used radiometric quantities are termed as follows.

Radiant flux (ϕ) is the rate of energy transfer by electromagnetic radiation. The units are Joules per second or watts. The Sun's radiant flux is 3.9×10^{26} W.

Monochromatic intensity (I_λ): The energy transferred by electromagnetic radiation in a specified direction per unit area per unit time normal to the direction considered at a given wavelength is known as monochromatic intensity, as defined by Wallace and Hobbs (2006). The integral of monochromatic intensity over a finite area of the electromagnetic spectrum is known as intensity or radiance with a unit of $\mathrm{Wm^{-2}\ sr^{-1}}$.

$$I = \int_{\lambda 1}^{\lambda 2} I_\lambda \mathrm{d}\lambda = \int_{\nu 1}^{\nu 2} I_\nu \mathrm{d}\nu \tag{2.2}$$

Both I_λ and I_ν are called monochromatic intensity, even though they are expressed in different units. They have the relation as

$$I_\nu = \lambda^2 I_\lambda \tag{2.3}$$

Monochromatic irradiance (E_λ) is the rate of energy transfer per unit area by radiation with a given wavelength through a plane surface. Irradiance covers an infinitesimal wavelength interval of the electromagnetic spectrum. The unit used for the measurement of irradiance is $Wm^{-2} nm^{-1}$.

Let the radiation be incident normal to a horizontal plane, then the monochromatic intensity can be:

$$E_\lambda = \int_{2\pi} I_\lambda \cos\theta d\omega \qquad (2.4)$$

where π indicates that the integration extends over the entire hemisphere, θ is the zenith angle, and $d\omega$ represents an elemental arc of solid angle.

The radiant flux divided by the area through which it passes is known as the *irradiance (E)* and is defined as

$$E = \int_0^\infty E_\lambda d\lambda \qquad (2.5)$$

Sometimes when referring to the Planck's function, $B_\lambda(T)$ may be used in place of E_λ. In general, the irradiance upon an element of surface may consist of contributions which come from infinite different directions.

2.2.4 Blackbody Radiation

Blackbody radiation refers to an object or system which absorbs all radiation incident upon it and reradiates energy which is characteristic of the radiating system only, and not dependent upon the type of radiation which is incident on it.

Physically a blackbody is an ideal perfect radiator and perfect absorber. All incident radiation falling on the blackbody is completely absorbed and hence known as black. Maximum possible emission occurs at all wavelengths and in all directions, thus the radiance is independent of direction, known as *isotropic radiation*.

2.2.4.1 Radiation Laws Governing Black Bodies

All objects will emit radiation when they are given suffcient energy in the form of heat. The emission of this radiation, or electromagentic radiation, is one of the ways in which objects can lose energy to their lower-temperature surroundings. The radiation is emitted as a continuous spectrum of all frequencies, but the distribution of energy among the frequencies of the spectrum of a blackbody depends only upon the temperature of the body. The basic laws of radiation governing the relationship between the temperature of the blackbody and the frequency spectrum of emission are briefly mentioned below.

1. Planck's Law
The emission of radiation of an ideal blackbody is defined by Planck's radiation law. It states that the emitted radiation power increases with increasing temperature

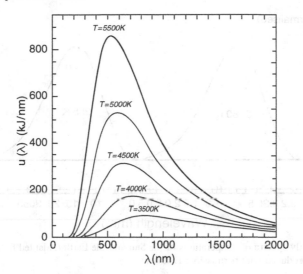

Fig. 2.2 Blackbody spectrum showing that the temperature is a function of wavelength based on the Planck's law of radiation (Courtesy: Wikipedia)

and that the maximum of the radiated spectrum is shifted toward shorter wavelengths with rising temperature. Planck's law provides the unique description of the monochromatic irradiance (E_λ) of radiation from a blackbody as a function of its temperature T:

$$E_\lambda = B_\lambda(T) = \frac{c_1}{\lambda^5[\exp(c_2)/\lambda T] - 1} \tag{2.6}$$

where c_1 and c_2 are constants. The shapes of the Planck's function for different temperatures are shown in Fig. 2.2. It can be seen that the wavelength of the peak of the blackbody radiation curve decreases in a linear fashion as the temperature increases. The monochromatic intensity shows a sharp cutoff at shorter wavelengths and increases steeply to a maximum. Thereafter, the intensity decreases rather slowly with increasing wavelengths. It explains why hotter bodies attain higher intensity peaks at shorter wavelengths. As the temperature of the blackbody decreases the emission intensity also decreases.

2. Wien's Displacement Law

Wien's displacement law states that there is an inverse relationship between the wavelength of the peak of the emission of a blackbody and its temperature. Differentiating the Planck's function and setting the derivative equal to zero gives the wavelength of the peak emission as

$$\lambda_{max} = \frac{b}{T} \tag{2.7}$$

where λ_{max} is the peak wavelength in meters, T is the temperature of the blackbody in Kelvin (K), and b is the proportionality constant, called Wien's displacement constant and equals 2.8978×10^6 nm K.

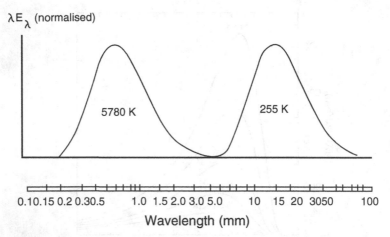

λE_λ (normalised)

5780 K   255 K

0.10.15 0.2 0.30.5   1.0 1.5 2.0 3.0 5.0   10   15 20 3050   100

Wavelength (mm)

Fig. 2.3 Blackbody spectra of temperatures of the Sun and the Earth (Adapted from Wallace and Hobbs 2006, Reproduced with permission: Elsevier)

Wien's law explains that objects of different temperature emit spectra that peak at different wavelengths. Hotter objects emit most of their radiation at shorter wavelengths; hence, they will appear to be bluer. Cooler objects emit most of their radiation at longer wavelengths; hence, they will appear redder. For example, the surface temperature of the Sun is 5,788 K, giving a peak at 502 nm, which is fairly in the middle of the visual spectrum. Due to the Rayleigh scattering of blue light by the atmosphere the white light is separated somewhat, resulting in a blue sky and a yellow Sun.

Figure 2.3 illustrates the normalized blackbody spectra of the Sun and the Earth. Over 99% of solar radiation is emitted at wavelengths shorter than 4,000 nm. By contrast, radiation emitted from objects such as the Earth lies mainly between 5,000 and 50,000 nm and peaks around 10,000–15,000 nm.

Furthermore, at any wavelength, a hotter object is more luminous than a cooler one. According to Wien's displacement law, the wavelength of maximum emission of Earth's atmosphere of temperature 255 K is 11,000 nm.

Light from the Sun and Moon: The surface temperature of the Sun is 5,778 K. Using Wien's law, this temperature corresponds to a peak emission at a wavelength of 2.89777 million nm K/5,778K = 502 nm. This wavelength is fairly in the middle of the most sensitive part of land animal visual spectrum acuity. Even nocturnal- and twilight-hunting animals must sense light from the waning day and from the Moon, which is reflected sunlight with this same wavelength distribution. Also, the average wavelength of starlight maximal power is in this region, due to the Sun being in the middle of a common temperature range of stars.

3. Rayleigh-Jeans Radiation Law

The Rayleigh-Jeans Law, attempts to describe the spectral intensity (I) of electromagnetic radiation at all wavelengths from a blackbody at a given temperature. For wavelength λ, it is

$$I_\lambda = \frac{2ckT}{\lambda^4} \tag{2.8}$$

where T is the temperature in Kelvin, and k is Boltzmann's constant.

In the microwave region of the electromagnetic spectrum, when the wavelength is at 5×10^6 nm and T is at terrestrial temperatures, we find that $hc/\lambda \ll kT$, so the Planck's function may be approximated by

$$I_\lambda = \frac{c_1 T}{c_2 \lambda^4} \tag{2.9}$$

In this Rayleigh-Jeans region, the Planck monochromatic radiance E_λ is a linear function of temperature T.

4. Stefan-Boltzmann Law

The Stefan-Boltzmann law, also known as Stefan's law, states that the total energy radiated per unit surface area of a blackbody in unit time, E, is directly proportional to the fourth power of the thermodynamic temperature T of the blackbody.

$$E = \sigma T^4 \tag{2.10}$$

The irradiance E has dimensions of power density and is measured in W m^{-2}. T is the absolute temperature and σ is the constant of proportionality known as Stefan-Boltzmann constant, which is nonfundamental in the sense that it is an agglomeration of other constants of nature. The value of the constant is

$$\sigma = \frac{2\pi^5 k^4}{15c^2 h^3} = 5.6704 \times 10^{-8}\,\mathrm{Wm^{-2}K^{-4}} \tag{2.11}$$

where k is Boltzmann's constant, h is Planck's constant, and c is the speed of light.

The surface of the Sun is 21 times as hot as that of the Earth, and therefore it emits 190,000 times as much energy per square meter. The distance from the Sun to the Earth is 215 times the radius of the Sun, reducing the energy per square meter by a factor of 46,000. Taking into account that the cross section of a sphere is one fourth of its surface area, we see that there is equilibrium of approximately 342 W m^{-2} surface area, or 1370 W m^{-2} cross-sectional area. This shows roughly why the temperature $T \sim 300$ K is approximately that of the Earth.

If we assume that the Earth is a perfect blackbody and include the effect of the Earth's albedo as 30% then the Earth's average surface temperature can be estimated as 255 K. But the actual measured value of Earth's surface temperature is 288 K. The difference of 33 K between such calculated value and the actual measured one is due to the effect of greenhouse gases, namely water vapor, carbon dioxide, and methane.

2.2.5 Atmospheric Scattering

Scattering is the process by which a particle in the path of an electromagnetic wave abstracts energy from the incident wave and reradiates that energy in all directions. In scattering, no energy transformation takes place, but only a change in the spatial distribution of the energy. Sunlight coming into the atmosphere can be scattered in any direction as it passes through the air. This diffuses the light and spreads it out in all directions so that it does not appear just as a single, straight beam.

In the atmosphere, the particles responsible for scattering cover the sizes from gas molecules (10^{-8} cm) to large raindrops and hail particles of about 1 cm. The relative intensity of the scattering pattern depends mainly on the ratio of particle size to wavelength of incident wave. There are three different types of scattering: *Rayleigh scattering*, *Mie scattering*, and *nonselective scattering*. Rayleigh scattering mainly consists of scattering from atmospheric gases. This occurs when the particles causing the scattering are smaller in size than the wavelengths of radiation in contact with them. This type of scattering is therefore wavelength-dependent. As the wavelength decreases, the amount of scattering increases. Because of Rayleigh scattering, the sky appears blue. This is because blue light is scattered around four times as much as red light, and UV light is scattered about 16 times as much as red light.

Mie scattering is caused by pollen, dust, smoke, water droplets, and other particles in the lower portion of the atmosphere. It occurs when the particles causing the scattering are larger than the wavelengths of radiation in contact with them. Mie scattering is responsible for the white appearance of the clouds. The last type of scattering is nonselective scattering. It occurs in the lower portion of the atmosphere when the particles are much larger than the incident radiation. This type of scattering is not wavelength-dependent and is the primary cause of haze.

2.2.6 Absorption and Emission

The maximum amount of radiation that can be emitted by any object at a given temperature is determined by Planck's law. A real object, which is not a perfect blackbody, will emit less than blackbody radiation. For a given wavelength λ, the emissivity (ε) is defined as the ratio of actual emitted radiance I_λ to the blackbody radiance B_λ:

$$\varepsilon_\lambda = \frac{I_\lambda}{B_\lambda} \tag{2.12}$$

The emissivity ranges from 0 to 1 for real objects and is a measure of how strongly a body radiates at a given wavelength. If the emissivity is independent of wavelength, then the emitter is called a gray body. Clouds and gases have emissivities which are a strong function of wavelength. The emissivity of the sea surface is close to 1 for visible wavelengths.

The emissivity of a perfect blackbody is unity, i.e., $\varepsilon_\lambda = 1$, whereas for a gray body it is constant. In the case of a selective radiation, the emissivity is a function of wavelength. A high emissivity, near to the maximum of 1, indicates an object that absorbs and radiates a large proportion of the incident energy and a low emissivity (closer to zero) indicates an object that absorbs and radiates a small proportion of the incident energy. The majority of natural objects, excluding water, are selective radiators and their emissivity is wavelength-dependent.

In a similar manner, we can define the monochromatic absorptivity a_λ, reflectivity r_λ, and transmissivity t_λ as the fraction of the incident monochromatic intensity S that a blackbody absorbs, reflects, and transmits, respectively, as

$$\text{Absorptivity } a_\lambda = \frac{I_\lambda(\text{absorbed})}{I_\lambda(\text{incident})} \tag{2.13}$$

$$\text{Reflectivity } r_\lambda = \frac{I_\lambda(\text{reflected})}{I_\lambda(\text{incident})} \tag{2.14}$$

$$\text{Transmissivity } t_\lambda = \frac{I_\lambda(\text{transmitted})}{I_\lambda(\text{incident})} \tag{2.15}$$

1. Kirchhoff's Law

A body in local thermodynamic equilibrium will emit the same amount of energy that it absorbs. Therefore the body does not heat up or cool down. Consider a body which is able to absorb and emit radiation. If I_λ is the incident spectral radiance, then the emitted spectral radiance E_λ is

$$E_\lambda = \varepsilon_\lambda = a_\lambda I_\lambda \tag{2.16}$$

where $a_\lambda I_\lambda$ is the absorbed spectral radiance and a_λ is the absorbance at a given wavelength.

For thermal equilibrium the emitted and absorbed radiation is the same; therefore, in the case of a blackbody:

$$I_\lambda = B_\lambda \tag{2.17}$$

so

$$E_\lambda = a_\lambda \tag{2.18}$$

This is known as Kirchhoff's law, which means that at a given wavelength, weak absorbers are weak emitters, and conversely, strong absorbers are strong emitters. Kirchhoff's law is a general statement equating emission and absorption in heated objects, following from general considerations of thermodynamic equilibrium.

An object at some nonzero temperature radiates electromagnetic energy. If it is a perfect blackbody, absorbing all light that strikes it, it radiates energy according to the blackbody radiation formula. More generally, if it is a gray body that radiates with some emissivity multiplied by the blackbody formula, Kirchhoff's law states that at thermal equilibrium, the emissivity of a body equals its absorptivity.

The absorptivity (or *absorbance*) is the fraction of incident light that is absorbed by the body/surface. In the most general form of the theorem, this power must be integrated over all wavelengths and angles. In some cases, however, emissivity and absorption may be defined as depending on wavelength and angle.

Kirchhoff's law has a corollary: by conservation of energy, the emissivity/absorptivity cannot exceed 1, so it is not possible to thermally radiate more energy than a blackbody, at equilibrium. In negative luminescence the angle and wavelength integrated absorption exceeds the material's emission. However, such systems are powered by an external source and are therefore not in thermal equilibrium.

2.2.7 Reflectivity and Transmissivity

Consider a spectral radiance I_λ incident on a slab of an absorbing material which is in local thermodynamic equilibrium (LTE). Figure 2.4 summarizes the radiant energy processes taking place including reflection, absorption, re-emission, and transmission.

In Fig. 2.4, we can see that the reflected radiation is $r_\lambda I_\lambda$, the transmitted radiation is $t_\lambda I_\lambda$, the absorbed radiation is $a_\lambda I_\lambda$, and the emitted radiation is $\varepsilon_\lambda B_\lambda$ which is a function of temperature, T.

For a blackbody, closely approximated by the Earth's surface or a dense cloud, $t_\lambda = I_\lambda = 0$, so the absorptivity is 1. For a body in LTE conservation of energy means that the absorbed radiation equals the incident radiation minus reflection and transmission contributions, so that:

$$a_\lambda I_\lambda = I_\lambda - r_\lambda I_\lambda - t_\lambda I_\lambda \tag{2.19}$$

or

$$a_\lambda + r_\lambda + t_\lambda = 1 \tag{2.20}$$

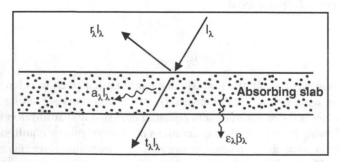

Fig. 2.4 Radiative transfer process for monochromatic radiance incident on a slab of absorbing medium (Adapted from Wallace and Hobbs 2006, Reproduced with permission: Elsevier)

The latter equation says that the processes of absorption, reflection, and transmission account for all the incident radiation falling on a body in LTE. In any atmospheric window region for which

$$t_\lambda = 1, \text{ then } a_\lambda = 0 \text{ and } r_\lambda = 0 \tag{2.21}$$

Monochromatic radiation incident upon an opaque surface with $t_\lambda = 0$ is either absorbed or reflected, so

$$a_\lambda + r_\lambda + t_\lambda = 1 \tag{2.22}$$

At any given wavelength strong reflectors are weak absorbers and weak reflectors are strong absorbers. For the atmosphere, the reflection of IR radiation is negligible, since the radiation wavelength is large compared to the size of molecules. Therefore, for the atmosphere

$$\varepsilon = 1 - t_\lambda \tag{2.23}$$

2.2.8 Radiance Temperature

Radiance temperature, sometimes called brightness temperature or blackbody temperature, is the temperature measured by a satellite instrument, usually detected in terms of radiance, but converted into a temperature through Planck's function at a given wavelength. The temperature T can be obtained from Planck's function as

$$T = \frac{c_2}{\lambda \, ln[c_1/\lambda^5 E_\lambda] + 1} \tag{2.24}$$

where T is called the brightness temperature/radiance temperature/equivalent blackbody temperature, λ is the wavelength, E_λ is the emitted radiation at wavelength λ, and c_1 and c_2 are empirical constants.

2.2.9 Solar Constant

The solar energy reaching the periphery of the Earth's atmosphere is known as the solar constant, and is considered to be constant for all practical purposes. The solar constant is defined as the amount of solar radiation incident, per unit area and time, on a surface which is perpendicular to the radiation and is situated at the outer limit of the atmosphere, the Earth being at its mean distance from the Sun.

The solar constant consists of all types of solar radiation, and is not just confined to visible light. Because of the difficulty in obtaining accurate measurements, the exact value of the solar constant is not known with certainty. It is measured by satellite to be roughly 1,366 W m^{-2}. The actual value of the energy varies with several factors, the most important being the Earth's distance from the Sun, which changes because of the Earth's elliptical orbit. It fluctuates by about 6.9% during a year from 1,412 W m^{-2} in early January to 1,321 W m^{-2} in early July, due to the Earth's varying distance from the Sun, and by a few parts per thousand from day to day. Thus, for the whole Earth, with a cross section of 127,400,000 km^2, the power is 1.740 × 10^{17} W, plus or minus 3.5%. The solar constant is not quite constant over long time periods either. The value 1,366 W m^{-2} is equivalent to 1.96 calories per minute per square centimeter, which can also be expressed as 1.96 langleys per minute (Wallace and Hobbs 2006).

The Earth receives a total amount of radiation determined by its cross section (πR^2), but as the planet rotates, this energy is distributed across the entire surface area ($4\pi R^2$). Hence, the average incoming solar radiation, taking into account the half of the planet not receiving any solar radiation at all, is one fourth the solar constant or ∼342 W m^{-2}. At any given location and time, the amount received at the surface depends on the state of the atmosphere and the latitude.

2.2.10 Albedo

Albedo is known as surface reflectivity of solar radiation. It is defined as the ratio of reflected to incident electromagnetic radiation. Albedo is a measure of surface diffuse reflectivity, which has a nondimensional value, normally represented in percent. The term has its origins from a Latin word *albus*, meaning white. It is quantified as the percentage of solar radiation of all wavelengths reflected by a body or surface to the amount incident upon it. An ideal white body has an albedo of 100%, whereas the albedo is 0% in the case of an ideal blackbody. The albedo of the Earth's surface varies with the type of material that covers it. The typical amounts of solar radiation reflected from various objects are shown in Table 2.2. Albedo values can

Table 2.2 Albedo of various surfaces

Surface	Details	Albedo (in %)
Soil	Wet–dry	4–40
Sand	—	15–45
Grass	Long–short	16–26
Agricultural crops	—	18–25
Tundra	—	18–25
Forests	Desiduous	15–20
	Coniferous	5–15
Water	Small zenith angle	3–10
	Large zenith angle	10–100
Snow	Old/fresh	40–95
Ice	Sea	30–45
	Glacier	20–40
Clouds	Thick	60–90
	Thin	30–50

range from 3% for water at small zenith angles to over 95% for fresh snow. On average the Earth and its atmosphere typically reflect about 30%, a number highly dependent on the vegetation, surface characteristics, forest cover, cloud cover and its distribution, etc (Ahrens 2006).

Surface reflectance values show large geographic variation. Mean annual albedo values differ considerably between the equator and the poles, largely due to the presence of snow and ice-covered surfaces along with cloudy skies in high latitudes, which greatly increases albedo values in those areas. Atmospheric reflectance principally varies with dust concentration, the zenith angle of the Sun, and the type and/or amount of cloud cover. Well-developed convective clouds reflect up to 90% of incident solar energy, making thick clouds appear bright from space. The reflectance properties of the surface change from one season to another. Throughout the high latitudes, snow cover and ice extent reach maximum values during the cold seasons, significantly increasing the surface reflectance values. Melting in the spring exposes bare soils that absorb a significantly greater portion of the incoming solar radiation, decreasing the albedo values.

2.2.11 The Greenhouse Effect

The greenhouse effect is the process in which the emission of infrared radiation by the atmosphere warms a planet's surface. The name comes from an analogy with the warming of air inside a greenhouse compared to the air outside the greenhouse. The Earth's average surface temperature is about 33°C warmer than it would be without the greenhouse effect. In addition to the Earth, Mars and especially Venus have greenhouse effects.

A schematic representation of the exchange of energy between outerspace, Earth's atmosphere, and its surface is shown in Fig. 2.5. Warming takes place in the atmosphere due to the fact that incoming visible radiation can penetrate to the ground with relatively little absorption, while much of the outgoing longwave infrared radiation is trapped by the atmosphere and emitted back to the ground. In order to satisfy the radiation balance of the Earth atmosphere system, the surface must compensate by emitting more radiation than it would in the absence of such an atmosphere. Thus greenhouse gases such as water vapor, carbon dioxide, methane, etc. act as an effective blanket around the Earth.

The Earth receives energy from the Sun in the form of radiation. The Earth and its atmosphere reflect about 30% of the incoming solar radiation. The remaining 70% is absorbed, warming the land, atmosphere, and oceans. For the Earth's temperature to be in steady state, this absorbed solar radiation must be very nearly balanced by energy radiated back to space in the infrared wavelengths. Since the intensity of infrared radiation increases with increasing temperature, one can think of the Earth's temperature as being determined by the infrared flux needed to balance the absorbed solar flux (Bengtson et al. 1999). The visible solar radiation mostly heats

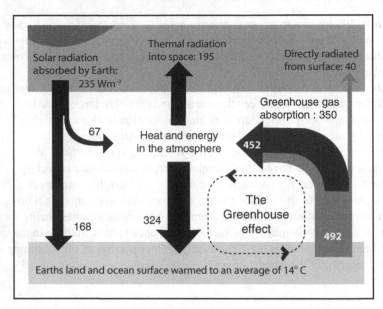

Fig. 2.5 Schematic representation of the exchanges of energy between outer space, and the Earth's atmosphere, and its surface (Courtesy: R. A. Rhode (a), Global Warming Art Project)

the surface, not the atmosphere, whereas most of the infrared radiation escaping to space is emitted from the upper atmosphere, not the surface. The infrared photons emitted by the surface are mostly absorbed in the atmosphere by greenhouse gases and clouds and do not escape directly to space.

The cause of the surface warming can be easily understood by starting with a simplified model of a purely radiative greenhouse effect that ignores energy transfer in the atmosphere by convection (sensible heat transport) and by the evaporation and condensation of water vapor (latent heat transport). In this purely radiative case, one can think of the atmosphere as emitting infrared radiation both upward and downward. The upward infrared flux emitted by the surface must balance not only the absorbed solar flux but also this downward infrared flux emitted by the atmosphere. The surface temperature will rise until it generates thermal radiation equivalent to the sum of the incoming solar and infrared radiation.

The opacity of the atmosphere to infrared radiation determines the height in the atmosphere from which most of the photons are emitted into space. If the atmosphere is more opaque, the typical photon escaping to space will be emitted from higher in the atmosphere. Since the emission of infrared radiation is a function of temperature, it is the temperature of the atmosphere at this emission level that is effectively determined by the requirement that the emitted flux balance the absorbed solar flux.

2.3 Radiative Transfer

Both the solar and terrestrial radiations passing through the atmosphere attenuates due to scattering by aerosols and absorption by gas molecules present in the atmosphere. The attenuation of the radiation depends on: (i) the intensity of radiation at that point, (ii) the local concentration of the gases and aerosol particles that are responsible for absorption and scattering, and (iii) the effectiveness of the absorbers and scatterers.

2.3.1 Beer's Law

Beer's law, also known as Bouguer's law or Lambert's law, states that the monochromatic intensity I_λ decreases monotonically with path length as the radiation passes the layer. The law can be demonstrated as follows.

Consider a parallel beam of radiation, I_λ, passing through an infinitesimally thin layer of the atmosphere containing absorbing gases and aerosols along a specific path (see Fig. 2.6). After passing through the layer, the monochromatic intensity of radiation is decreased by

$$dI_\lambda = -I_\lambda \rho r k_\lambda \sigma ds \qquad (2.25)$$

where ρ is the density of air, r is the mass of the absorbing gas per unit mass of air, and k_λ is the mass absorption or scattering coefficient.

As shown in Fig. 2.6, the differential path length along the ray path of the incident radiation is $ds = sec\theta dz$. Substituting this value for ds in Eq. (2.25), we get

$$dI_\lambda = -I_\lambda \rho r k_\lambda \sigma sec\theta dz \qquad (2.26)$$

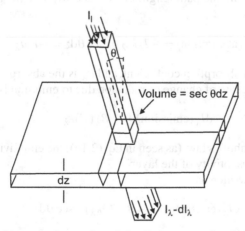

Fig. 2.6 Depletion of incoming beam of parallel radiation on passing through a slab of absorbing material (Adapted from Wallace and Hobbs 2006, Reproduced with permission: Elsevier)

Now let us consider the case of depletion of radiation due to absorption and scattering from the top of the atmosphere ($z = \infty$) down to any level (z). Integrating Eq. (2.26) gives

$$\ln I_{\lambda\infty} - \ln I_\lambda = \sec\theta \int_z^\infty k_\lambda \rho r dz \qquad (2.27)$$

Taking the antilog on both sides, we get

$$I_\lambda = I_{\lambda\infty} \exp(-\tau_\lambda \sec\theta) = I_{\lambda\infty} t_\lambda \qquad (2.28)$$

where $\tau_\lambda = \int_z^\infty k_\lambda \rho r dz$ is a dimensionless quantity referred to as the normal optical depth or optical thickness. It is a measure of the cumulative depletion that a beam of radiation directed straight downward with zero zenith angle would experience in passing through the layer, and $t_\lambda = \exp(-\tau_\lambda \sec\theta)$ is the transmissivity of the layer.

In the absence of scattering, the monochromatic absorptivity

$$a_\lambda = 1 - t_\lambda = 1 - \exp(-\tau_\lambda \sec\theta) \qquad (2.29)$$

approaches unity exponentially with increasing optical depth.

In the above cases, we have dealt with the scattering and absorption of solar radiation in the atmosphere in the absence of emission. Now we will see the absorption and emission of infrared radiation in the absence of scattering.

2.3.2 Schwarszchild's Equation

Let us derive the equation that governs the transfer of the infrared radiation through a gaseous medium. The rate of change of the monochromatic intensity of outgoing terrestrial radiation along the path length of $ds = \sec\theta dz$, due to absorption within the layer, is written as

$$dI_\lambda(\text{absorption}) = -I_\lambda k_{a\lambda} \rho r \sec\theta dz = -I_\lambda a_\lambda \qquad (2.30)$$

where $k_{a\lambda}$ is the mass absorption coefficient, and a_λ is the absorptivity of the layer.

The corresponding rate of change of radiation due to emission is

$$dI_\lambda(\text{emission}) = -B_\lambda(T)\varepsilon_\lambda \qquad (2.31)$$

According to Kirchhoff's law, (as seen in Eq. (2.16)) the emissivity ε_λ of a blackbody is equal to its absoptivity of the layer.

Eq. (2.31) can be written as

$$dI_\lambda(\text{emission}) = -B_\lambda(T)k_\lambda \rho r \sec\theta dz \qquad (2.32)$$

The attenuation of monochromatic intensity due to both emission and absorption of the layer can be obtained by adding Eq. (2.30) and Eq. (2.31) and we obtain

$$dI_\lambda = -\left[I_\lambda a_\lambda - B_\lambda(T)\right] k_\lambda \rho r \sec\theta dz \tag{2.33}$$

This equation is known as Schwarzschild's equation. It states that as the radiation passes through an isothermal layer, its monochromatic intensity exponentially approaches that of a blackbody radiation corresponding to the temperature of the layer (Wallace and Hobbs 2006).

2.3.3 Solar Absorption and Atmospheric Heating

The absorption of solar radiation in the atmospheric layers can be estimated by using the Beer Lambert law. According to Beer Lambert law, the intensity of solar radiation of a given wavelength I_λ absorbed by a thin layer of the atmosphere of thickness dz in terms of molecular number density can be represented as:

$$dI_\lambda = -I_\lambda \sigma_{a\lambda} n \sec\theta dz \tag{2.34}$$

where, $\sigma_{a\lambda} n = k_a$ is the absorption coefficient, in which $\sigma_{a\lambda}$ is the absorption cross section and n is the molecular number density.

Integration of this absorption over a path through the atmosphere gives

$$I_\lambda = -I_{0\lambda} \exp\left[-\int \sigma_{a\lambda} n(z) \sec\theta dz\right] \tag{2.35}$$

The optical thickness is presented as

$$\tau_a = \int \sigma_{a\lambda} n(z) \sec\theta dz \tag{2.36}$$

The transmission t through the path s is

$$t_\lambda(z) = \exp[-\tau_{a\lambda}(z) n(s) \sec\theta ds] \tag{2.37}$$

Solar radiation penetrates the atmosphere at angle of incidence, which depends on the latitude, season, and local time. By spherical geometry, the cosine of the solar zenith angle (θ) can be written as

$$\cos\theta = \cos\phi \cos\delta \cos(HA) + \sin\phi \sin\delta \tag{2.38}$$

where ϕ = latitude, δ = solar declination angle (23.5° for solstice, 0° equinox), HA = hour angle, for which 0° is local noon.

Let us assume a parallel path atmosphere, $ds = \sec\theta dz$. This approximation is good until we have $\theta > 75°$. At that solar zenith angle, we need to take Earth's curvature into account.

The molecular number density $n(z)$ often varies exponentially with altitude. The number density represents the concentration of molecules, and at any height z can be represented by the relation

$$n(z) = n(z_0)\exp(-z/H)\tag{2.39}$$

where z_0 is the number density at height $z = 0$, and H is the scale height. Thus,

$$I(z) = I_\infty\exp\left[-\sec\theta\int_z^\infty \sigma_a n_0\exp(-z/H)dz\right]\tag{2.40}$$

$$I(z) = I_\infty\exp[-\sec\theta\sigma_a n_0 H\exp(-z/H)]\tag{2.41}$$

where I_∞ is the solar intensity outside the Earth's atmosphere.

The rate of energy disposition by absorption is proportional to the rates of formation of ions, the rates of photodissociation and production of heat. They are all directly linked to the rate of energy deposition in the atmosphere by absorption. If there is any fluorescence, then the rate of energy deposition will be somewhat less.

$$r = -\frac{dI}{\sec\theta dz}\tag{2.42}$$

$$r = \sigma_a n_0 I(\infty)\exp\left[-\frac{z}{H}\right] + \tau_0\exp\left[\frac{z}{H}\right]\tag{2.43}$$

where

$$\tau_0 = \sigma_a n_0 H\sec\theta\tag{2.44}$$

The altitude with the maximum absorption, which is found by differentiating the expression r and setting it equal to zero and then solving for z, is:

$$z_m = H\ln(\tau_0)\tag{2.45}$$

$$z_m = H\ln(\sigma_a n_0 H\sec\theta)\tag{2.46}$$

For an overhead Sun, the altitude of maximum absorption is:

$$z_m = H\ln(\sigma_a n_0 H)z_0\tag{2.47}$$

so that the altitude of maximum energy absorption at any other solar zenith angle is:

$$z_m = z_0 + H\ln(\sec\theta)\tag{2.48}$$

The rate of energy deposition at the maximum is given by:

$$r_m = \sigma_a n_0 I_\infty\cos\theta\exp\left[-1-\frac{z_0}{H}\right]\tag{2.49}$$

The variation of the rate of energy deposition is given by the equation:

$$\frac{r}{r_0} = \exp\left[1 - Z - \sec\theta\exp(-Z)\right] \tag{2.50}$$

where $Z = (z - z_0)/H$, and r_0 is the rate of energy absorption for $\theta = 0$.

$$r_0 = \sigma_a n_0 I(\infty)\exp\left[-1 - \frac{z_0}{H}\right] \tag{2.51}$$

Figure 2.7 depicts the maximum deposition rate with altitude and shows the effect of solar zenith angle. It is evident from Fig. 2.7 that the height of maximum deposition rate z_m is independent of the intensity, but strongly dependent on (i) the nature and distribution of the absorbing gas, (ii) the solar zenith angle, and (iii) the wavelength of the radiation.

Since the solar spectrum and absorption occur over a range of wavelengths, several strongly absorbing layers will occur at different altitudes.

Oxygen and ozone absorb strongly at different wavelengths and are illustrated in Fig. 2.8.

$$\tau_a(\lambda, z, \theta) = \sec\theta\left[\int_z^\infty \sigma_a(O_2, \lambda)(O_2(z))dz + \left(\int_z^\infty \sigma_a(O_3, \lambda)(O_3(z))dz\right)\right] \tag{2.52}$$

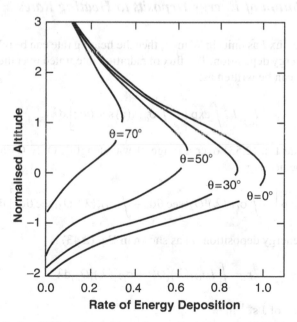

Fig. 2.7 Effect of solar zenith angle on the maximum deposition rate with altitude (Brasseur and Solomon 2005, Courtesy: Springer)

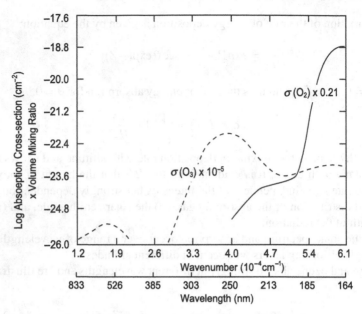

Fig. 2.8 Absorption of solar radiation by molecular oxygen and ozone (Brasseur and Solomon 2005, Courtesy: Springer)

2.3.4 Translation of Energy Deposits to Heating Rates

If we denote the flux I as units in W m^{-2}, then the heating rate can be related directly to the rate of energy deposition. The flux of radiation integrated over the entire range of wavelengths can be written as:

$$I = I_\infty \int \exp\left[-\int \sigma_{a\lambda} n(z) \sec \theta dz\right] d\lambda \qquad (2.53)$$

Now integrate Eq. (2.53) over a range of wavelengths. Only O_2 and O_3 really contribute to the heating:

$$I = \int_\lambda I_\infty \exp\left[-\int \sigma_{O_2}(\lambda)[O_2]\sec\theta dz - \int \sigma_{O_3}(\lambda)[O_3]\sec\theta dz\right] d\lambda \qquad (2.54)$$

The rate of energy deposition (r) as shown in Eq. (2.43) is

$$r = \int_\lambda I\sigma_{O_2}(\lambda)[O_2] + \sigma_{O_3}(\lambda)[O_3] d\lambda \qquad (2.55)$$

where r has units of J s^{-1} m^{-3}.

If all of this energy goes into heating, then the rate of energy deposition equals the rate of heating. By the First Law of Thermodynamics:

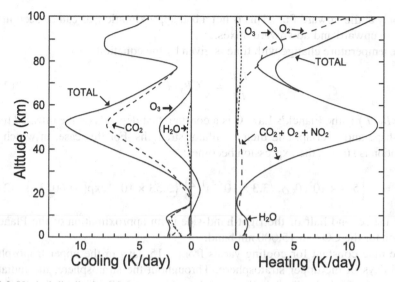

Fig. 2.9 Vertical distribution of shortwave heating rates and longwave cooling rates (Brassur and Solomon 2005, Courtesy: Springer)

$$\frac{dU}{dt} = \dot{Q} \tag{2.56}$$

which can be represented as

$$\rho C_p \frac{DT}{Dt} = r \tag{2.57}$$

$$\frac{DT}{Dt} = \frac{1}{\rho C_p} \int_\lambda I \left\{ \sigma_{O_2}(\lambda)[O_2] + \sigma_{O_3}(\lambda)[O_3] \right\} d\lambda \tag{2.58}$$

where ρ is the density, and C_p is the specific heat, or specific enthalpy (1,005 J kg^{-1} K^{-1}). The solar shortwave heating rates by O_3, O_2, NO_2, H_2O, CO_2 and the terrestrial longwave cooling rates by CO_2, O_3, and H_2O with height extending from the surface to 100 km are depicted in Fig. 2.9.

2.3.5 Infrared Heating and Cooling

Infrared heating plays a relatively small role in stratospheric heating. The main infrared heating is by absorption of the 2,700 nm and 4,300 nm bands of CO_2. The 9,600 nm band of O_3 provides some heating near the tropopause.

Infrared cooling is primarily due to the 15,000 nm band of CO_2. The second most important infrared cooling is in the 9,600 nm band of O_3, which is most important near the stratopause. We can use the cooling-to-space approximation to estimate the cooling rate. In this approximation, we ignore the downward flux of infrared and look only at the upward flux. This is not a bad approximation, as can be seen in the figure on upward and downward fluxes.

The temperature change with time is given by the equation:

$$\frac{DT}{Dt} = -CB_v(T) \tag{2.59}$$

where $B_v(T)$ is the Planck's Law. C is a constant that depends on the infrared lines, their shape and overlap, and their oscillator strengths. For this case, in which the absorption is strong, the expression becomes:

$$\frac{DT}{Dt} = -\left\{5.4 \times 10^5 (\theta_{CO_2}/3.3 \times 10^{-4})^{1/2}\right\}\left\{3.53 \times 10^{-4}\exp(-960/T)\right\} \tag{2.60}$$

where the second half of the right-hand-side is an approximation of the Planck's function for the CO_2, in 15,000 nm band.

The time constant for cooling varies from $\sim$15 days in the upper troposphere to 3–5 days in the upper stratosphere. Throughout the stratosphere, the radiative lifetimes are much shorter than the typical transport at midlatitudes.

2.3.6 Radiative Heating due to Absorption

In order to determine the heating rate of the layer between altitude z and $z + dz$, the energy balance at each boundary of the layer must be identified. As the thickness dz approaches zero, the energy absorbed per unit volume is given by the net flux divergence DF/dz. Therefore, the variation of the temperature in the layer per unit time is given by

$$\frac{DT}{Dt} = -\frac{1}{\rho C_p}\frac{DF}{Dz} \tag{2.61}$$

$$\frac{DT}{Dt} = -\frac{g}{C_p}\frac{DF}{Dp} \tag{2.62}$$

where C_p is the specific heat at constant pressure, ρ is the total air density, g is the acceleration due to gravity, and p is the pressure.

2.3.7 Vertical Profiles of Radiative Heating

The rate of change of temperature due to absorption or emission of radiation within the atmospheric layer is given by

$$\rho C_p\frac{DT}{Dt} = \frac{DF(z)}{Dz} \tag{2.63}$$

where $F = F\uparrow + F\downarrow$ is the net flux and ρ is the total density of the air. The rate of heating per unit wave number v is given as

$$\left[\frac{DT}{Dt}\right]_v = -\frac{1}{\rho C_p}\frac{DF(z)}{Dz} \tag{2.64}$$

$$= -\frac{1}{\rho C_p}\frac{D}{Dz}\left[\int_{4\pi}I_v\mu d\omega\right] \tag{2.65}$$

$$= -\frac{1}{\rho C_p}\int_{4\pi}\frac{D}{Dz}I_v d\omega \tag{2.66}$$

$$= -\frac{2\pi}{\rho C_p}\int_{-1}^{1}\frac{D}{Dz}I_v d\mu \tag{2.67}$$

where $\mu = \cos\theta$, ω is the solid angle, and $ds = dz/\mu = \sec\theta dz$. Substituting for dI_v/ds from Schwarszchild's equation, we get

$$\left[\frac{DT}{Dt}\right]_v = \frac{2\pi}{\rho C_p}\int_{-1}^{1}[k_v I_v - B_v]d\mu \tag{2.68}$$

This equation is generally used in estimating the infrared radiative heating rates.

The three most important radiatively active greenhouse gases in the atmosphere are carbon dioxide, water vapor, and ozone. In the troposphere, all the three constituents produce radiative cooling in the longwave part of the spectrum. Water vapor is the dominant contributor, but decreases with height. Upper troposphere experiences a net radiative cooling due to the presence of greenhouse gases.

On the contrary, the stratosphere is nearly in radiative equilibrium state. Radiative heating due to the absorption of solar radiation in the ultraviolet part by ozone molecules exactly balances the longwave cooling to space by carbon dioxide, water vapor, and ozone, in which carbon dioxide is the most significant contributor to the long wave cooling in the stratosphere.

2.4 Solar Radiation and Earth's Atmosphere

Solar radiation is the dominant direct energy source for the stratospheric ozone photochemistry, global atmospheric circulation, tropospheric weather systems, and affects all physical, chemical, and biological processes. The Sun provides a natural influence on the Earth's atmosphere and climate. In order to understand stratosphere–troposphere interaction processes, the changes in solar radiation on passing through the Earth's atmosphere must be first understood.

2.4.1 Absorption of Solar Radiation

The vast amount of energy continuously emitted by the Sun is dispersed into outer space in all directions. Only a small fraction of this energy is intercepted by the Earth and other solar planets.

Solar radiation occurs over a wide range of wavelengths. With a surface temperature of 5,780 K, the energy flux at the surface of the Sun is approximately 63×10^6 W m^{-2}. The production of the radiation by the Sun depends on the physical and chemical characteristics of the solar atmosphere. However, the energy of solar radiation is not divided evenly over all wavelength regions but rather sharply centered on the wavelength band of 200–2,000 nm.

Figure 2.10 shows solar radiation at the top of the atmosphere and the actual radiation at sea level which has been reduced due to absorption by atmospheric gases. In passing through outer space, which is characterized by vacuum, the different types of solar energy remain intact and are not modified until the radiation reaches the top of the Earth's atmosphere. In outer space, gamma ray, X-ray, ultraviolet, and infrared radiations are present. Table 2.3 illustrates attenuation of solar radiation at its various spectral regions due to the absorption of atmospheric constituents.

Electromagnetic radiation coming from the Sun travels at a speed of 3×10^8 m s^{-1}. This radiation comprises wavelengths that vary from the very short gamma rays to the very long microwaves. About 43% of the total radiant energy emitted from the Sun is in the visible parts of the spectrum. The bulk of the remainder lies in the near-infrared (49%) and ultraviolet (7%) sections. Less than 1% of solar radiation is emitted as X-rays, gamma waves, and radio waves.

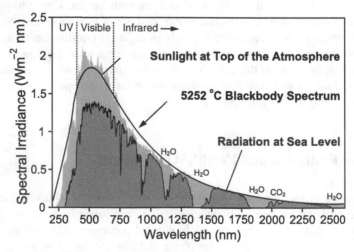

Fig. 2.10 Absorption of solar radiation by the Earth's atmosphere (Courtesy: Robert A. Rhode (b), Global warming Art)

Table 2.3 Solar radiation and its abosorption in the Earth's atmosphere

Band	Wavelength (nm)	Atmospheric effects
Gamma ray	<0.03	Completely absorbed by the upper atmosphere
X-ray	0.03–3	Completely absorbed by the upper atmosphere
Ultraviolet (UV)	3–300	
UV_c	200–280	Completely absorbed by oxygen, nitrogen, and ozone in the upper atmosphere
UV_b	280–320	Mainly absorbed by ozone in the lower stratosphere
UV_a	320–400	Transmitted through the atmosphere, but atmospheric scattering is severe
Visible	400–700	Transmitted through the atmosphere, with moderate scattering of the shorter waves
Infrared (IR)	700–14,000	
Reflected IR	700–3000	Mostly reflected radiation
Thermal IR	3,000–14,000	Absorption at specific wavelengths by carbon dioxide, ozone, and water vapor, with two major atmospheric windows

In addition to gamma rays and X-rays, which are absorbed high in the atmosphere, ultraviolet (UV) radiation in the atmosphere is divided into three spectra: UV_a, UV_b, and UV_c, as shown in Fig. 2.11.

UV_a falls right below visible light, with wavelengths that vary from 320 to 400 nm. Although it is not absorbed by ozone, UV_a is the least energetic and the least damaging of all UV radiation.

UV_b radiation, which ranges in wavelength from 280 to 320 nm, is more energetic than UV_a and is thought to be harmful to the biosphere. It exists in lesser amounts and is largely absorbed by ozone.

UV_c, at 200–280 nm, which is the most energetic and most damaging but least prevalent of the UV radiation types, is totally absorbed by ozone and normal diatomic oxygen high in the atmosphere.

Ozone is the most effective in absorbing radiation at the 250 nm wavelength. In fact, it is 100 times more efficient at 250 nm than it is at 350 nm. After ozone absorbs this shortwave radiation, it reradiates at generally longer wavelengths which initially

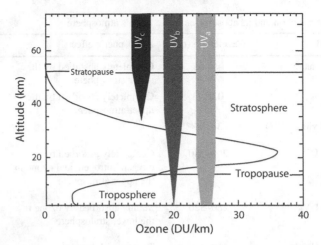

Fig. 2.11 Penetration of UV spectral bands in the troposphere and stratosphere (Source: WMO 2002)

goes in all directions. Some part of the radiation is reabsorbed by other atmospheric constituents, some make it to Earth's surface, and some return to space. The net effect, however, is an increase in temperature in the upper stratosphere.

Depletion in solar radiation on reaching Earth's surface is illustrated in Fig. 2.12. Many factors affect the amount of UV radiation that reaches Earth's surface. In addition to the amount of ozone in the stratosphere, the angle of the Sun, length of daylight hours, and path length of radiation through the atmosphere are all determined by latitude and time of year. Solar output, and the type and thickness of clouds are also important factors.

As the Sun's energy spreads through space its spectral characteristics do not change because space contains almost no interfering matter. However, the energy

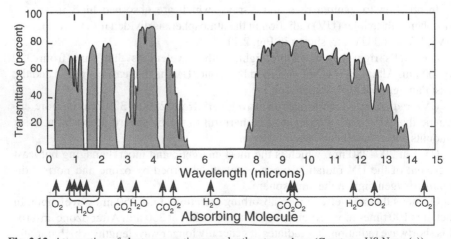

Fig. 2.12 Attenuation of electromagnetic waves by the atmosphere (Courtesy: US Navy (a))

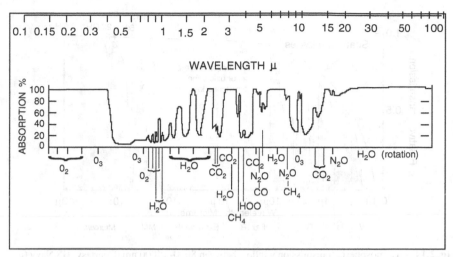

Fig. 2.13 Atmospheric absorption for terrestrial radiation and solar radiation at ground level (Adapted from Wallace and Hobbs 2006, Reproduced with permission: Elsevier)

flux drops monotonically as the same total radiated energy spreads over the surface. As the radiation reaches the outer limit of the Earth's atmosphere, the radiative flux is approximately 1,360 W m^{-2}.

Solar radiation plays two important roles in the atmosphere. One is the heating and cooling of the atmosphere. The infrared part of the solar spectrum can heat and cool the atmospheric layers, whereas the UV/visible part will heat certain layers of the atmosphere depending upon its absorption characteristics. Generally, heating is by solar UV absorption by O_2 in the mesosphere and thermosphere and by O_3 in the stratosphere. The cooling is generally by infrared absorption by CO_2 (15,000 nm), O_3 (9,600 nm), and H_2O (80,000 nm) in the lower and middle atmosphere. Atmospheric absorption of solar and terrestrial radiation for a zenith angle at 50° at ground level is depicted in Fig. 2.13.

Photochemical dissociation is the other significant effect of the solar radiation on the atmosphere, caused mainly due to the stratospheric ozone and ions in the ionospheric region. Even though only less than 1% of the solar radiation is in the UV, it is responsible for most of the photochemical processes, such as ionization, dissociation, etc., taking place in the Earth's atmosphere.

2.4.2 Atmospheric Window

The atmospheric window refers to the region of the electromagnetic spectrum that is relatively free from the effects of atmospheric attenuation. The atmospheric window lies approximately at wavelengths of infrared radiation between 8,000 and 14,000 nm (Fig. 2.14).

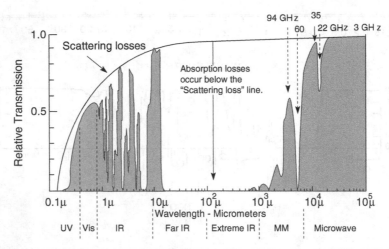

Fig. 2.14 An atmospheric transmission window between 8000–14000 nm (Courtesy: US Navy (b))

The absorptions of the principal natural greenhouse gases are concentrated in two belts. Gases such as carbon dioxide and methane, along with less abundant hydrocarbons, absorb at wavelengths longer than the atmospheric window region due to the presence of relatively long C–H and carbonyl bonds. The bonds of water vapor and ammonia absorb at wavelengths shorter than 8,000 nm. Except for the bonds in ozone, no bonds between carbon, hydrogen, oxygen, and nitrogen atoms absorb between these two ranges. This means that radiation in the atmospheric window is almost completely reflected from the Earth into space. Without this window, the Earth would become much too warm to support life, and possibly so warm that it would lose its water as Venus did early in solar system history. Thus, the existence of a window in the electromagnetic spectrum is critical to Earth remaining a habitable planet.

In recent decades, the existence of the atmospheric window has become threatened by the development of highly unreactive gases containing bonds between fluorine and either carbon or sulphur. The stretching frequencies of bonds between fluorine and other light nonmetals are such that strong absorption in the atmospheric window will always be characteristic of compounds containing such bonds. Bonds to other halogens also absorb in the atmospheric window, though much less strongly.

Moreover, the unreactive nature of such compounds that makes them so valuable for many industrial purposes means that they are not removable in the natural circulation of the Earth's atmosphere. It is estimated, for instance, that perfluorocarbons (CF_4, C_2F_6, C_3F_8) can stay in the atmosphere for over 50,000 years, which may be an underestimate given the absence of natural sources of these gases.

This means that such compounds have an enormous global warming potential. One kilogram of sulphur hexafluoride will, for example, cause as much warming as 23 tons of carbon dioxide over 100 years. Perfluorocarbons are similar in this respect, and even carbon tetrachloride (CCl_4) has a global warming potential of 1,800 compared to carbon dioxide.

2.4.3 Attenuation of Solar Radiation in the Stratosphere and Troposphere

The penetration of solar radiation into the Earth's lower atmosphere depends on the absorption of each constituent present in the stratosphere and troposphere.

The stratosphere controls the amount of solar radiation reaching the surface of the Earth, and the troposphere regulates the amount of radiation escaping from the Earth's surface into space. The absorption bands of O_3, O_2 (rich in the stratosphere), and O_2, H_2O, and CO_2 (abundant in the troposphere) are critical to maintaining the radiation balance of the Earth's atmosphere.

Figure 2.15 shows that O_3 absorbs ultraviolet in the stratosphere region. Molecular oxygen absorbs ultraviolet as well as some visible and infrared light. Water vapor (H_2O) in the troposphere absorbs highly in the range of 400–900 nm and again above 1,200 nm. The other important greenhouse gas, CO_2, has high absorption around 1,400 nm and above.

When molecules absorb energy, the absorbed energy may cause a chemical change or it may be re-emitted. Often molecules re-emit energy at wavelengths longer than that at which it was absorbed. Thus when molecules such as H_2O and CO_2 absorb visible or infrared light, they often re-emit it as longer wavelength infrared.

Table 2.4 illustrates significant absorption bands of solar radiation in the stratosphere and troposphere. Solar radiation below 100 nm is almost completely

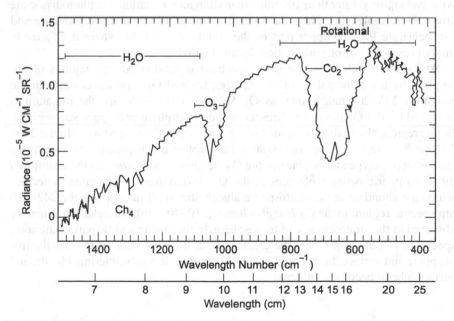

Fig. 2.15 Absorption of terrestrial radiation by greenhouse gases (Courtesy: Jet Propulsion Laboratory, NASA)

Table 2.4 Significant absorbtion bands of solar radiation in the stratosphere and troposphere (Brasseur and Solomon 2005)

Wavelength (nm)	Absorbtion band	Absorbtion characteristics
175–200	O_2 Schumann Runge bands	Absorption by mesospheric and upper stratospheric oxygen molecules; ozone effects important only in the stratosphere
200–242	O_2 Herzberg continuum O_3 Hartley band	Absorption by stratospheric oxygen molecules, absorption by stratospheric ozone
242–310	O_3 Hartley band	Absorption by stratospheric ozone, leading to $O(^1D)$
310–400	O_3 Huggins bands	Absorption by stratospheric and tropospheric ozone, leading to $O(^3P)$
400–850	O_3 Chappuis bands	Absorption by tropospheric ozone causes photodissociation even at the surface

absorbed above 100 km by molecular and atomic oxygen and molecular nitrogen. At wavelengths greater than 100 nm, solar ultraviolet radiation can photodissociate atmospheric molecules. The solar Lyman-α line at 121.6 nm is very intense and can penetrate into the upper part of the middle atmosphere, where it effectively dissociates water vapor, carbon dioxide, and methane.

At longer wavelengths, the solar spectrum is subdivided into regions of absorption by the principal absorbing species, O_2 and O_3. For wavelengths in the region of 175–200 nm, known as O_2 *Schumann Runge bands*, the radiation is absorbed by the O_2 molecules present in the mesosphere and upper stratosphere. In this region, absorption by ozone takes place only in the stratosphere. In the 200–242 nm band region, there are two strong absorption bands present. Absorption by stratospheric oxygen takes place in the O_2 *Herzberg continuum* and the absorption by stratospheric ozone molecules in the O_3 *Hartley band*. The ozone molecules which are abundant in the stratosphere absorb almost all radiation in the 242–310 nm spectral region. In the O_3 *Huggins bands* (310–400 nm), the radiation is mainly absorbed in the stratosphere and troposphere. In the visible and IR part of the solar spectrum, i.e., above 400 nm, the greater part of the solar photons reaches the troposphere and surface. In this region, the effects of molecular scattering, clouds, and surface albedo become significant.

2.5 The Atmosphere and Its Radiative Processes

Radiative processes constitute the ultimate source and sink of energy in the atmospheric system. Radiative processes have an enormous influence on the behavior of the atmosphere because they govern the amount of energy entering and leaving the Earth atmosphere system and hence the amount of energy available to heat the air, evaporate water to produce moisture, and cause the movement of air masses. This energy initially enters the atmosphere as shortwave radiation from the Sun, but it is transferred to the troposphere and stratosphere in different ways, giving these two layers of the atmosphere very different structures and characteristics (Fig. 2.16).

The contrasting temperature patterns in the troposphere and stratosphere are the result of differences in the way radiative energy is transferred to the atmosphere. The stratosphere is heated from the top downward as intense ultraviolet radiation is absorbed by oxygen and ozone. The troposphere is heated from the bottom upward as the Earth's surface warmed by incoming solar radiation emits longwave infrared radiation, which is then absorbed and re-emitted by greenhouse gases in the air above. A small amount of heat is also transferred directly to the air by direct contact with the surface and by the evaporation and condensation of moisture.

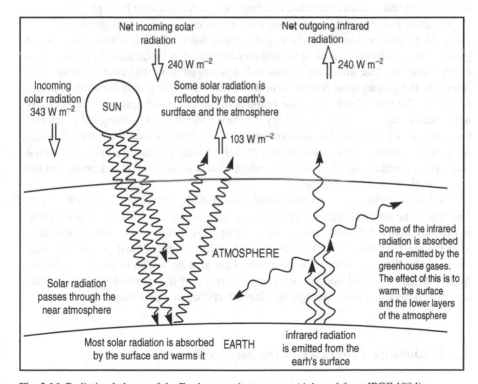

Fig. 2.16 Radiation balance of the Earth atmosphere system (Adapted from IPCC 1994)

2.5.1 Radiative Processes in the Troposphere

In the troposphere, very little of the incoming solar radiation is absorbed directly by the atmosphere. Instead, the shortwave radiation warms the Earth's surface, which then transfers heat energy to the atmosphere in a variety of ways – partly through direct contact between the surface and the air, partly through the evaporation and condensation of moisture, but mostly through the emission of longwave infrared radiation, which is absorbed by water vapor and other greenhouse gases in the air such as carbon dioxide, methane, nitrous oxide, and ozone. By re-emitting some of this longwave radiation back toward the Earth's surface, these gases retain heat at the bottom of the atmosphere and help to make it warmer. As a result of this greenhouse effect, the Earth's average temperature is some $33°$ C warmer than it would otherwise be and the planet is able to support life.

Because it is heated in this way, air in the troposphere is generally warmest at the surface and becomes cooler with increasing altitude. Since warm air is less dense than cool air, the warm air rises and cooler air moves in at the surface to take its place. This simple convective flow is modified by the Earth's rotation, by surface features, and by temperature differences between the equator and the poles. The end result is a rather turbulent layer of the atmosphere in which air circulates in complicated and variable patterns, moving energy and moisture from place to place.

Radiative processes in the troposphere are generally well known, particularly the clear-sky longwave transfer, including the absorption properties of most greenhouse gases. Their treatment in atmospheric general circulation models depends on several approximations. The fluxes are computed as averaged quantities over a few spectral intervals, the propagation is limited to the vertical upwelling or downwelling directions, and the role of subgrid scale features of the clouds or aerosols is essentially neglected. Although this methodology is believed to have only a marginal impact on the accuracy of computed longwave fluxes, the analysis of satellite, ground-based, or aircraft measurements (Ramanathan et al. 1995) has generated a concern that the radiative algorithms used in climate models could significantly underestimate the atmospheric shortwave absorption.

Modeling studies have demonstrated the sensitivity of the simulated climate to changes in the atmospheric absorption. As a radiative forcing, anomalous absorption is fundamentally different from water vapor or CO_2 in that it does not significantly alter the Earth's net radiation budget. Instead, it shifts some of the deposition of solar energy from the ground to the atmosphere (Li et al. 1997), with implications for the hydrological cycle and vertical temperature profile of the atmosphere. Anomalous absorption may not, however, appreciably affect climate sensitivity.

2.5.2 Radiative Processes in the Stratosphere

Stratosphere consists of only about 10–20% of the total atmospheric mass, but the changes in the stratospheric dynamics and chemistry are important because of their effect on the tropspheric weather systems. There is increasing evidence that stratospheric effects can have a detectable and perhaps significant influence on tropospheric climate (Randel and Wu 1999).

The stratosphere is heated from the top downward and is therefore warmer at the top than it is at the bottom. Consequently, the densest air in the stratosphere is at the bottom, there is little vertical mixing of the air, and the stratosphere is very stable. Heat is added to the stratosphere when strong UV_c radiation from the Sun is absorbed by oxygen molecules and causes them to split. One of the results of this process is the production of ozone and the formation of the ozone layer in the stratosphere. More warming occurs when the ozone molecules intercept and are destroyed by intense but slightly less powerful UV_b radiation. A beneficial byproduct of these processes is that most of the ultraviolet radiation that is harmful to plant and animal life on Earth is filtered out in the stratosphere and does not reach the Earth's surface. Some additional heating of the stratosphere also occurs because ozone absorbs infrared radiation emitted by the Earth's surface.

Solar radiative heating of the stratosphere is mainly from absorption of UV and visible radiation by ozone, along with contributions due to the near-infrared absorption by carbon dioxide and water vapor. Depletion of the direct and diffuse solar beams arises from scattering by molecules, aerosols, clouds, and surface.

The longwave process consists of absorption and emission of infrared radiation, principally by carbon dioxide, methane, nitrous oxide, ozone, water vapor, and halocarbons (CFCs, HFCs, HCFCs, etc.). The timescale for the radiative adjustment of stratospheric temperatures is less than about 50–100 days. For CO_2, part of the main 15,000 nm band is saturated over quite short vertical distances, so that some of the upwelling radiation reaching the lower stratosphere originates from the cold upper troposphere. When the CO_2 concentration is increased, the increase in absorbed radiation is quite small and increased emission leads to a cooling at all heights in the stratosphere. But for gases such as the CFCs, whose absorption bands are generally in the 8,000–13,000 nm atmospheric window, much of the upwelling radiation originates from the warm lower troposphere, and a warming of the lower stratosphere results, although there are exceptions. Methane and nitrous oxide are in between. In the upper stratosphere, increases in all well-mixed gases lead to a cooling as the increased emission becomes greater than the increased absorption. Equivalent CO_2 is the amount of CO_2 used in a model calculation that results in the same radiative forcing of the surface-troposphere system as a mixture of greenhouse gases (see, e.g., IPCC 1996) but does not work well for stratospheric temperature changes (WMO 1999).

An ozone loss leads to a reduction in the solar heating, while the major longwave radiative effects from the 9,600 and 14,000 nm bands produce a cooling tendency in the lower stratosphere and a positive radiative change above (Ramaswamy et al. 1996). Following volcanic eruptions, large amount of aerosols are loaded in the stratosphere, leading to an increase in heating due to the longwave absorption. For the solar beam, aerosols enhance the planetary albedo while the interactions in the near-infrared spectrum yield a heating which is about one third of the sum of the total solar heating and longwave heating. In addition, ozone losses can result from heterogeneous chemistry occurring on or within sulphate aerosols, and those changes produce a radiative cooling.

The decreases in ozone in the Antarctic ozone hole area are very likely to contribute a negative radiative forcing of the troposphere surface that offsets as much as one half of the positive radiative forcing attributable to the increases in CO_2 and other greenhouse gases. It appears that most of the observed decreases in upper-tropospheric and lower-stratospheric temperatures were due to ozone decreases rather than increased CO_2 (Ramaswamy et al. 1996; Bengtson et al. 1999).

The subject of solar effects on climate and weather has a new beginning, thanks to the observational studies (Labitzke and van Loon 1997), and modeling studies that suggest viable mechanisms involving the stratosphere. As solar irradiance changes, proportionally much greater changes are found in the ultraviolet, which leads to photochemically induced ozone changes. The altered UV radiation changes the stratospheric heating rates per amount of ozone present. The stratosphere thus provides a potential response to the tropospheric weather systems (Shindell et al. 1999).

2.6 Stratospheric Cooling

Cooling of the stratosphere is favored by the ozone reduction. But the main cause of stratospheric cooling is the release of carbon dioxide in the troposphere. Therefore, global warming (i.e., tropospheric warming) and stratospheric cooling are parallel effects. Further cooling of the stratosphere may have an impact on the future development of the ozone layer, because a cold stratosphere is necessary for ozone depletion. We may therefore keep in mind that releasing more CO_2 may favor the ozone hole formation. To understand all the factors contributing to the stratospheric cooling, however, is rather complicated.

2.6.1 Causes of Stratospheric Cooling

There are several causes for the stratosphere cooling. The two better known reasons are: (i) the depletion of stratospheric ozone and (ii) the increase in atmospheric carbon dioxide.

2.6.1.1 Cooling Due to Ozone Depletion

In the case of less ozone in the stratosphere, absorption of solar UV radiation is reduced. As a result, lesser amount of solar energy is transformed into heat in the stratosphere. The stratospheric heating is thereby reduced because of the reduced absorption of solar UV light. Ozone in the lower stratosphere also acts as a greenhouse gas and absorbs infrared radiation. At about 20 km altitude the effects of UV light and infrared radiation effect are nearly equal.

The lower stratosphere seems to be cooling by about 0.5° C per decade. This general trend is interrupted by heavy volcanic eruptions (Kodera 1994), which lead to a temporary warming of the stratosphere for 1–2 years. Afterward, temperatures again begin the increasing trend. Calculations by several research groups (Kiehl et al. 1995; Kiehl and Trenberth 1997; Chanin and Ramaswamy 1999) generally find a larger cooling trend for the recent two decades (1979–2000), compared to the previous period of 1958–1978.

2.6.1.2 Cooling Due to the Greenhouse Effect

Greenhouse gases (CO_2, O_3, CFC) generally absorb and emit the infrared heat radiation at a certain wavelength. If this absorption is very strong as in the 15,000 nm absorption band of CO_2, the greenhouse gas can block most of the outgoing infrared radiation already close to the Earth's surface. Nearly no radiation from the surface can, therefore, reach the CO_2 residing in the upper troposphere or lower stratosphere. On the other hand, CO_2 emits heat radiation to space. In the stratosphere this emission becomes larger than the energy received from below by absorption. In total, CO_2 in the lower stratosphere and upper troposphere loses energy to space. It cools these regions of the atmosphere. Other greenhouse gases, such as ozone and chlorofluorocarbons (CFCs), have a weaker impact, because their absorption in the troposphere is smaller. They do not entirely block the radiation from the ground in their wavelength regimes and can still absorb energy in the stratosphere and heat this region of the atmosphere (Ramaswamy et al. 2001).

Figure 2.17 shows the time series of global mean temperature anomalies in the lower stratosphere (16–21 km), tropopause layer (9.5–16 km), mid-troposphere (1.5–9.5 km), and surface. The years of major volcanic eruptions are indicated at the bottom of the figure. The figure shows that the stratosphere has cooled significantly since about 1980, mostly as a result of ozone loss but also because of the accumulation of greenhouse gases in the troposphere. Prior to 1980, the tropopause shows a warming trend, which turns into a cooling trend in the later years. In both the surface and mid-troposphere levels, cooling was observed between 1962 and 1976, except for the year 1973. Afterward, a warming trend is generally observed in these layers.

2.6.2 Stratospheric Cooling Rates

Figure 2.18 shows how water vapor, carbon dioxide, and ozone contribute to long-wave cooling in the stratosphere. Especially for CO_2 it is obvious, that there is no cooling in the troposphere, but a strong cooling effect in the stratosphere. Ozone, on the other hand, cools the upper stratosphere, but warms the lower stratosphere (Ramaswamy et al. 2006).

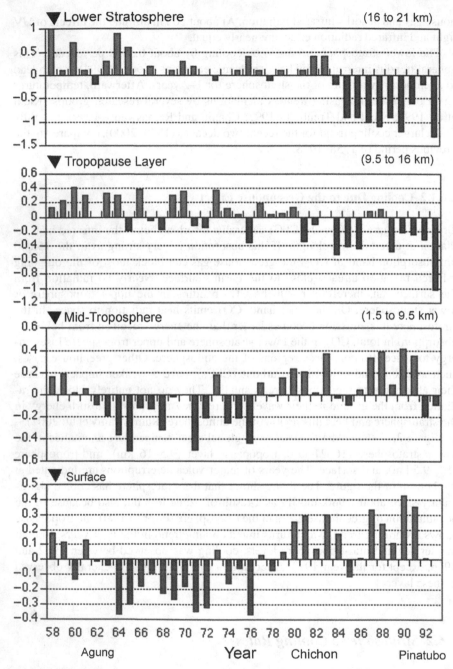

Fig. 2.17 Global mean temperature anomalies in the lower stratosphere (16–21 km), tropopause (9.5–16 km), mid-troposphere (1.5–9.5 km) and surface layers (Adapted from National Weather Service, NOAA)

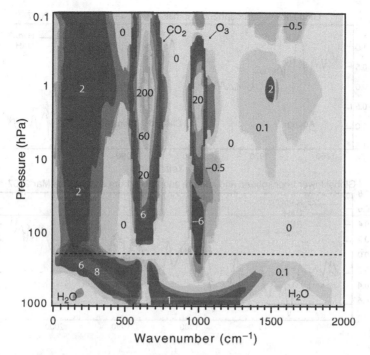

Fig. 2.18 Effect of CO_2 and O_3 on stratospheric cooling at different layers (The values are in units of 10^{-3} K (day cm^{-1}) (Adapted from P. Haynes and T. Shepherd, Courtesy: SPARC)

The impact of ozone decrease is more relevant in the lower stratosphere in the region around 20 km of altitude. The impact of cooling due to increase of carbon dioxide is high in the upper stratosphere between 40 and 50 km (see Fig. 2.18). Due to these different effects, cooling is not homogeneous over the whole stratosphere.

2.6.3 Other Influences

Surface warming due to the greenhouse effect might also change the heating of the Arctic stratosphere by changing planetary waves and/or their propagation. These waves are triggered by the surface structure of the northern hemisphere (mountain ranges like the Himalayas, alternation of land and sea). Recent studies also show that an increase in the stratospheric water vapor concentration could have a strong cooling effect, comparable in its extent to the effect of ozone loss.

Figure 2.19 illustrates the time series of global mean temperatures in the lower troposphere and lower stratosphere from 1958 to 2007. The increase in tropospheric temperature is noted in the mid-1960s and is continuing. The mean lower troposphere has increased nearly about 0.8° C during the last 40 years. In the lower stratosphere, the decrease in temperature trend exists for the last five decades.

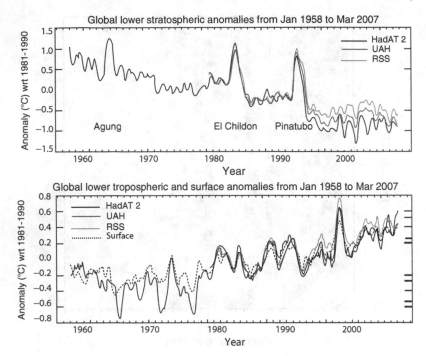

Fig. 2.19 Time series of global mean temperature anomalies in the lower troposphere and lower stratosphere for the last 50 years (WMO 2007)

During the last 50 years the global mean lower stratospheric temperature has decreased by about 2° C.

Nevertheless, cooling is observed and expected due to increasing carbon dioxide and decreasing ozone. Further stratospheric cooling would make the formation of an Arctic ozone hole more likely. We have to keep in mind that CO_2 emissions do not only lead to tropospheric warming but also to stratospheric cooling.

Another important feature of the stratosphere is the cold pool of air that forms at high latitudes during the winter. This cold air is centered in the lower stratosphere at about 25 km. During the southern hemisphere winter air can reach temperatures colder than $-90°C$ near the south pole. In the northern hemisphere, the lowest temperatures reach about $-65°C$.

As a result, a zone of strong westerly winds (or vortex) forms and surrounds each pole. Because the temperature contrast is greatest in the vicinity of the south pole, the vortex that forms there during the southern hemisphere winter is considerably stronger than the vortex that forms during the northern hemisphere winter.

Stratospheric cooling and tropospheric warming are intimately connected, not only through radiative processes, but also through dynamical processes, such as the formation, propagation, and absorption of planetary waves. At present not all causes of the observed stratospheric cooling are completely understood.

2.7 Effects of Solar Activity on Stratosphere and Troposphere

The subject of solar effects on tropospheric weather and climate has enjoyed a resurgence in the late 1990s, partly because of observational studies (Labitzke and van Loon 1997), and also due to modeling studies that suggest viable mechanisms involving the stratosphere. As solar irradiance changes, proportionally much greater changes are found in the ultraviolet which lead to photochemically induced ozone changes, and the altered UV radiation changes the stratospheric heating rates per amount of ozone present (Haigh 1996; Shindell et al. 1999). Including the altered ozone concentrations gave an enhanced tropospheric response provided the stratosphere was adequately resolved.

Solar activity in the 11-year cycle is considered to be one of the driving forces behind periodic changes in temperatures and pressure heights of the Earth's lower stratosphere from pole to pole. The Sun's output has varied about 0.1% over one solar cycle during the past several decades. Over centuries, however, larger variations may occur. For example, an extended quiet period on the Sun may have chilled the Earth during the Little Ice Age between the mid-1550s and mid-1800s. During the long severe winters and short wet summers of that period Alpine glaciers advanced down river canyons, Dutch canals froze over, and farming became difficult farther north.

Influence of the 11-year cycle on middle atmospheric temperature showed a negative response in the stratosphere and positive association in the mesosphere with the changes in solar activity (Mohanakumar 1995), as illustrated in Fig. 2.20. The temperature decreases by 2–3% from its mean value in the stratosphere and increases by 46% in the mesosphere for an increase in 100 units of solar radio flux. Atmospheric pressure is found to be more sensitive to solar changes. An average solar maximum condition enhances the pressure in the stratosphere by 5% and in the upper mesosphere by 16–18% compared to the respective mean values. Increase in the solar radio flux tends to strengthen winter westerlies in the upper stratosphere over the midlatitude and summer easterlies in the middle stratosphere over tropics. Larger variability in the zonal wind is noted near stratopause height.

Ten-to-twelve-year oscillation in the stratosphere of the northern hemisphere corresponded to four 11-year solar cycles, beginning in 1958 (Labitzke 2002). A strong correlation exists between the solar cycle and the 10–12-year oscillation of the lower stratosphere's mean temperatures and constant pressure heights above sea level.

The annual mean temperatures of the lower stratosphere are well correlated with the solar cycle in the summer of either hemisphere, but only weakly correlated in winter. That is, during the summer months in either hemisphere, the average temperatures of the lower stratosphere rise and fall with the waxing and waning of the Sun's energy output over its 11-year cycle.

Effect of the 11-year solar cycle on the activity of the October/November Southern Annular Mode (SAM) was examined by the analysis of the ERA40 reanalysis data from 1968 to 2001. It was found that year-to-year variability of October–November mean SAM signal significantly differed according to the solar activity. In the high solar years, signal of the SAM extends up to the upper

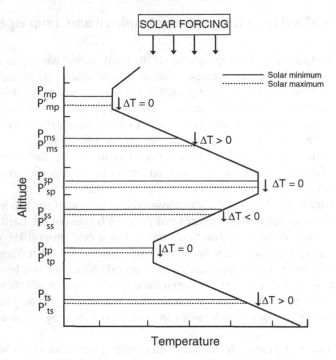

Fig. 2.20 Illustration of the influence of the solar forcing on the thermal structure of the Earth's neutral atmosphere (P – pressure during solar maximum; P' – pressure during solar minimum; ΔT – change in temperature from solar minimum to maximum; mp – mesopause, ms – mesosphere; sp – stratopause; ss – stratosphere; tp – tropopause; and ts – troposphere) (Mohanakumar 1995, Courtesy: Springer)

stratosphere during October–December and the activity below lower the stratosphere lasts until the following early winter, whereas in the low solar years, signal of the SAM is confined only inside the troposphere from October to December and it breaks off in January.

A possible mechanism of the solar influence on climate through dynamical processes is suggested by Kodera (2004). Variation of the solar ultraviolet (UV) heating rate resulting from changes in the solar irradiance and ozone concentration produces a small anomalous zonal wind in the subtropical stratopause region. This initial solar effect in the stratopause region is amplified through interaction with planetary waves propagating from the winter troposphere. There are two possible mechanisms for the downward penetration of the solar effect. One is that the zonal wind anomalies produced in the subtropical stratopause region shifts poleward and downward in the westerly jet in the winter hemisphere similar to the polar night jet oscillation. When zonal wind anomaly reaches the lower stratosphere, it induces changes in meridional propagation of tropospheric waves. Tropospheric wave change creates a circulation change similar to the Arctic or Antarctic oscillation. Another possibility is that the modification of the wave mean–flow interaction in

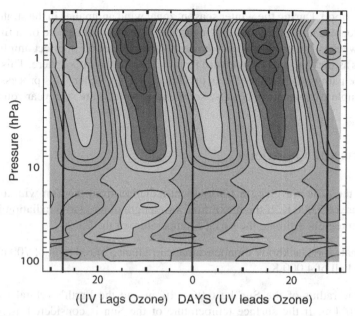

Fig. 2.21 Response of lower stratospheric ozone to solar variability (Adapted from Hirooka and Kuroda 2004, Courtesy: American Geophysical Union)

the upper stratosphere induces also change in stratospheric meridional circulation. The stratospheric circulation further modulates the Hadley circulation in the equatorial troposphere. This process may be responsible for the solar influence on the summer monsoon and the El Niño/Southern Oscillation (ENSO).

Figure 2.21 shows a correlation study between solar flux and ozone concentrations in the middle atmosphere, based on satellite observations by Hirooka and Kuroda (2004). They reported that solar ultraviolet (UV) flux changes by 6–8% near 200 nm between extremes in the 11-year solar cycle. However, the total insolation change is negligibly small. Such UV flux changes would bring about ozone changes in the middle atmosphere, which might directly influence tidal wave activity there. On the other hand, the ozone changes are considered to bring about changes of temperature and wind fields in the middle atmosphere. Hence, meridional planetary wave propagation could be influenced according to the change of resultant refractive properties of the atmosphere. In addition, some investigators have suggested the existence of an 11-year component in the Hadley circulation in the troposphere, which might influence planetary wave activity through convective activity changes. Evidences supporting this speculation have been obtained, but whether some of these relations to the 11-year solar cycle are true or accidental has not been verified.

Baldwin and Dunkerton (2005) reported that the atmosphere is approximately twice as sensitive to solar variability as would be expected from radiative calculations. This suggests that processes amplify an atmospheric solar signal, which is seen in the stratosphere, to affect surface climate. However, the mechanism is not well understood. During the winter season, zonal wind anomalies in the stratosphere tend to progress northward and downward through the stratosphere on a timescale of a few weeks. The process of wave mean–flow interaction in effect amplifies the anomalies as they progress downward from ~ 1 hPa to the troposphere. This creates a plausible pathway for solar effects on climate during winter. This process cannot be the complete explanation because solar climate effects are seen year-round.

Problems and Questions

2.1. Find the peak wavelength for incident radiation from the Sun with a surface temperature of 6,000 K. If the maximum wavelength of terrestrial radiation is in the 11,400 nm region, estimate the surface temperature of the Earth.

2.2. Estimate the blackbody monochromatic irradiance of red light of 700 nm from a hot iron ball of 4,000 K.

2.3. Let the radius of the Sun be 6.96×10^6 m and the Earth's orbital radius be 1.495×10^8 km. If the surface temperature of the Sun is considered as 5,780 K, estimate the value of solar constant at the top of the Earth's atmosphere.

2.4. In a dry atmosphere, the temperature of a parcel of air is 8°C at 700 m. Find the potential temperature of the air parcel.

2.5. The axis of the Earth is tilted at an angle of $23°27'$ from the vertical. Discuss the seasonal variations in temperature and zonal wind, if the axis would be tilted 15° and also the case of 30° from the vertical.

2.6. Imagine the case that the atmosphere allows all terrestrial radiation to pass through, without absorption, reflection, or scattering. In such a case, explain the diurnal variation of surface temperature. Discuss the variation of surface radiation budget.

2.7. Consider a parallel beam of radiation is passing an atmospheric layer of thickness 200 m containing an absorbing gas of density 0.2 kg m^{-3}. Let the incident beam pass through the layer at an angle of 30° from the normal. Calculate the optical thickness and absorptivity of the layer at a given wavelength region of absorption coefficient of 0.1 m^2 kg^{-1}.

2.8. Estimate the transmissivity of a layer of 500 m thickness containing gas of average density 0.1 kg m^{-3} in which a parallel beam of monochromatic radiation passes at an angle of 60° relative to the normal layer. Assume the absorption coefficient at the wavelength region is 0.05 m^2 kg^{-1}.

2.9. Estimate how much sea level would rise if the entire Antarctic ice cap melts, given that the approximate Antarctic ice cap thickness is 3,000 m, area of Antarctica is 14×10^6 km^2, and surface area of world ocean is 361×10^6 km^2.

2.10. Discuss the three mechanisms by which heat energy can be transferred. Mention which of the mechanisms play a significant role in maintaining Earth's global energy budget.

2.11. What changes would you expect in the amount of absorbed visible radiation, if all the permanent ice in the Arctic Ocean is removed (given: Albedo of the open ocean is 0.20 and that of ice is 0.35).

2.12. Cameras in weather satellites can take pictures using infrared radiation. What is the source of the infrared radiation? Explain why high cold clouds in infrared satellite photos look bright, and the warm surface of the Earth looks dark.

2.13. Estimate the solid angle subtended by the Earth as seen by a geosynchronous satellite. For simplicity, assume that the satellite sees the entire disc of the Earth (hints: the radius of the Earth is 6,370 km, and the radius of a geosynchronous orbit is 42,000 km).

2.14. Consider two places at the same latitude and elevation above sea level. Both places have clear skies and the same amount of water vapor in the air, but one place has snow on the ground and the other place has a grassy field. Which place is more likely to have a colder daytime high temperature? Explain.

2.15. On cloudy afternoon days in summer it feels very hot during the night hours. Also when the air is dry, the Earth's surface can become very cold during the night. Why?

2.16. In some of the cars, people put dark plastic film on the windows to keep their cars cooler in summer. Based on radiation principles, explain how a car with dark windows could be cooler than a car with clear windows.

2.17. In the afternoon of some summer days many cumulus clouds develop because the Sun heats the Earth's surface. After the Sun sets, the source of surface heating is removed, but sometimes the clouds destabilize the atmosphere and severe thunderstorms occur even after sunset. How?

2.18. Satellites have cooling systems to regulate their temperatures. One method of temperature control is to have painted plate hinged to the side of a satellite. When the plate is flat against the side of the satellite, the temperature is maximum. When one end of the plate is swung out from the side like an awning, the satellite is cooler. Explain how this works, even if the color of the plate may be the same as the color of the satellite and even though no air circulates through a satellite in space.

2.19. Diffuse gases like those in the Earth's atmosphere absorb and emit radiation at a series of fairly precise frequencies. Why?

References

Ahrens CD (2006) Meteorology Today, An Introduction to Weather, Climate and Environment, Eighth edition, Thompson Brooks/Cole (USA)

Andrews DJ, Holton JR, Leovy CB (1987) Middle Atmosphere Dynamics, Academic, New York

Baldwin MP, Dunkerton TJ (1999): Propagation of the Arctic Oscillation from the stratosphere to the troposphere, J Geophys Res, 104: 30937–30946

Baldwin MP, Dunkerton TJ (2005) The solar cycle and stratosphere-troposphere dynamical coupling, J Atmos Solar Terr Phys, 67: 71–82

Bengtson L, Roeckner E, Stendel M (1999) Why is the global warming proceeding much slower than expected? J Geophys Res, 104: 3865–3876

Brasseur G, Solomon S (2005): Aeronomy of the Middle Atmosphere, 3rd edition, Springer, The Netherlands

Chandrasekhar S (1950). Radiative Transfer. Dover, New York

Chanin M-L, Ramaswamy V (1999) Trends in stratospheric temperatures, Scientific Assessment of Ozone Depletion: 1998, WMO Report No 44, Geneva

Goody RM (1964) Atmospheric Radiation I: Theoretical Basis, Oxford University Press (Clarendon), London and New York

Haigh JD (1996) The impact of solar variability on climate, Science, 272: 981–983

Haynes P, Shepherd T (2001) Report on the SPARC Tropopause Workshop, Bad Tolz, Germany, April 17–21, 2001, SPARC Newsletter (http://www.aero.jussieu.fr/~sparc/News17/ReportTropopWorkshopApril2001/Shepherd_figure1.jpg)

Hirooka T, Kuroda Y (2004) Plausible solar influences on wave activities in the middle atmosphere, Fall Meeting, American Geophysical Union

IPCC (1996) Climate Change 1995. The IPCC Second Scientific Assessment Report

Jet Propulsion Laboratory, NASA (http://origins.jpl.nasa.gov/library/exnps/f4-3.gif)

Kiehl JT, Trenberth K (1997) Earth's annual global mean energy budget. Bull Am Met Soc, 78: 197–208

Kiehl JT, Hack JJ, Zhang MH, Cess RD (1995) Sensitivity of a GCM climate to enhanced shortwave absorption, J Atmos Sci, 8: 2200–2212

Kodera K (1994) Influence of volcanic eruptions on the troposphere through stratospheric dynamical processes in the Northern Hemisphere winter, J Geophys Res, 99: 1273–1282

Kodera K (2004) Solar influence on the Indian Ocean monsoon through dynamical processes, Geophys Res Letts, 31: L24209, doi:10.1029/ 2004GL020928

Kondratyev KYa (1972) Radiation Processes in the Atmosphere, WMO Publication No 39, Geneva

Labitzke K (2002) The solar signal of the 11-year sunspot cycle in the stratosphere differences between the northern and southern summers, J Met Soc Japan, 80: 963–971

Labitzke K, van Loon H (1997) The signal of the 11-year sunspot cycle in the upper troposphere-lower stratosphere, Space Sci Rev, 80: 393–410

Li Z, Moreau L, Arking A (1997) On solar energy disposition: A perspective from observation and modeling, Bull Am Met Soc, 78: 53–70

Liou KN (1980) An Introduction to Atmospheric Radiation, Academic, New York

Mohanakumar K (1995) Solar activity forcing of the middle atmosphere, Ann. Geophys, 13: 879–885

Paltridge GW, Platt CMR (1976) Radiative Processes in Meteorology and Climatology, Elsevier Scientific Pub Co, Amsterdam, The Netherlands

Ramanathan V, Subasilar B, Zhang G, Conant W, Cess RD, Kiehl JT, Grassl H, Shi L (1995) Warm Pool heat budget and shortwave cloud forcing – A missing physics, Science, 267: 499–503

Ramaswamy V, Schwarzkopf MD, Randel WJ (1996) Fingerprint of ozone depletion in the spatial and temporal pattern of recent lower-stratospheric cooling, Nature, 382: 616–618

Randel WJ, Wu F (1999) Cooling of the Arctic and Antarctic polar stratosphere due to ozone depletion, J Clim, 12: 1467–1479

Rhode RA (a), Greenhouse Effect, Global warming art (http://www.globalwarmingart.com/wiki/Image:Greenhouse_Effect.png)

Rhode RA (b) Solar Radiation Spectrum, Global warming Art (http://globalwarmingart. com/wiki/Image:Solar Spectrum.png)

Shindell DT, Miller RL, Schmidt GA, Pandolfo L (1999) Simulation of recent northern winter climate trends by greenhouse-gas forcing, Nature, 399: 452–455

Smith WL (1985) Satellites. In Handbook of Applied Meteorology, edited by D. D. Houghton, Wiley and Sons, New York

University of Colorado, The Electromagnetic Spectrum, Chart by LASP, Boulder (http://lasp.colorado.edu/cassini/images/Electromagnetic%20Spectrum_noUVIS.jpg)

US Navy (a) Atmospheric absorption, U.S. Federal Government (http://ewhdbks.mugu.navy.mil/ atmospheric_absorption.png)

US Navy (b) Atmospheric transmittance, U.S. Federal Government (http://ewhdbks.mugu. navy.mil/transmit.gif)

Wallace JM, Hobbs PV (2006) Atmospheric Science – An Introductory Survey. Second Edition, Elsevier, New York

Wikipedia, Planck's law of black body radiation (http://en.wikipedia.org/wiki/Planck's_ law_of_black-body_radiation)

WMO (1999) Scientific assessment of ozone depletion : 1998, Global Ozone Research and Monitoring Project, WMO Report No. 44, Geneva

WMO (2007) Scientific Assessment of Ozone Depletion : 2006, Global Ozone Research and Monitoring Project, WMO Report No. 50, Geneva

Chapter 3
Dynamics of the Troposphere and Stratosphere

3.1 Introduction

Dynamics of the troposphere and stratosphere are fundamentally inseparable. However, the mechanism involved in the generation and maintenance of the circulation in the troposphere and stratosphere are different. Large-scale circulation in the troposphere is primarily driven by the differential absorption of solar energy at the surface, whereas in the stratosphere, eddies are as fundamental to the circulation as the differential solar heating.

Waves and eddies play a major role in maintaining the global atmospheric circulation. The waves generated in the troposphere propagate into the stratosphere with absorption, so that stratospheric changes depend upon where and how they are absorbed, and in turn these changes can affect the troposphere. Without eddies the zonal mean temperature would match the radiative equilibrium temperature with a ∼10–20-day lag, and would vary annually. The downward propagation of zonal mean anomalies provides a purely dynamical stratosphere–troposphere link. Also, the flow would be only the zonal mean flow, which is determined by the meridional temperature gradient and the thermal wind balance; and there would be no stratosphere–troposphere exchange (Holton et al. 1995). Heating and cooling patterns result from eddies driving the stratosphere away from radiative equilibrium.

In this chapter, the basics of atmospheric dynamics and the important quantities necessary for understanding the stratosphere–troposphere interactions are explained. Detailed description and derivation of equations in atmospheric dynamics can be obtained from Heiss (1959), Andrews et al. (1987), Pedlosky (1987), Holton (2004), Martin (2006), etc.

3.2 Basic Quantities of Atmospheric Dynamics

Earth's atmosphere can be considered as a continuous fluid medium or continuum. Various physical quantities, such as pressure, temperature, and density, characterize the state of the atmosphere and are assumed to have unique values at each point in the atmosphere as continuous functions of space and time.

3.2.1 Equation of State

Equation of state, also known as the ideal gas equation, is defined as an equation relating temperature, pressure, and specific volume of a system in thermodynamic equilibrium. These field variables are related to each other by the equation of state for an ideal gas. The equation of state for dry air (ideal gas law) is given by

$$p\alpha = RT \tag{3.1}$$

or

$$p = \rho RT \tag{3.2}$$

where p is the pressure; α is the specific volume; R is the specific gas constant ($R = 287 \text{ J kg}^{-1}\text{K}^{-1}$); ρ is the density; and T is the absolute temperature.

3.2.2 Hydrostatic Equation

In the absence of atmospheric motion, the force of gravity must be exactly balanced by the vertical pressure gradient force. This state of the atmosphere is known as under *hydrostatic equilibrium*. The hydrostatic equation is the formal expression of pure hydrostatic equilibrium, in which the vertical pressure gradient force is balanced by the gravity force.

Consider an air mass lying between heights z and $z + dz$ as shown in Fig. 3.1. The vertical pressure gradient acting on the air mass is $\partial p / \partial z$. The weight of the air mass per unit area is ρg. Under hydrostatic equilibrium, the vertical pressure gradient force (also known as the buoyancy force) is balanced by the body force (gravity force). The hydrostatic equation can be written as

$$\frac{\partial p}{\partial z} = -\rho g \tag{3.3}$$

where p is the pressure, ρ the density, g the acceleration due to gravity, and z the geometric height. For cyclonic-scale motions the error committed in applying the hydrostatic equation to the atmosphere is less than 0.01%. Strong vertical accelerations in thunderstorms and mountain waves may be 1% of gravity or more in extreme situations.

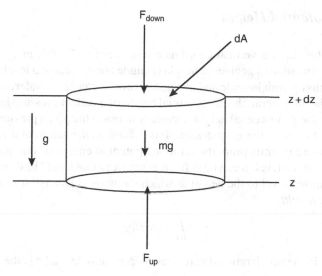

Fig. 3.1 Balance between upward buoyancy force and downward gravity force maintaining the hydrostatic equilibrium

In isobaric coordinate systems, hydrostatic equation can be written as

$$\frac{\partial \Phi}{\partial z} = \frac{RT}{g} \tag{3.4}$$

where $RT/g = H$ is known as the *scale height*.

Combining the equation of state, the hydrostatic equation can be represented as

$$\frac{\partial \Phi}{\partial z} = -\alpha = -\frac{RT}{p} \tag{3.5}$$

where Φ is the geopotential and α the specific volume.

From the hydrostatic equation, it is apparent that there exists a single-valued monotonic relationship between pressure and height in the atmosphere. Therefore, we can use pressure as the independent vertical coordinate and geopotential height as a dependent variable, $Z(x, y, p)$. The vertical coordinate can be written as

$$Z = -H\left(\frac{p}{p_s}\right) \tag{3.6}$$

where p_s is the standard reference pressure (1,000 hPa) and H is the scale height with T_s as the global average temperature.

For an isothermal atmosphere at temperature T_s, the coordinate z is equal to geometric height, and the density profile is given by the reference density

$$\rho_0(z) = \rho_s \exp(-z/H) \tag{3.7}$$

3.2.3 Geopotential Height

Geopotential height is a vertical coordinate referenced to Earth's mean sea level, which is a modification to geometric height (altitude above mean sea level) using the variation of gravity with latitude and elevation. Thus, it can be considered a gravity-adjusted height. In general, the geopotential height of a certain pressure level would correspond to the geopotential height necessary to reach the given pressure.

At an elevation of h, the *geopotential* Φ is defined as the height of a given point in the atmosphere in units proportional to the potential energy of unit mass at this height relative to sea level. It can be defined as the work that would be done in lifting unit mass from sea level to the height at which the mass is located. It is also called a the *dynamic height*.

$$\Phi = \int_0^h g(\phi, z) \mathrm{d}z \tag{3.8}$$

where $g(\phi, z)$ is the acceleration due to gravity, ϕ is latitude, and z is the geometric elevation.

Thus, it is the gravitational potential energy per unit mass at that level. The *geopotential height* is

$$Z_g = \frac{\Phi}{g_0} \tag{3.9}$$

where g_0 is the standard gravity at mean sea level, which is used instead of height for most forms of meteorological upper air data.

Atmospheric scientists often use geopotential height rather than geometric height because doing so in many cases makes analytical calculations more convenient. For example, the primitive equations which weather forecast models solve are more easily expressed in terms of geopotential than geometric height. Using the geopotential height will eliminate centrifugal force and air density, which is very difficult to incorporate in the equations.

At higher levels from the surface of the Earth, the geopotential height underestimates the geometric height; but this difference is generally small (a few meters or less) in the lower atmosphere, so that the difference is often ignored. The mention of *height* on a meteorological upper air sounding or chart more likely means geopotential height than actual height. This assumption is not applicable above the tropopause.

3.2.4 Hypsometric Equation

The thickness of a layer between two constant pressure surfaces which is proportional to the temperature is known as *hypsometric* equation. The expression relating the thickness, between two isobaric surfaces to the mean temperature of the layer is represented as

$$h = z_1 - z_2 = \frac{R\overline{T}}{g} \ln\left(\frac{p_1}{p_2}\right) \tag{3.10}$$

where z_1 and z_2 are geometric heights at pressure levels p_1 and p_2, respectively; R is the gas constant for dry air; $\overline{T}$ is the mean temperature of the layer; and g is gravity.

Thus the thickness of a layer between two pressure levels is proportional to the mean temperature of the layer. Pressure decreases more rapidly with height in a cold layer than a warm layer. The pressure decreases exponentially with geopotential height by a factor of e^{-1} per scale height.

$$p(Z) = p(0) \exp\left(\frac{-Z}{H}\right) \tag{3.11}$$

3.3 Fundamental Laws of Conservation

Fundamental laws of fluid dynamics and thermodynamics are applicable to the atmospheric motions, in terms of field variables (pressure, density, and temperature) as dependent variables and space and time as independent variables. Basically atmospheric motions are governed by three fundamental physical principles: laws of conservation of momentum (Newton's second law of motion), mass (continuity), and energy (first law of thermodynamics).

3.3.1 Equations of Motion (Conservation of Momentum)

Newton's second law of motion states that the rate of change of momentum (for unit mass, it is acceleration) of an object relative to coordinates fixed in space (inertial frame of reference) equals the sum of all forces acting on it.

$$\frac{D(m\vec{V})}{Dt} = \sum \vec{F} \tag{3.12}$$

The fundamental forces in atmospheric motions are the pressure gradient force , gravitational force, and frictional force.

$$\frac{D(m\vec{V})}{Dt} = -\frac{1}{\rho}\vec{\nabla}p + \vec{g}^* + \vec{F}_r \tag{3.13}$$

where $\vec{F}_r$ is the frictional force per unit mass. Since the motion is referred to as the rotating coordinate system of the Earth, the apparent forces, the centrifugal force, and the Coriolis force arising due to the rotation of the Earth should also be included in the equation to validate Newton's law of motion applicable to the acceleration of the terrestrial coordinate system.

$$\frac{D\vec{V}}{Dt} = -\frac{1}{\rho}\vec{\nabla}p - 2\vec{\Omega}\times\vec{V} + \vec{g}^* + \Omega^2\vec{R} + \vec{F}_r \tag{3.14}$$

where $2\vec{\Omega}\times\vec{V}$ is the Coriolis term, $\vec{g}^*$ is the gravitational force per unit mass, and $\Omega^2\vec{R}$ is the centrifugal force. The gravitational force acts always toward the center of the Earth and the centrifugal force acts radially outward from the axis of rotation. The gravity force, or simply the *gravity*, is the resultant of these two forces.

The total acceleration on an individual fluid particle is equated to the sum of the forces acting on the particle within the fluid. Written for a unit mass of fluid in motion in a coordinate system fixed with respect to the Earth, the vector equation of motion for the atmosphere is

$$\frac{D(\vec{V})}{Dt} = -\frac{1}{\rho}\vec{\nabla}p - 2\vec{\Omega}\times\vec{V} + \vec{g} + \vec{F}_r \tag{3.15}$$

where $\vec{g} = \vec{g}^* + \Omega^2\vec{R}$ is the acceleration due to gravity (gravity force), $\vec{V}$ is the three-dimensional velocity vector, $\vec{\Omega}$ the angular velocity of the Earth, ρ is the density, p the pressure, and $\vec{F}_r$ is the frictional force per unit mass. The usual form for the scalar equations of motion in spherical coordinates (λ, ϕ, r), with λ as the longitude, ϕ the latitude, and r the distance from the center of the Earth, is as follows:

$$\frac{\partial u}{\partial t} + u\frac{\partial u}{r\cos\phi\partial\lambda} + v\frac{\partial u}{r\partial\phi} + w\frac{\partial u}{\partial z} - \left(2\Omega + \frac{u}{r\cos\phi}\right)(v\sin\phi - w\cos\phi) + \frac{1}{\rho r\cos\phi}\frac{\partial p}{\partial\lambda} = F_\lambda \tag{3.16}$$

$$\frac{\partial v}{\partial t} + u\frac{\partial v}{r\cos\phi\partial\lambda} + v\frac{\partial v}{r\partial\phi} + w\frac{\partial v}{\partial z} - \left(2\Omega + \frac{u}{r\cos\phi}\right)(u\sin\phi) + \frac{vw}{r} + \frac{1}{\rho}\frac{\partial p}{r\partial\phi} = F_\phi \tag{3.17}$$

$$\frac{\partial w}{\partial t} + u\frac{\partial w}{r\cos\phi\partial\lambda} + v\frac{\partial w}{r\partial\phi} + w\frac{\partial w}{\partial z} - \left(2\Omega + \frac{u}{r\cos\phi}\right)(u\cos\phi) - \frac{v}{r} + \frac{1}{\rho}\frac{\partial p}{\partial z} + g = F_z \tag{3.18}$$

Since the atmosphere is very shallow compared to the radius of the Earth, we can approximate $r = a + z$ by $r \approx a$, where a is the mean radius of the Earth and z is the height above mean sea level.

The three components of the momentum equation in spherical coordinate system can be written as

$$\frac{Du}{Dt} - 2\Omega v\sin\phi + 2\Omega w\cos\phi + \frac{uw}{a} - \frac{uv}{a}\tan\phi + \frac{1}{\rho}\frac{\partial p}{\partial x} = F_x \tag{3.19}$$

$$\frac{Dv}{Dt} + 2\Omega u\cos\phi + \frac{vw}{a} + \frac{u^2}{a}\tan\phi + \frac{1}{\rho}\frac{\partial p}{\partial y} = F_y \tag{3.20}$$

$$\frac{Dw}{Dt} - 2\Omega u\cos\phi - \left(\frac{u^2 + v^2}{a}\right) + g + \frac{1}{\rho}\partial p\partial z = F_z \tag{3.21}$$

where the total derivative

$$\frac{D}{Dt} = \frac{\partial}{\partial t} + u\frac{\partial}{\partial x} + v\frac{\partial}{\partial y} + w\frac{\partial}{\partial z} \qquad (3.22)$$

in spherical coordinate $\partial x = a\cos\phi\,\partial\lambda$ and $\partial y = a\,\partial\phi$.

Most global numerical weather prediction models and general circulation models use an approximate version of the above nonhydrostatic primitive equations. This version involves the approximation of the vertical equation of motion by the hydro-static equation (see section 3.2.2). These approximations result in the quasi-static primitive equations.

The above equation can be simplified by applying *scale analysis*. It is a powerful tool used for the simplification of equations, especially in atmospheric motions containing many terms with varying degrees of magnitudes. By applying this technique it is possible to determine the terms in the equations which have negligible contribution for meteorological problem and can be eliminated. Simplified horizontal components of the equation of motion (Eqs. (3.19) and (3.20)) are represented as

$$\frac{Du}{Dt} - fv + \frac{\partial\Phi}{\partial x} = F_x \qquad (3.23)$$

$$\frac{Dv}{Dt} + fu + \frac{\partial\Phi}{\partial y} = F_y \qquad (3.24)$$

where the first term in Eqs. (3.23) and (3.24) is the total derivative of u and v, respectively; the second term is the Coriolis forcing; the third term is the pressure gradient forcing; and F_x and F_y are zonal and meridional components of drag forcing due to small eddies. Simplification of the vertical component of the equation of motion (Eq. 3.21) gives the hydrostatic equation given in section 3.2.2.

3.3.2 Equation of Continuity (Conservation of Mass)

Equation of continuity (continuity equation) is a hydrodynamical equation that expresses the principle of the conservation of mass in a fluid. It equates the increase in mass in a hypothetical fluid volume to the net flow of mass into the volume. The equation of continuity is usually written in two alternative forms: *mass divergence* form, based on Eulerian control volume, and *velocity divergence* form, derived from the Lagrangian control volume.

3.3.2.1 Mass Divergence Form

Consider an infinitesimal cube having sides δx, δy, δz fixed in space, through which air flows. The air flow along the x direction of the cube is schematically shown in Fig. 3.2.

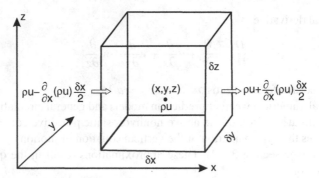

Fig. 3.2 Mass flow through a fixed control volume along the x-direction

The mass flux at the center of the cube is given by ρu. Applying Taylor series on the center point, the rate of mass inflow through the side A of the cube is given by

$$\left[\rho u - \frac{\partial(\rho u)}{\partial x}\frac{\delta x}{2}\right]\delta y \delta z \tag{3.25}$$

The rate of mass outflow through the side B of the cube is

$$\left[\rho u + \frac{\partial(\rho u)}{\partial x}\frac{\delta x}{2}\right]\delta y \delta z \tag{3.26}$$

The rate of accumulation of mass along the x direction flow inside the infinitesimal cube must be equal to the difference between the inflow and outflow. This can be expressed as

$$\frac{\partial M_x}{\partial t} = \left[\rho u - \frac{\partial(\rho u)}{\partial x}\frac{\delta x}{2}\right]\delta y \delta z - \left[\rho u + \frac{\partial(\rho u)}{\partial x}\frac{\delta x}{2}\right]\delta y \delta z \tag{3.27}$$

$$\frac{\partial M_x}{\partial t} = -\frac{\partial(\rho u)}{\partial x}\delta x \delta y \delta z \tag{3.28}$$

where M_x is the accumulation of mass in the cube along the x direction. Similarly, we can represent the rates of mass accumulation in the y and z directions

$$\frac{\partial M_y}{\partial t} = -\frac{\partial(\rho v)}{\partial y}\delta x \delta y \delta z \tag{3.29}$$

$$\frac{\partial M_z}{\partial t} = -\frac{\partial(\rho w)}{\partial z}\delta x \delta y \delta z \tag{3.30}$$

Now the net rate of mass accumulation in the cube is

$$\frac{\partial M}{\partial t} = -\left[\frac{\partial(\rho u)}{\partial x} + \frac{\partial(\rho v)}{\partial y} + \frac{\partial(\rho w)}{\partial z}\right]\delta x \delta y \delta z \tag{3.31}$$

By definition, the net rate of mass accumulation per unit volume is equal to the rate of change of density.

Net rate of accumulation of mass per unit volume is

$$\frac{\partial \rho}{\partial t} = -\left[\frac{\partial(\rho u)}{\partial x} + \frac{\partial(\rho v)}{\partial y} + \frac{\partial(\rho w)}{\partial z}\right] = -\vec{\nabla}.\rho\vec{V} \qquad (3.32)$$

$$\frac{\partial \rho}{\partial t} + \vec{\nabla}.(\rho\vec{V}) = 0 \qquad (3.33)$$

This expression is known as the *mass divergence* form of the equation of continuity.

3.3.2.2 Velocity Divergence Form

An alternate form of the equation of continuity can be derived as follows. Consider a cube of fixed mass, $\delta M = \rho\, \delta x\, \delta y\, \delta z$, but varying dimensions of δx, δy, δz. In this case, the mass is fixed, then $D(\delta M)/Dt = 0$. By applying the chain rule,

$$\frac{D(\rho\delta x\delta y\delta z)}{Dt} = \frac{D(\rho)}{Dt}\delta x\delta y\delta z + \rho\frac{D(\delta x)}{Dt}\delta y\delta z + \rho\frac{D(\delta y)}{Dt}\delta x\delta z + \rho\frac{D(\delta z)}{Dt}\delta x\delta y = 0 \qquad (3.34)$$

Now,

$$\frac{Lt}{\delta x \to 0}\frac{D(\delta x)}{Dt} = \partial u; \quad \frac{Lt}{\delta y \to 0}\frac{D(\delta y)}{Dt} = \partial v; \quad \frac{Lt}{\delta z \to 0}\frac{D(\delta z)}{Dt} = \partial w \qquad (3.35)$$

Dividing throughout by $\partial x\, \partial y\, \partial z$, we get

$$\frac{D\rho}{Dt} + \rho\frac{\partial u}{\partial x} + \rho\frac{\partial v}{\partial y} + \rho\frac{\partial w}{\partial z} = 0 \qquad (3.36)$$

$$\frac{D\rho}{Dt} + \rho(\vec{\nabla}.\vec{V}) = 0 \qquad (3.37)$$

where ρ is the fluid density and $\vec{V}$ is the velocity vector. This expression is called the equation of continuity in velocity divergence form.

In the case of an incompressible fluid, $D\rho/Dt = 0$, so the equation of continuity is simplified as $\vec{\nabla}.\vec{V} = 0$.

It can be shown that the equation of continuity in mass divergence form is the same as the velocity divergence form.

Recall the vector identity

$$\vec{\nabla}.(\rho\vec{V}) = \rho\vec{\nabla}.\vec{V} + \vec{V}.\vec{\nabla}\rho \qquad (3.38)$$

Substitute this identity in Eq. (3.33)

$$\frac{\partial \rho}{\partial t} + \rho \vec{\nabla}.\vec{V} + \vec{V}.\vec{\nabla}\rho = 0 \tag{3.39}$$

$$\frac{1}{\rho}\frac{D\rho}{Dt} + \vec{\nabla}.\vec{V} = 0 \tag{3.40}$$

which is same as the equation in velocity divergence form.

3.3.2.3 Continuity Equation in Isobaric Coordinate System

Consider the Lagrangian control volume $\delta V = \delta x\, \delta y\, \delta z$ and applying the hydrostatic equation, $\delta p = -\rho g \delta z$. The mass of this fluid element is conserved following the motion, then $\delta M = -\rho \delta x \delta y \delta p / g$.

$$\delta V = -\frac{\delta x \delta y \delta p}{\rho g} \tag{3.41}$$

Under the quasi-static approximation with pressure as vertical coordinate, the equation takes the form

$$\frac{1}{\delta M}\frac{D\delta M}{Dt} = \frac{g}{\delta x \delta y \delta p}\frac{D}{Dt}\left[\frac{\delta x \delta y \delta p}{g}\right] = 0 \tag{3.42}$$

Applying chain rule differentiation

$$\frac{1}{\delta x}\frac{D\delta x}{Dt} + \frac{1}{\delta y}\frac{D\delta y}{Dt} + \frac{1}{\delta z}\frac{D\delta p}{Dt} = 0 \tag{3.43}$$

$$\frac{\delta u}{\delta x} + \frac{\delta v}{\delta y} + \frac{\delta \omega}{\delta p} = 0 \tag{3.44}$$

As the limit of $\delta x,\ \delta y,\ \delta z \to 0$, we obtain the continuity in equation

$$\frac{\partial u}{\partial x} + \frac{\partial v}{\partial y} + \frac{\partial \omega}{\partial p} = 0 \tag{3.45}$$

or in vectorial form

$$\frac{\partial \omega}{\partial p} = -\vec{\nabla}.\vec{V}_H \tag{3.46}$$

where p is the pressure, $\vec{V}_H$ is the horizontal velocity, $\omega = Dp/Dt$ is the vertical velocity in isobaric surface. Horizontal divergence ($\vec{\nabla}.\vec{V}_H > 0$) is accompanied by vertical compression ($\partial \omega/\partial p < 0$), and horizontal convergence is accompanied by vertical stretching.

The above equation does not contain density field and its time derivative, which is the advantage of the isobaric coordinate system. Since the equation does not contain time derivative it is not a *prognostic* (predictive) equation.

3.3.3 Conservation of Energy

The law of conservation of energy states that the sum of all energy in the universe is constant. The Sun is the primary source of energy for the Earth and its atmosphere. When solar radiation is absorbed at the Earth's surface and its atmosphere, it is termed as internal energy. Conversion of internal energy into various other forms of energy is a challenging problem in atmospheric sciences (Martin 2006).

In order to derive an expression for the energy equation, let us consider the three components of the momentum equation in Cartesian coordinate system as

$$\frac{Du}{Dt} = -\frac{1}{\rho}\frac{\partial p}{\partial x} + 2\Omega v \sin\phi - 2\Omega w \cos\phi - \frac{uw}{a} + F_x \tag{3.47}$$

$$\frac{Dv}{Dt} = -\frac{1}{\rho}\frac{\partial p}{\partial y} - 2\Omega u \sin\phi - \frac{vw}{a} + F_y \tag{3.48}$$

$$\frac{Dw}{Dt} = -\frac{1}{\rho}\frac{\partial p}{\partial z} + 2\Omega u \cos\phi - g - \frac{u^2+v^2}{a} + F_z \tag{3.49}$$

Multiplying the x component of the equation (Eq. 3.47) by u, y component (Eq. 3.48) by v, and z component (Eq. 3.49) by w, and added together, we get:

$$u\frac{Du}{Dt} + v\frac{Dv}{Dt} + w\frac{Dw}{Dt} = -\frac{1}{\rho}\left[u\frac{\partial p}{\partial x} + v\frac{\partial p}{\partial y} + w\frac{\partial p}{\partial z}\right] - gw + [uF_x + vF_y + wF_z] \tag{3.50}$$

$$\frac{D}{Dt}\left[\frac{u^2+v^2+w^2}{2}\right] = -\frac{1}{\rho}\vec{V}.\vec{\nabla}p - gw + \vec{V}.\vec{F_r} \tag{3.51}$$

where the frictional force $\vec{F_r} = \hat{i}F_x + \hat{j}F_y + \hat{k}F_z$.

Note that in this expression, Coriolis and curvature terms are absent. This indicates that the rotation effect and the curvature terms have no role in controlling the energy of the Earth atmosphere system.

In the above equation, we know that

$$-gw = -g\frac{Dz}{Dt} = \frac{D\Phi}{Dt} \tag{3.52}$$

Substituting in Eq. 3.51, we get

$$\frac{D}{Dt}\left[\frac{u^2+v^2+w^2}{2} + \Phi\right] = -\frac{1}{\rho}\vec{V}.\vec{\nabla}p + \vec{V}.\vec{F_r} \tag{3.53}$$

This equation is known as the *mechanical energy equation*, because it deals with the mechanical forms of energy. The sum of the kinetic energy and gravitational potential energy is called the mechanical energy. This expression states that following the motion, the rate of change of mechanical energy per unit volume is equal to the rate at which the work is done by the pressure gradient force.

In order to include the thermal energy in the energy equation, we have to incorporate the first law of thermodynamics, which is represented as

$$\dot{Q} = C_v \frac{DT}{Dt} + p \frac{D\alpha}{Dt} \tag{3.54}$$

where $\dot{Q}$ is the diabatic heating rate, C_v is the specific heat at contant volume, α is the specific volume. $\dot{Q}$ represents the absorption of solar radiation, which converts to both internal energy in the form of temperature changes and the mechanical energy made apparent in doing work, $D\alpha/Dt$.

Equation (3.53) can be rearranged as

$$0 = \frac{D}{Dt} \left[\frac{u^2 + v^2 + w^2}{2} + \Phi \right] + \frac{1}{\rho} \vec{V}.\vec{\nabla}p - \vec{V}.\vec{F_r} \tag{3.55}$$

This zero valued expression in Eq. (3.55) can be added to Eq. (3.54), so that

$$\dot{Q} = C_v \frac{DT}{Dt} + \rho \frac{D\alpha}{Dt} + \frac{D}{Dt} \left[\frac{u^2 + v^2 + w^2}{2} + \Phi \right] + \frac{1}{\rho} \vec{V}.\vec{\nabla}p - \vec{V}.\vec{F_r} \tag{3.56}$$

Noting that

$$\frac{1}{\rho} \vec{V}.\vec{\nabla}p = \alpha \left[\frac{Dp}{Dt} - \frac{\partial p}{\partial t} \right] \tag{3.57}$$

$$p \frac{D\alpha}{Dt} + \alpha \frac{Dp}{Dt} = \frac{D(p\alpha)}{Dt} \tag{3.58}$$

Rewriting Eq. (3.56) we get

$$\dot{Q} = \frac{D}{Dt} \left[\frac{u^2 + v^2 + w^2}{2} + \Phi + C_v T + p\alpha \right] - \alpha \frac{\partial p}{\partial t} - \vec{V}.\vec{F_r} \tag{3.59}$$

which is known as the energy equation. If the flow is adiabatic, steady state, and frictionless, then the quantity:

$$\frac{(u^2 + v^2 + w^2)}{2} + \Phi + C_v T + p\alpha = \text{constant} \tag{3.60}$$

This equation is known as Bernoulli's equation for an incompressible flow. In this case

$$\frac{u^2 + v^2 + w^2}{2} + \Phi + p\alpha = \text{constant} \tag{3.61}$$

This relationship suggests that for an atmosphere at rest, an increase in elevation produces a decrease in hydrostatic pressure.

3.4 Thermodynamics of Dry Atmosphere

Differentiating the equation of state (Eq. 3.1) with respect to time yields

$$p\frac{D\alpha}{Dt} + \alpha\frac{Dp}{Dt} = R\frac{DT}{Dt} \tag{3.62}$$

Substituting for $pD\alpha/Dt$ from Eq. (3.54) and using $C_p = C_v + R$, we get

$$C_p\frac{DT}{Dt} - \alpha\frac{Dp}{Dt} = \dot{Q} \tag{3.63}$$

Dividing throughout by T, and noting that $\alpha/T = R/p$ by equation of state, we get

$$C_p\frac{D\ln T}{Dt} - R\frac{D\ln p}{Dt} = \frac{\dot{Q}}{T} \tag{3.64}$$

where the term $\dot{Q}/T$ is called *entropy*. An isentropic process is one in which the entropy is constant with time, so that

$$C_p\frac{D\ln T}{Dt} - R\frac{D\ln p}{Dt} = 0 \tag{3.65}$$

3.4.1 Potential Temperature

Potential temperature θ is the temperature a parcel of air would have if it were adiabatically compressed (or expanded) from its original pressure, p, to a standard pressure, p_0 (usually 1,000 hPa). Lines of constant θ are often referred to as isentropes, and the flow along surfaces of constant potential temperature is known as isentropic flow.

Now integrate Eq. (3.65) from a given pressure p and temperature T, to a reference pressure p_0 (say mean sea level pressure) and to the potential temperature θ,

$$\int_T^\theta C_p D\ln T = \int_p^{p_0} R D\ln p \tag{3.66}$$

which gives as,

$$C_p(\ln\theta - \ln T) = R(\ln p_0 - \ln p) \tag{3.67}$$

Taking antilogarithm, after rearranging the expression, we get

$$\theta = T\left[\frac{p_0}{p}\right]^{R/C_p} \tag{3.68}$$

The above expression is known as *Poisson's equation*. Where, T and p are the initial temperature and pressure of the parcel of air, p_o is the standard pressure level (1,000 hPa), R is the universal gas constant, and C_p is the specific heat constant.

Potential temperature is conserved when there are no diabatic effects. An air parcel will tend to move on an isentropic surface, or surface of constant potential temperature.

Potential temperature is closely related to a quantity called *entropy*. The two-dimensional surfaces of constant potential temperature in the atmosphere (which are roughly parallel to the land surface) are known as isentropic surfaces (surfaces of constant entropy). Air parcels in which no heat is added or lost move on isentropic surfaces, so that potential temperature is conserved along the air parcel trajectory. It turns out that this assumption of no heat added or lost is a fairly good one for periods of 5–10 days. By using isentropic surfaces, atmospheric scientists are able to reduce the problem of tracking air parcel motion from a three-dimensional (latitude, longitude, altitude) problem to a two-dimensional (latitude, longitude) one on an isentropic surface.

The vertical gradient of potential temperature determines the stratification of the air. If potential temperature rises with altitude, the air is said to be stably stratified. If potential temperature falls with altitude, the air is said to be negatively stratified. If potential temperature is unchanged, the air is said to be neutrally stratified. In mathematical terms, where potential temperature is denoted by θ we have

$$D\theta/Dz > 0 \text{ for stable air}$$
$$D\theta/Dz = 0 \text{ for neutral air}$$
$$D\theta/Dz < 0 \text{ for unstable air}$$

The stratosphere is the layer of the atmosphere where the potential temperature increases with altitude; hence, it is a stably stratified region. The static stability of the stratosphere acts as a sort of cap on the weather, which is confined to the troposphere.

3.4.2 Atmospheric Stability

Stability is a measure of the degree to which the atmosphere resists turbulence and vertical motion. It describes the state of the atmosphere when a parcel of air returns to its original position, when it is displaced up or down. This will depend on whether the temperature of the parcel of air is related with that of its new surrounding. Stability criterion is an important factor in the troposphere.

Let us consider the equation for potential temperature (θ) and differentiate Eq. (3.68) with respect to the height, z, as

$$\frac{\partial ln\theta}{\partial z} = \frac{\partial lnT}{\partial z} + \frac{R}{C_p}\left[\frac{\partial p_0}{\partial z} - \frac{\partial p}{\partial z}\right] \tag{3.69}$$

This equation can be written as

$$\frac{1}{\theta}\frac{\partial \theta}{\partial z} = \frac{1}{T}\frac{\partial T}{\partial z} - \frac{R}{C_p p}\frac{\partial p}{\partial z} \qquad (3.70)$$

Since $\partial p/\partial z = -\rho g$ by hydrostatic equation, and applying the equation of state (Eq. 3.1), we can modify the above equation as

$$\frac{T}{\theta}\frac{\partial \theta}{\partial z} = \frac{\partial T}{\partial z} + \frac{g}{C_p} \qquad (3.71)$$

where $g/C_p = \Gamma_d$ is the dry adiabatic lapse rate, and $\partial T/\partial z = -\Gamma$ is the environment lapse rate
That is,

$$\Gamma = \Gamma_d - \frac{T}{\theta}\frac{\partial \theta}{\partial z} \qquad (3.72)$$

The stability of the atmosphere can be assessed by the three conditions. If $(\partial\theta/\partial z) > 0$, then $\Gamma < \Gamma_d$, so the atmosphere is statically stable. In this case an adiabatic ascent of a parcel of air, which is cooler than the environment, will return to its original position. When $(\partial\theta/\partial z) = 0$, then $\Gamma = \Gamma_d$, and the atmosphere is neutrally stable. The lifted air parcel of a dry air will occupy a new position, because the temperature of the parcel and the surrounding are the same. In the third condition, if $(\partial\theta/\partial z) < 0$, then $\Gamma > \Gamma_d$, which corresponds to an absolutely unstable stratification. In this case a dry air parcel lifted will always be warmer than the surroundings and move away from its original position, and is freely convective.

3.4.3 Brunt Vaisala Frequency

In a statically stable atmosphere, if a parcel is colder than its environment upon being lifted, it will be forced to move back downward to its original level once the impulse force is exhausted. In such cases a series of oscillations will result about its original level. The frequency of these buoyancy oscillations is termed Brunt-Vaisala frequency, which depends on the restoring force (the product of gravity and the density difference between the parcel and the environment) that acts on them.

Let δz be the vertical displacement of an air parcel about its initial level, then according to Newton's second law

$$\frac{F_z}{\text{Mass}} = \frac{Dw}{Dt} = \frac{D^2}{Dt^2}(\delta z) \qquad (3.73)$$

Let ρ and T are the density and temperature of the parcel and ρ' and T' of that of the environment, then the restoring force per unit mass for the displaced parcel can be written as

$$\frac{F_z}{\text{Mass}} = -\left[\frac{\rho' - \rho}{\rho'}\right] g \qquad (3.74)$$

Applying equation of state, Eq. (3.72) can be modified as

$$\frac{F_z}{\text{Mass}} = -\left[\frac{1}{T'} - \frac{1}{T}\right] gT' \tag{3.75}$$

$$\frac{F_z}{\text{Mass}} = -g\left[\frac{T - T'}{T}\right] \tag{3.76}$$

In this expression $(T - T')$ can be replaced by $(\Gamma_d - \Gamma)\delta z$, because the dry parcel cools at dry adiabatic lapse rate (Γ_d) and can be compared to the environment whose temperature changes at a rate described by the environmental lapse rate (Γ). Thus, the restoring force per unit mass

$$\frac{F_z}{\text{Mass}} = -\frac{g}{T}(\Gamma_d - \Gamma)\delta z \tag{3.77}$$

Equations (3.73) and (3.77) turns into second-order differential equation.

$$\frac{D^2(\delta z)}{Dt^2} + \frac{g}{T}(\Gamma_d - \Gamma)\delta z = 0 \tag{3.78}$$

The solution of the above equation describes the buoyancy oscillation with a period $2\pi/N$, where

$$N^2 = \left[\frac{g}{T}(\Gamma_d - \Gamma)\right] \tag{3.79}$$

which can be written as

$$N = \left[\frac{g}{\theta}\frac{\partial\theta}{\partial z}\right]^{1/2} \tag{3.80}$$

where N is known as Brunt-Vaisala frequency and its unit is (s^{-1}). For average tropospheric conditions, $N \sim 1.2 \times 10^{-1}\ s^{-1}$ so that the period of buoyancy oscillation is about 8 minutes. Similary the value of N in the stratosphere is $\sim 2.2 \times 10^{-2}\ s^{-1}$, the period of buoyancy oscillation is about 40 minutes.

For a statically stable atmosphere (i.e., $\partial\theta/\partial z > 0$), $N > 0$ and buoyancy oscillations are generated. For the absolutely unstable case ($\partial\theta/\partial z < 0$), N is imaginary and corresponds to a growing disturbance. In the neutral case ($\partial\theta/\partial z = 0$), and $N = 0$, there is no buoyancy oscillations. Air parcels undergo buoyancy oscillations in association with atmospheric gravity waves (Nappo 2002).

3.4.4 Thermodynamic Energy Equation

For a system in thermodynamic equilibrium, the development of the weather system is controlled by thermodynamic processes corresponding to the first law of thermodynamics. The first law of thermodynamics is a simple form of prognostic equation relating to the rate of change of temperature of an air parcel as it moves through the atmosphere. The changes in temperature affect the thickness pattern which determines the distribution of geopotenial Φ on pressure surfaces.

The first law of thermodynamics applies to a moving fluid parcel and can be represented in the form

$$C_p \frac{DT}{Dt} - \alpha \frac{Dp}{Dt} = \dot{Q} \tag{3.81}$$

where $\dot{Q}$ is the diabatic heating. Substituting the value of α from the equation of state and the vertical velocity $\omega = Dp/Dt$ in Eq. (3.63), dividing throughout by C_p, and after rearranging the equation, we get

$$\frac{\partial T}{\partial t} + u\frac{\partial T}{\partial x} + v\frac{\partial T}{\partial y} + \omega \left[\frac{\partial T}{\partial p} + \frac{RT}{pC_p} \right] = \frac{\dot{Q}}{C_p} \tag{3.82}$$

$$\frac{DT}{Dt} = \frac{kT}{p}\omega + \frac{\dot{Q}}{C_p} \tag{3.83}$$

where $k = R/C_p$.

The rate of change of total thermodynamic energy is equal to the rate of diabatic heating plus the rate at which work is done on the parcel by external forces.

In Eq. (3.83), the first term on the right-hand side corresponds to the rate of change of temperature due to adiabatic compression or expansion. The second term represents the effect of diabatic heat sources and sinks, due to absorption of solar radiation and absorption and emission of longwave radiation and latent release. In the stratosphere, heat absorbed and released due to chemical and photochemical reactions are significant. In the troposphere, especially in the tropical region, the heat added to or removed from the air parcel and its environment due to unresolved scales of motion like convective plumes are also to be considered as a part of the diabatic heating.

The local rate of change of temperature can be obtained from Eq. (3.82) as

$$\frac{\partial T}{\partial t} = -\vec{V}_H \cdot \vec{\nabla} T + \left[\frac{\kappa T}{\rho} - \frac{\partial T}{\partial p} \right] \omega + \frac{\dot{Q}}{C_p} \tag{3.84}$$

The first term on the right-hand side represents the horizontal advection of temperature. The second term is the combined effect of adiabatic compression and vertical advection. In a stably stratified atmosphere, the term $\partial T/\partial p$ is less than the adiabatic lapse rate, so that the term within the parentheses is positive. It indicates that sinking (ascending) air motion always favors local warming (cooling). For adiabatic conditions, the last term vanishes ($\dot{Q} = 0$), and the air parcel conserves potential temperature as it moves along its three-dimensional trajectories.

3.5 Primitive Equations

The primitive equations are a version of the Navier-Stoke's equations which describe hydrodynamical flow on the sphere under the assumptions that vertical motion is much smaller than horizontal motion (hydrostatics) and that the fluid layer depth is small compared to the radius of the sphere. Thus, they are a good approximation of global atmospheric flow and are used in most meteorological models.

In general, nearly all forms of the primitive equations relate the five variables (u, v, ω, T, Φ) and their evolution over space and time.

3.5.1 Forms of the Primitive Equations

The precise form of the primitive equations depends on the vertical coordinate system chosen, such as pressure coordinates, log pressure coordinates, or sigma coordinates. Furthermore, the velocity, temperature, and geopotential variables may be decomposed into mean and perturbation components using Reynolds decomposition.

In this form pressure is selected as the vertical coordinate and the horizontal coordinates are written for the Cartesian tangential plane (i.e., a plane tangent to some point on the surface of the Earth). This form does not take the curvature of the Earth into account, but is useful for visualizing some of the physical processes involved in formulating the equations due to its relative simplicity. Primitive equations are as follows.

The horizontal momentum equation is given as

$$\frac{D\vec{V}_H}{Dt} = -\frac{1}{\rho}\vec{\nabla}\phi - f\hat{k} \times \vec{V}_H \tag{3.85}$$

The x and y components of the geostrophic momentum equation can be written as

$$\frac{Du}{Dt} - fv = -\frac{\partial \Phi}{\partial x} \tag{3.86}$$

$$\frac{Dv}{Dt} + fu = -\frac{\partial \Phi}{\partial y} \tag{3.87}$$

The hydrostatic equation, a special case of the vertical momentum equation in which there is no vertical acceleration, is:

$$0 = -\frac{\partial \Phi}{\partial p} - \frac{RT}{p} \tag{3.88}$$

The continuity equation, connecting horizontal divergence/convergence to vertical motion under the hydrostatic approximation $(\partial p = -\rho \partial \Phi)$ is:

$$\frac{\partial u}{\partial x} + \frac{\partial v}{\partial y} + \frac{\partial \omega}{\partial p} = 0 \qquad (3.89)$$

The thermodynamic energy equation, a consequence of the first law of thermodynamics is:

$$\frac{\partial T}{\partial t} + u\frac{\partial T}{\partial x} + v\frac{\partial T}{\partial y} + \omega\left[\frac{\partial T}{\partial p} + \frac{RT}{pC_p}\right] = \frac{\dot{Q}}{C_p} \qquad (3.90)$$

These five fundamental equations along with the statement of the conservation of water vapor substance form the basis for any numerical weather prediction scheme.

3.5.2 Horizontal Equation of Motion

The equations of motion are a complicated set of expressions of different scale interactions. Applying scale analysis, the two significant terms of the horizontal equation of motions are the pressure gradient force and the Coriolis force. As a first approximation to the full equations of motion we can consider the horizontal pressure gradient force and the Coriolis force terms to be in approximate balance with one another as

$$\vec{\nabla}\phi = -f\hat{k} \times \vec{V}_H \qquad (3.91)$$

3.6 Balanced Wind Flow

In the atmosphere, the air naturally moves from areas of high pressure to areas of low pressure, due to the pressure gradient force. As soon as the air starts to move the Coriolis force deflects it due to the rotation of the Earth. As the air moves from the high pressure area, its speed increases, and so does the deflection from the Coriolis force. The deflection increases until the Coriolis and pressure gradient forces are in balance, at which point the air is no longer moving from high to low pressure, but instead moves along an isobar, a line of equal pressure. It is assumed that the atmosphere starts in a geostrophically unbalanced state and describes how such a state would evolve into a balanced flow. In practice, the flow is nearly always balanced.

3.6.1 Geostrophic Wind

Geostrophic wind (V_g) is defined as the wind which would exist in the atmosphere, if the flow is horizontal and hydrostatic without acceleration and friction. In other words, the geostrophic wind is the wind resulting from the balance between Coriolis force and pressure gradient force.

$$\vec{V}_g = \frac{1}{f}(\hat{k} \times \vec{\nabla}\phi)$$
(3.92)

where

$$\vec{V}_g = \hat{i}u_g + \hat{j}v_g \text{ and } \vec{\nabla}\phi = \hat{i}\frac{\partial\phi}{\partial x} + \hat{j}\frac{\partial\phi}{\partial y}$$
(3.93)

Horizontal components are:

$$u_g = -\frac{1}{f}\frac{\partial\Phi}{\partial y}$$
(3.94)

$$v_g = \frac{1}{f}\frac{\partial\Phi}{\partial x}$$
(3.95)

where Φ is the geopotential height and f is the Coriolis parameter.

In natural coordinates, the geostrophic wind equation can be written as

$$V_g = -\frac{1}{f}\frac{\partial\Phi}{\partial n}$$
(3.96)

where $\partial\phi/\partial n$ is the horizontal geopotential field distribution in the normal plane.

Geostrophic wind blows parallel to straight isobars or contours (Fig. 3.3). Frictional effects are neglected, which is usually a very good approximation for the synoptic scale instantaneous flow in the midlatitude mid-troposphere. However, although ageostrophic terms are relatively small, they are important for the time evolution of the flow.

In either hemisphere, the geostrophic wind circulates cyclonically around the center of a low pressure area. The tighter the spacing of the isobars or geopotential height contours, the stronger the Coriolis force required to balance the pressure gradient force and hence the higher the geostrophic wind speed.

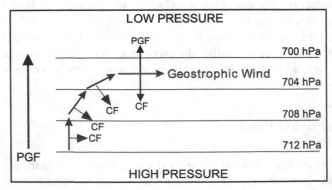

Fig. 3.3 Geostrophic wind flow parallel to straight isobars, which is maintaining the balance between horizontal pressure gradient force (PGF) and the Coriolis Force (CF) (Adapted from M. Pidwirny 2006)

The validity of the geostrophic approximation is dependent on the local Rossby number. It is invalid at the equator where the Coriolis parameter f is zero, and therefore generally not used in the tropics.

3.6.2 Ageostrophic Wind

The geostrophic wind is strictly valid only in regions of zero wind acceleration. Since the wind is a vector quantity, with magnitude and direction, if either of these properties is changed over time, the wind is accelerated. Thus the geostrophic condition is violated when there is a change in the wind speed along the direction, and/or wind direction changes along the flow. The along-flow speed changes are most prominent in the vicinity of the local wind maxima, especially in the jet core regions. Also along-flow direction changes are seen in the region of troughs and ridges. These regions are usually associated with sensible weather in the form of circulation systems, clouds, and precipitation.

Ageostrophic wind (V_{ag}) is defined as the departure from the geostrophic wind. It is the difference between the actual wind at a location and the calculated geostrophic wind at that point.

$$\vec{V}_{ag} = \vec{V} - \vec{V}_g \tag{3.97}$$

3.6.3 Gradient Wind

Wind above the Earth's surface does not always travel in a straight line. In many cases, the winds flow around the curved isobars of a high (anticyclone) or low (cyclone) pressure center. A wind that blows around curved isobars above the level of friction is called a gradient wind. Gradient wind results from a three-way balance between pressure gradient force, Coriolis force, and centrifugal force.

The gradient wind flows parallel to curved isobars. The centrifugal accelerations observed in association with the curvature of the trajectories of air parcels tend to be much larger than those associated with the speeding up or slowing down of air parcels as they move downstream. Hence when the horizontal acceleration (DV/Dt) is large, its scalar magnitude can be approximated by the centrifugel acceleration ($-V^2/R$), where R is the local radius of curvature of the trajectory.

$$\frac{V^2}{R} = -\vec{\nabla}\Phi - f\hat{k} \times \vec{V}_H \tag{3.98}$$

The equation for the gradient wind (V_G) is

$$V_G = -\frac{fR}{2} \pm \sqrt{\left(\frac{f^2R^2}{4} - R\frac{\partial\Phi}{\partial n} \right)} \tag{3.99}$$

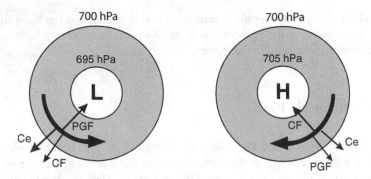

Fig. 3.4 Gradient wind flow around circular isobars. It is a three-way balance between the pressure gradient force (PGF), Coriolis force (CF), and centrifugal force (Ce) (Adapted from M. Pidwirny 2006)

$$\text{or } V_G = -\frac{fR}{2} \pm \sqrt{\left(\frac{f^2R^2}{4} + fRV_g\right)} \qquad (3.100)$$

The gradient wind equation is a representation of the entire normal component of the equation of motion. There is a balance between all the forces present in the normal component of the equation in natural coordinates: (i) the centrifugal force $-V^2/R$; (ii) the pressure gradient force $-\partial\Phi/\partial n$; and (iii) the Coriolis force $-fv$.

The gradient wind is very much like the geostrophic wind, in that it is a frictionless wind which allows for flow that is parallel to the isobars. The one difference between the geostrophic wind and the gradient wind is that the gradient wind includes the centrifugal force, thereby allowing curvature in the flow field. At first glance, this may seem to be simply an addition to the geostrophic wind equation, since the two flows are very similar to one another. For example, if the curvature is equal to zero, then geostrophic flow is obviously the result. However, the addition of the centrifugal force complicates the equation tremendously and creates several different solutions to the gradient wind equation.

Figure 3.4 describes the forces that produce gradient winds around high- and low-pressure centers. Around a low-pressure center, the gradient wind consists of the Coriolis force and centrifugal force acting away from the center of rotation, while the pressure gradient force acts towards the center. In a high pressure center, the Coriolis forces is directed toward the center of the high-pressure region, while the pressure gradient and centrifugal forces are directed outward.

3.6.4 Relation between Geostrophic Wind and Gradient Wind

When comparing gradient wind speeds with geostrophic wind speeds for the same pressure gradient force, there are some differences. Gradient flow around a high-pressure system will be faster than the geostrophic flow if the pressure gradient force is constant. The opposite is true when considering low-pressure systems. In this case, the gradient wind around the low-pressure system is less than the geostrophic wind if the pressure gradient force is constant. The formula below defines a ratio of geostrophic flow to gradient flow:

$$\frac{\vec{V}_g}{\vec{V}_G} = 1 + \frac{V_g}{fR} \tag{3.101}$$

where V_g is geostrophic wind speed, V_G is gradient wind speed, and R_T is the radius of curvature of the trajectory.

Thus, if the value of the ratio is greater than one, then the flow is cyclonic. If the value is less than one, the flow is anti-cyclonic.

3.6.5 Thermal Wind

Thermal wind (V_T) is defined as the vertical shear of geostrophic wind in hydrostatic equilibrium and it arises from a horizontal temperature gradient.

Thermal wind equation may be written in the form

$$\vec{V}_T - \frac{1}{f}\hat{k} \times \vec{\nabla}(\Phi_1 - \Phi_0) \tag{3.102}$$

where the Φ_1, Φ_0 are geopotential height fields with $\Phi_1 > \Phi_0$, and f is the Coriolis parameter. The equation also makes obvious that the thermal wind does not apply in the tropics: f is zero at the equator, and generally small in the tropics.

In terms of geostrophic wind (V_g), the thermal wind can be written as:

$$\frac{\partial \vec{V}_g}{\partial \ln p} = -\frac{R}{f}\hat{k} \times \vec{\nabla}_p T \tag{3.103}$$

where $\vec{V}_g$ is the geostrophic wind vector, p the pressure, R the gas constant for air, and f the Coriolis parameter. Thermal wind in the atmopshere is depicted in Fig. 3.5.

The thermal wind is not actually a wind, but a wind difference between two pressure levels. It is only present in an atmosphere with horizontal gradients of temperature, i.e., baroclinicity. In a barotropic atmosphere the geostrophic wind is independent of height. The name stems from the fact that this wind flows around areas of low (high) temperature in the same manner as the geostrophic wind flows around areas of low (high) pressure.

It is a theoretical wind that blows parallel to the thickness lines, for the layer considered, analogous to how the geostrophic wind blows parallel to the height contours. The closer the thickness isopleths, the stronger the thermal wind. Cold air is always located to the left of the thermal wind (as you face downstream) and warm air is located on the right. Since thickness contours are tighter on the cold side of thermal wind, the lower thickness values will be found on the left side of the thermal

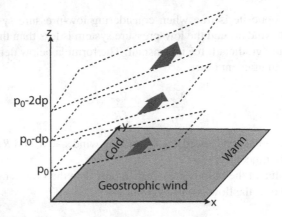

Fig. 3.5 Thermal wind showing the relationship between the vertical shear of the geostrophic wind and horizontal thickness gradient. The dotted line encloses isobaric surfaces which tilt with increasing height

wind. The speed and direction of the thermal wind are determined by vector geometry where the geostrophic wind at the upper level is subtracted from the geostrophic wind at the lower level.

The zonal and meridional components of the thermal wind equation are:

$$\frac{\partial u_g}{\partial \ln p} = \frac{R}{f} \left[\frac{\partial T}{\partial y} \right]_p \tag{3.104}$$

$$\frac{\partial v_g}{\partial \ln p} = -\frac{R}{f} \left[\frac{\partial T}{\partial x} \right]_p \tag{3.105}$$

where the subscript p implies that the differentiation occurs while the pressure is constant.

3.6.6 Applications of Thermal Wind

The thermal wind relationship has important dynamical consequences for the general circulation of the atmosphere. The thermal wind concept forms the cornerstone of modern dynamic meteorology as well as the first-order balance for the midlatitude air flows. Some of the important applications are briefly mentioned as follows.

3.6.6.1 Increasing Westerlies with Height

In the northern hemisphere, the polar region is cold while the equatorial region is warm. Because the thermal wind circles the area of low temperature in the same manner as the geostrophic wind flows, the thermal wind in the northern hemisphere midlatitude is westerly. This can be seen by looking at a globe from above the north pole: a westerly current flows counterclockwise around the globe. This applies similarly to the southern hemisphere, and because f is negative there, it yields the same result.

Based on the thermal wind concept, the vertical wind profile in the midlatitude northern hemisphere troposphere can be explained. If the thermal wind is westerly, the atmospheric flow will become more westerly with height, as the thermal wind describes the wind change with height. Therefore, if at a certain level, say at the top of the boundary layer, the wind speed is close to zero, the wind speed will have a strong eastward component at higher levels.

3.6.6.2 Formation of Jet Streams

The thermal wind concept basically describes the jet stream, a westerly current of air with maximum wind speeds close to the tropopause which is fundamentally the result of the temperature contrast between equator and pole.

3.6.6.3 Advection of Warm or Cold Air

If the geostrophic wind at a level advects (i.e., transports) warm or cold air, the thermal wind causes a turning of wind direction with height. A similar argument as in the other example with regard to how the thermal wind is related to the temperature distribution can be made.

3.6.6.4 Veering and backing

The geostrophic wind that advects warm air into a region of colder air causes the wind to turn right (clockwise, veering) with height, while cold air advection into a region of warmer air results in the wind turning left (counterclockwise, backing) in the northern hemisphere.

3.6.7 Barotropic and Baroclinic Atmospheres

A barotropic atmosphere is one in which the density depends only on the pressure, so that isobaric surfaces are also surfaces of constant density. The isobaric surfaces will also be isothermal; hence from the thermal wind equation, the geostrophic wind is independent of height. Therefore, the motions of a rotating barotropic fluid are strongly constrained. Thermal wind does not exist in a barotropic atmosphere.

In the case of baroclinic atmosphere, the density depends on both the temperature and the pressure $\rho = \rho(p, T)$. In a baroclinic atmosphere, the geostrophic wind generally has vertical shear and is related to the horizontal temperature gradient by the thermal wind equation. Thus thermal wind exists in a baroclinic atmosphere. Baroclinicity is a measure of the stratification in a fluid.

Baroclinity is proportional to

$$\vec{\nabla} p \times \vec{\nabla} \rho \qquad (3.106)$$

which again is proportional to the angle between surfaces of constant pressure and surfaces of constant density. Thus, in a barotropic fluid these surfaces are parallel. Areas of high atmospheric baroclinity are characterized by the frequent formation of cyclones.

3.7 Circulation, Vorticity, and Divergence

Circulation and vorticity are the two primary measures of rotation in a fluid. The atmosphere is characterized by the omnipresence of a variety of fluid eddies and swirling winds.

3.7.1 Circulation

Circulation is the flow or motion of fluid through a given area or volume. In atmospheric science, it is used to explain the flow of air as it moves around a pressure system in the atmosphere. It describes smaller patterns in semipermanent pressure systems as well as the relatively permanent global circulation of air.

Circulation is a macroscopic measure of the rotation of the fluid and is a scalar quantity. Circulation about a closed contour in a fluid is defined as the line integral of the velocity vector that is locally tangent to the contour (see Fig. 3.6).

$$C = \oint \vec{V}.\vec{dl} = \oint |V|\cos\alpha.\vec{dl} \qquad (3.107)$$

where C is the circulation and $\vec{V}$ is the velocity vector, and $\vec{dl}$ represents the displacement vector along the edge of the fluid element.

In the northern hemisphere, the circulation is taken as positive, when $C > 0$ for counterclockwise (cyclonic) rotation around the contour, and is negative when $C < 0$ for clockwise (anticyclonic) rotation.

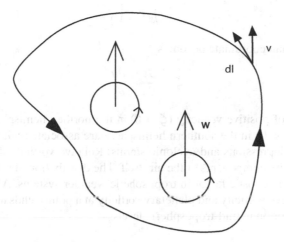

Fig. 3.6 Circulation around a closed fluid contour

3.7.2 Vorticity

Vorticity is the microscopic measure of the rotation of the air or water (any fluid) considered at a point, and is a vector field. The vorticity is obtained by taking the curl of velocity. In atmospheric science, we are primarily concerned with the rotational motion about an axis that is perpendicular to the Earth's surface. Thus we are interested only in the vertical component of the absolute and relative vorticity, which are designated as η and ζ, respectively:

$$\eta = \hat{k}.(\vec{\nabla} \times \vec{V}_a) \tag{3.108}$$

$$\zeta = \hat{k}.(\vec{\nabla} \times \vec{V}) \tag{3.109}$$

where $\vec{V}_a$ is the absolute velocity and $\vec{V}$ is the relative velovity.

The difference between absolute vorticity and relative vorticity is the planetary vorticity, which is the local vertical component of the vorticity of the Earth due to its rotation, and is simply the Coriolis parameter, f.

3.7.3 Relative Vorticity

Relative vorticity (ζ) is the curl of the relative velocity vector and is a measure of the rotation of air in the atmosphere. Both shear and curvature effects can produce vorticity in the atmosphere. For large-scale atmospheric flows, the vertical component of the relative vorticity is important. It depends only on the horizontal velocity and is given as

$$\zeta = \hat{k}(\vec{\nabla} \times \vec{V}) \tag{3.110}$$

which in Cartesian coordinate becomes

$$\zeta = \frac{\partial v}{\partial x} - \frac{\partial u}{\partial y} \tag{3.111}$$

The regions of positive vorticity ($\zeta > 0$) in the northen hemisphere, and negative vorticity ($\zeta < 0$) in the southern hemisphere are associated with cyclonic disturbances, like depressions and cyclonic storms. Relative vorticity depends on the horizontal velocity components of the air itself. The distribution of relative vorticity is an important diagnostic factor in tropospheric weather systems. Absolute vorticity (sum of relative vorticity and planetary vorticity at a point) tends to be conserved following the motion in mid-tropospheric levels.

3.7.4 Vorticity in Natural Coordinate (t, n, z) System

Vorticity in a natural coordinate (t, n, z) system can be represented as

$$\zeta = -\frac{\partial V}{\partial n} + \frac{V}{R_s} \tag{3.112}$$

where R_s is the radius of curvature and V is the horizontal wind speed.

It shows that the net vertical vorticity component consists of two parts: (i) rate of change of zonal wind speed normal to the direction of flow ($-\partial V/\partial n$), known as the shear vorticity; and (ii) turning of the wind along a streamline (V/R_s), called the curvature vorticity. It may be noted that even straight flow may have vorticity, if the speed changes normal to the direction of flow. On the other hand, curved flow may have zero vorticity, provided that the shear vorticity is equal and opposite to curvature vorticity (for details see Holton 2004).

3.7.5 Planetary Vorticity

The spin imparted to an object by the spinning Earth about a local vertical is known as planetary vorticity, which is same as the Coriolis parameter, f. It is always positive in the northern hemisphere.

The Earth rotates around its axis, which goes through the poles. From Fig. 3.7 it can be seen that Earth's spin effect is minimum at the equator and maximum at poles. The planetary vorticity is zero at equator and maximum at poles.

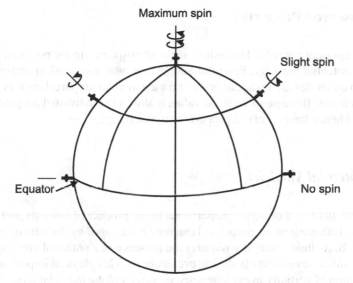

Fig. 3.7 Spinning effect of Earth's rotation

3.7.6 Absolute Vorticity

Absolute vorticity (η) is the sum of planetary vorticity (f) and the relative vorticity (ζ).

$$\eta = \zeta + f \qquad (3.113)$$

3.7.7 Divergence

The flow of wind resulting in a horizontal outflow of air from a region is known as divergence. The opposite of divergence is convergence, which is always associated with ascending air masses. Mathematically it is represented as $\vec{\nabla}.\vec{V}$. The horizontal divergence (D) is represented as

$$D = \frac{\partial u}{\partial x} + \frac{\partial v}{\partial y} \qquad (3.114)$$

Horizontal divergence (convergence) plays a major role in vorticity of the atmosphere. Convergence spins up cyclonic vorticity, whereas divergence spins up anticyclonic vorticity. In surface weather charts, we can see that the area of horizontal divergence is associated with descending motion which produces clear weather conditions. On the other hand, the area of convergence is the region of upward motion with strong convective clouds.

3.8 Conserved Properties

The two important conserved quantities in the atmosphere are the potential temperature and potential vorticity. Potential temperature was discussed in section 3.4.1. These two quantities are invariant or fixed for a particular air parcel even as the parcel moves about. Because of its fixed value, it allows the individual air parcel to be traced, and hence the conserved property acts as a tracer.

3.8.1 Potential Vorticity

Potential vorticity is a quantity proportional to the product of vorticity and stratification that, following an air parcel and can only be changed by diabatic or frictional processes. Baroclinic instability requires the presence of a potential vorticity gradient along which waves amplify during cyclogenesis. This plays an important role in the generation of vorticity in cyclogenesis, especially along the polar front. It is also very useful in tracing intrusions of stratospheric air deep into the troposphere in the vicinity of jet streaks (Wallace and Hobbs 2006).

In general, potential vorticity is a measure of the absolute vorticity to the effective depth of the vortex. Potential vorticity in a homogeneous incompressible fluid can be represented as

$$\text{Potential vorticity} = \frac{\zeta + f}{\delta z} \tag{3.115}$$

where δz is the depth of the parcel.

3.8.2 Ertel's Potential Vorticity

Ertel's Potential vorticity (PV) is defined as the product of the vertical component of absolute vorticity and static stability. It is measured in units of $\text{K kg}^{-1} \text{ m}^2 \text{ s}^{-1}$.

$$PV = (\zeta + f) \left[-g \frac{\partial \theta}{\partial p} \right] \tag{3.116}$$

where $\zeta = \hat{k}(\vec{\nabla} \times \vec{V})$ is the relative vorticity, which can be either shear vorticity or curvature vorticity and f is the planetary vorticity.

Potential vorticity is a combination of absolute vorticity and the gradient of potential temperature into a scalar quantity (i.e., a quantity that has a magnitude but no direction) that is conserved under frictionless, adiabatic conditions. Because it contains both dynamic (vorticity) and thermodynamic (potential temperature) properties, the statement of its conservation is quite general.

Potential vorticity consists of both vorticity and a component due to vertical changes in potential temperature. It turns out that potential vorticity increases when

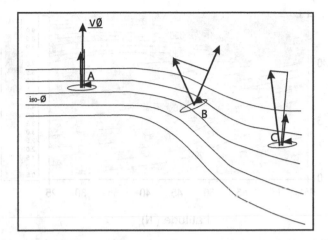

Fig. 3.8 Conservation of potential vorticity in a fluid flow

the stability (given by the vertical gradient of potential temperature) increases, as is the case when you go from the troposphere into the stratosphere. Under appropriate conditions, wind and temperature fields can be derived from potential vorticity. Thus, potential vorticity provides a very powerful dynamic tool.

The Ertel's potential vorticity is a conservative quantity during adiabatic evolution. The PV indicates that an air parcel may see its absolute vorticity increased when the static stability decreases, as is graphically shown in Fig. 3.8.

The potential vorticity is usually measured in potential vorticity units or pvu. A pvu is $10^{-6} m^2 \, Ks^{-1} \, kg^{-1}$. The PV is relatively low in the troposphere (smaller than 1 pvu) and the PV gradient is flat. There is, however, a tendency to have a slightly higher PV at high latitudes than near the equator because of the planetary component of the absolute vorticity.

At the tropopause, the PV gradient becomes very strong because of the very high values of the static stability in the stratosphere. Above the surface of 1.5 or 2 pvu, the PV increases very quickly in the stratospheric air. Figure 3.9 gives the PV in the stratosphere relative to the troposphere. The iso-potential vorticity surfaces of 1.5 or 2 pvu define the dynamical tropopause. This separates tropospheric air from stratospheric air in terms of PV. This is the classical definition of the thermal tropopause in terms of temperature gradients.

3.9 Vorticity Equation

The vorticity equation is an important prognostic equation in atmospheric sciences. It describes the total derivative (that is, the total change due to local change with time and advection) of vorticity, and thus can be stated in either relative or absolute form.

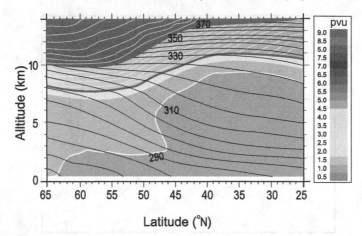

Fig. 3.9 Potentail vorticity distribution in the troposphere and lower stratosphere over midlatitudes (Adapted from Virtual Lab)

3.9.1 Vorticity Equation in x, y, p Coordinates

Starting from the horizontal equations of motion

$$\frac{Du}{Dt} = \frac{\partial u}{\partial t} + u\frac{\partial u}{\partial x} + v\frac{\partial u}{\partial y} + w\frac{\partial u}{\partial p} = fv - g\frac{\partial z}{\partial x} \tag{3.117}$$

$$\frac{Dv}{Dt} = \frac{\partial v}{\partial t} + u\frac{\partial v}{\partial x} + v\frac{\partial v}{\partial y} + w\frac{\partial v}{\partial p} = -fu - g\frac{\partial z}{\partial y} \tag{3.118}$$

we can write them as

$$\frac{\partial u}{\partial t} + u\frac{\partial u}{\partial x} + v\frac{\partial u}{\partial y} + w\frac{\partial u}{\partial p} - fv = -g\frac{\partial z}{\partial x} \tag{3.119}$$

$$\frac{\partial v}{\partial t} + u\frac{\partial v}{\partial x} + v\frac{\partial v}{\partial y} + w\frac{\partial v}{\partial p} + fu = -g\frac{\partial z}{\partial y} \tag{3.120}$$

Taking ∂ (Eq. 3.119)$/\partial x - \partial$ (Eq. 3.118)$/\partial y$ and using the definition $\zeta = (\partial v/\partial x - \partial u/\partial y)$, we get

$$\frac{\partial \zeta}{\partial t} + u\frac{\partial \zeta}{\partial x} + v\frac{\partial \zeta}{\partial y} + w\frac{\partial \zeta}{\partial p} + (\zeta + f)\left[\frac{\partial u}{\partial x} + \frac{\partial v}{\partial y}\right] + \left[\frac{\partial \omega}{\partial x}\frac{\partial v}{\partial p} - \frac{\partial \omega}{\partial y}\frac{\partial u}{\partial p}\right] + v\frac{\partial f}{\partial y} = 0 \tag{3.121}$$

or

$$\frac{D(\zeta + f)}{Dt} = -(\zeta + f)\left[\frac{\partial u}{\partial x} + \frac{\partial v}{\partial y}\right] - \left[\frac{\partial \omega}{\partial x}\frac{\partial v}{\partial p} - \frac{\partial \omega}{\partial y}\frac{\partial u}{\partial p}\right] \tag{3.122}$$

The left-hand side term of Eq. (3.121) represents, the total rate of change of $(\zeta + f)$, which is the absolute vorticity, including the relative vorticity ζ and the vorticity due to the Earth's rotation f. The first term on the right-hand side is the divergence term and the second term is the twisting or tilting term.

The divergence term states that the change of the absolute vorticity is proportional to the magnitude of divergence and the magnitude of the absolute vorticity. Since the absolute vorticity is positive in general, we can say that an increase in absolute vorticity produces convergence and the decrease in absolute vorticity provides divergence. This term is responsible for the formation of intense cyclones in the midlatitudes and hurricanes in the tropics (Gill 1982; Martin 2006).

The tilting term will tilt the vorticity in the horizontal direction due to vertical wind shear to the vorticity in the vertical direction. This term is generally ignored in synoptic meteorology because ω is generally small for synoptic scale motion. However, ω can be quite large in storms and this term is quite important in generating storm scale vorticity, which causes the storm rotation and ultimately leads to the formation of rapid rotation in tornadoes.

3.9.2 Simplified Vorticity Equation

The convergence–divergence is usually stronger near the surface and in the upper levels below the tropopause, and is usually weak in the middle levels around 450–550 hPa level. If a nondivergent level exists in a synoptic scale motion (tilting term negligible), then

$$\frac{D(\zeta + f)}{Dt} = 0 \tag{3.123}$$

or

$$\frac{\partial(\zeta + f)}{\partial t} \approx -u\frac{\partial(\zeta + f)}{\partial x} + v\frac{\partial(\zeta + f)}{\partial y} \tag{3.124}$$

The first equation states that the absolute vorticity is conserved if we follow the movement of an air parcel on the level of nondivergence. The second equation states that the local change of the absolute vorticity equals the horizontal advection of the absolute vorticity. In the natural coordinates, this equation can be written as

$$\frac{\partial(\zeta + f)}{\partial t} \approx -V\frac{\partial(\zeta + f)}{\partial s} \tag{3.125}$$

3.9.3 Quasi-geostrophic Vorticity Equation

Simplifying the vorticity equation by the following assumptions, viz. (i) neglecting the vertical advection and twisting terms; (ii) neglecting relative vorticity (ζ)

compared to the Coriolis parameter (f) in divergence term; (iii) approximating the horizontal velocity by the geostrophic wind in the advection term; and (iv) replacing the relative vorticity by its geostrophic value, we obtain the quasi-geostrophic vorticity equation

$$\frac{\partial \zeta_g}{\partial t} = -\vec{V}_g.\vec{\nabla}(\zeta_g + f) + f_0\frac{\partial \omega}{\partial p} \qquad (3.126)$$

where ζ_g is the relative vorticity and f is the planetary vorticity. The quasi-geostrophic vorticity equation states that the local rate of change of geostrophic vorticity is given by the sum of advection of absolute vorticity by geostrophic wind plus the concentration or dilution of vorticity by stretching or shrinking of fluid columns (the divergence effect).

By the definition of the geopotential tendency $\chi = \partial \Phi/\partial t$, the quasi-geostrophic vorticity equation can be modified as

$$\frac{1}{f_0}\nabla^2\chi = -\vec{V}_g.\vec{\nabla}\left(\frac{1}{f_0}\nabla^2\Phi + f\right) + f_0\frac{\partial \omega}{\partial p} \qquad (3.127)$$

It can be noted that the horizontal wind is not replaced by its geostrophic value in the divergence term. When the geostrophic wind is computed using a constant Coriolis parameter, it is just the small departures of the horizontal wind from geostrophy which account for the divergence. This divergence and the corresponding vertical motion field are dynamically necessary to keep the temperature changes hydrostatic and vorticity changes geostrophic in synoptic scale systems.

3.9.4 Quasi-geostrophic Potential Vorticity Equation

The geostrophic wind equations cannot be used to predict the evolution of the flow since they do not contain time derivative, and are thus diagnostic equations, but not prognostic (predictive) ones. However, by performing more detailed scale analysis, it is possible to derive a system of prognostic equations which still embody the geostrophic balance as the lowest-order approximation, but include in certain terms the possibility of departure from geostrophy which are known as quasi-geostrophic equations as mentioned above.

There is no single quasi-geostrophic system, since the form of the equations depends, among other things, on the horizontal scale of the motions of interest compared to the Earth's radius. A common feature of these equations is that they can be used to construct prognostic equations for the quasi-geostrophic potential vorticity as shown below.

$$\left[\frac{\partial}{\partial t} + \vec{V}_g.\vec{\nabla}\right]q = \frac{D_g q}{Dt} = 0 \qquad (3.128)$$

where q is the quasi-geostrophic potential vorticity defined by

$$q \equiv \left[\frac{1}{f_0} \nabla^2 \Phi + f + \frac{\partial}{\partial p} \left(\frac{f_0}{\sigma} \frac{\partial \Phi}{\partial p} \right) \right]$$ (3.129)

The quantity q is made up of contributions from the geostrophic relative vorticity, the planetary vorticity due to the Earth's rotation, and the stretching vorticity. As the parcel moves about the atmosphere the geostrophic relative vorticity, the planetary vorticity, and the stretching vorticity terms may change. However, their sum is conserved according to the geostrophic potential vorticity equation.

3.10 Mean Meridional Circulation of the Middle Atmosphere

Now we will see the main processes responsible for the observed mean meridional circulation in the stratosphere and mesosphere. The dynamical behavior of the stratosphere differs considerably from that of the troposphere. The presence of ozone in the stratosphere determines to a very large degree the radiative drive and to a certain extent the temperature structure of the middle atmosphere (Plumb 1982; Salby 1996).

A comparison between the radiatively determined temperature and observed zonal mean temperatures in the stratosphere and mesosphere shows that the observed temperatures are significantly warmer than radiatively determined ones throughout the winter polar region. Nevertheless, the observed temperatures are colder than the radiatively determined ones in the summer poles. The stratosphere has a warm summer pole and a relatively cold winter pole, while the mesosphere exhibits a cold summer pole and a relatively warm winter pole. This indicates that processes other than radiation play a significant role in order to maintain the observed structure of the middle atmosphere (Geller 1983; O'Neill 1997; Haynes 2005).

The observed seasonal cycle in mean zonal winds shows that in both the seasons, the presence of sunlight over the summer pole produces ozone and strong heating. However, due to the absence of sunlight no heating occurs in the winter polar region. This causes a winter pole to summer pole temperature gradient in the stratosphere. Due to the thermal wind balance, easterlies increase with height during summer and westerlies increase with height during winter in the middle atmosphere (Hamilton 1998). The latitudinal temperature gradient reverses above 65 km, causing the winds to decrease with altitude in both hemispheres.

The dynamical processes produced by eddies or wave motion in the stratosphere and mesosphere give rise to the meridional circulation in the middle atmosphere that closely balances over sufficiently long timescales, the relaxation force caused by the temperature deviation from the radiative equilibrium. Thus the middle atmosphere is a dynamically driven region compared to the troposphere, which is more sensitive to thermal driven heat engine. A schematic representation indicating the role of different aspects of the dynamics of the stratosphere and tropopshere is given in Fig. 3.10.

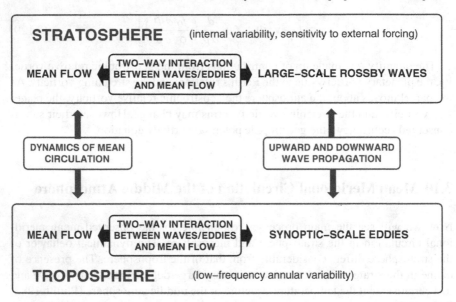

Fig. 3.10 Schematic representation of various aspects involved in maintaining the mean meridional circulation of the troposphere and stratosphere (Haynes 2005, Courtesy SPARC)

3.10.1 Zonally Averaged Circulation

First, let us look at the zonal mean circulation. We will use the log-pressure coordinate:

$$z = -H \ln \left(\frac{p}{p_0} \right) \tag{3.130}$$

where p_o is taken to be 1,000 hPa.

$$p(z) = p_0 \exp \left(\frac{-z}{H} \right) \tag{3.131}$$

where $p(z)$ is the pressure at height z and H is the scale height (g/RT).

To find the zonal means, which will be needed to examine the vertical and meridional flows, we must first write all the variables as the sum of a mean and a perturbed (or eddy) part. Thus,

$$u = \bar{u} + u' \tag{3.132}$$

The Eulerian means are found by taking the zonal averages for a fixed latitude, altitude, and time

$$\bar{u} = \frac{2}{\pi} \int_0^{2\pi} u(\lambda, \phi, z, t) d\lambda \tag{3.133}$$

Assume that the flow does not wander too far in the meridional direction, so that we can use the beta plane approximation.

$$f = f_0 + \beta y \tag{3.134}$$

where $f_0 = 2\Omega sin\phi_0$ and $\beta = (2\Omega/a)cos\phi_0$ and a is mean radius of the Earth = 6,371 km.

3.10.1.1 Zonal Mean Momentum Equation

Applying all the above assumptions, the resultant zonal mean momentum equation is

$$\frac{\partial \bar{u}}{\partial t} - f_0\bar{v} = -\frac{\partial(\overline{v'u'})}{\partial y} + \overline{X} \tag{3.135}$$

where $\partial \bar{u}/\partial t$ is the change in the mean zonal momentum with time, f_0v is the Coriolis forcing, $-\partial \overline{v'u'}/\partial y$ is the mean eddy momentum flux divergence, and $\overline{X}$ is the mean zonal eddy drag.

3.10.1.2 Thermodynamic Energy Equation

Thermodynamic energy equation can be written as:

$$\frac{\partial T}{\partial t} + N^2\frac{H}{R}\bar{w} = -\frac{\partial(\overline{v'T'})}{\partial y} + \frac{\overline{Q}}{C_p} \tag{3.136}$$

where $\partial \overline{T}/\partial t$ is the local change of T with time, $N^2(H/R)\bar{w}$ is the adiabatic cooling, $\partial(\overline{v'T'})/\partial y$ is the mean eddy heat flux divergence, and $\overline{Q}/C_p$ is the diabatic effects.

3.10.1.3 Static Stability Parameter

The most convenient measure of static stability for dynamical studies is the square of the buoyancy frequency (N^2), which is the frequency of adiabatic oscillation for a fluid parcel displaced vertically from its equilibrium level in a stably stratified atmosphere (see section 3.4.3). The lapse rate of temperature ($-DT/Dz$) is negative in the stratosphere and positive both in the troposphere and mesosphere. This indicates that the static stability in the stratosphere is much higher compared to that in the troposphere and mesosphere.

$$N^2 = \frac{R}{H}\left[\frac{kT_0}{H} + \frac{DT_0}{Dz}\right] = g\frac{D\ln\theta_0}{Dz} \tag{3.137}$$

where N is the buoyancy frequency, the atmosphere's natural resonance frequency for gravity-forced oscillations. We have neglected advection by ageostrophic mean meridional circulation and by vertical eddy flux divergences. When $N^2 > 0$, the atmosphere is statically stable. The static stability in the stratosphere is on the order of $N^2 \sim 5 \times 10^{-4} s^{-2}$, and in the mesosphere it is $N^2 \sim 3 \times 10^{-4} s^{-2}$. For modeling purposes, N^2 is often assumed to be constant throughout the stratosphere and mesosphere. But in the real atmosphere, it varies with altitude, latitude, longitude, and season.

The meridional momentum equation can be found by assuming geostrophic balance:

$$f_0 \bar{u} = -\frac{\partial \bar{\Phi}}{\partial y} \tag{3.138}$$

which, when combined with the hydrostatic relationship, gives the thermal wind relation:

$$f_0 \frac{\partial \bar{u}}{\partial z} + \frac{R}{H}\left[\frac{\partial \bar{T}}{\partial y}\right] = 0 \tag{3.139}$$

The ageostrophic mean meridional circulation is constrained by the thermal wind equation: the relationship between the zonal mean wind and the potential temperature distribution.

Without a mean meridional circulation, $\bar{v}$, $\bar{w}$, the eddy momentum flux divergence, and the eddy heat flux divergence, would tend to change the mean zonal wind and temperature fields so that thermal wind balance would be destroyed.

But small departures of the mean zonal wind from geostrophic balance cause a mean meridional circulation, thus maintaining the thermal wind balance. So a balance is established in the meridional direction so that the Coriolis force $f_0 u \sim$ the divergence of eddy momentum fluxes and the adiabatic cooling $\sim$ diabatic heating and convergence of eddy heat fluxes.

The temperature difference between the summer and winter poles in the 30–60 km region is less than expected from radiative equilibrium. Above 60 km, the gradient is even backwards. To understand these differences, the Eulerian mean circulation does not account for the tendency of eddy heat flux convergence and adiabatic cooling to cancel out each other, with the diabatic heating being a small factor.

Thus, an air parcel can rise only if its potential temperature is increased by diabatic heating. It is this small diabatic heating which gives rise to a residual meridional circulation, which in turn determines the mean meridional mass flow.

We can define the residual circulation as follows:

$$v^* = \bar{v} - \frac{R}{\rho_0 H}\left[\frac{\partial(\rho_0 \overline{v'T'}/N^2)}{\partial z}\right] \tag{3.140}$$

$$w^* = \bar{w} + \frac{R}{\rho_0 H}\left[\frac{\partial(\rho_0 \overline{v'T'}/N^2)}{\partial y}\right] \tag{3.141}$$

Now adiabatic motions and eddy thermal flux divergence and convergence are accounted for without assuming that they drive the circulation.

So, the zonal mean momentum equation and the thermodynamic energy equations have the form:

$$\frac{\partial \bar{u}}{\partial t} - f_0 \bar{v}^* = \frac{1}{\rho_0}\vec{\nabla}.\vec{F} + \bar{X} \equiv \bar{G} \tag{3.142}$$

$$\frac{\partial \overline{T}}{\partial t} + N^2 \frac{H}{R} \overline{w}^* = \frac{J}{C_p} \tag{3.143}$$

where $\vec{F}$ is the Eliassen-Palm (EP) flux (see Chapter 4, section 4.1), which arises from large-scale eddies, and X is the forcing from the small-scale eddies, like gravity wave drag. G is the total zonal drag force. $\vec{F} = \hat{j}F_y + \hat{k}F_z$, where the two components are given by

$$F_y = -\rho_0 \overline{u'v'} \tag{3.144}$$

$$F_z = \frac{\rho_0 f_0 \overline{R'_v T'}}{N^2 H} \tag{3.145}$$

The effects of nonzero EP flux divergence give rise to wave–mean flow interactions. F measures the flux of wave activity, which is a globally conserved quantity, if the flow is linear, steady, adiabatic, and nondissipative, i.e., in nonacceleration conditions. The drag on the mean flow will occur only when the EP flux divergence is nonzero, and then a meridional circulation is possible.

Now we can see how this all works by starting with a simple model and progressing to more complex models.

The air normally moves from the tropics to the poles, about $1/4$ of Earth's circumference, in periods of about 2 years. Thus, in steady-state condition, the value of $G \sim 10^{-4} s^{-1} \times 0.2 \, m \, s^{-1} = 2 \times 10^{-5} m \, s^{-2}$.

First assume there is no seasonal cycle. In this case, all time derivatives become zero, and the equations of motion become:

$$-f_0 \overline{v}^* = \frac{1}{\rho_0} \vec{\nabla}.\vec{F} + \overline{X} \equiv \overline{G} \tag{3.146}$$

$$N^2 \left(\frac{H}{R}\right) \overline{w}^* = \frac{\overline{Q}}{C_p} \tag{3.147}$$

The Coriolis force due to the residual meridional velocity just balances the eddies due to large and small eddies. The residual adiabatic cooling is just balanced by diabatic heating. If the eddy forcing did not exist, then the residual meridional velocity would be zero and the meridional drift would stop.

The two relationships are connected by the continuity equation, which gives the relationship:

$$-\frac{\partial \overline{G}}{\partial y} + \frac{f_0}{\rho_0} \left[\frac{\partial}{\partial z}(\rho_0 \overline{Q}k/N^2 H)\right] = 0 \tag{3.148}$$

If the eddy forcing does not exist, then the diabatic heating and the residual vertical velocity must also be zero. The simplest model of the atmosphere is thus one that is in radiative equilibrium, with the zonal mean temperature being equal to the steady-state radiative balance.

Next, we assume a diabatic heating that mimics the annual solar cycle. In this model, we parameterize the diabatic heating as the departure of the stratospheric temperatures from the radiative equilibrium value

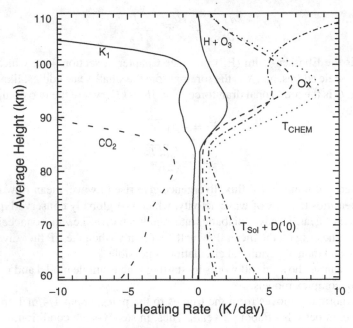

Fig. 3.11 Global mean heating rate calculated for solar minimum period (IPCC 2001)

$$\frac{\overline{Q}}{C_p} = \alpha_r \left[\overline{T} - T_r(y,z,t) \right] = -\alpha_r \delta \overline{T} \tag{3.149}$$

where α_r is the Newtonian cooling rate, and T_r is the radiative equilibrium temperature.

The stratosphere's temperature response will lag the temperature forcing because of the air's thermal capacity. This lag is small compared to the annual cycle because the thermal relaxation time is 5–20 days. Otherwise, this model produces an annually varying temperature distribution that looks very much like radiative equilibrium as shown in Fig. 3.11.

If we substitute this annually varying diabatic heating rate into the continuity equation, we find the equation:

$$\frac{1}{f_0} \frac{\partial \overline{G}}{\partial y} = -\frac{1}{\rho_0} \left[\frac{\partial}{\partial z} \left[\frac{\rho_0 R \alpha_r}{N^2 H} \delta \overline{T} \right] \right] \approx \frac{R \alpha_r}{N^2 H} \delta \overline{T} \tag{3.150}$$

where we assume that the density change is the most important change with altitude.

$$\delta \overline{T} \sim \tau_r \frac{N^2 H^2}{f_0 R} \frac{\partial \overline{G}}{\partial y} \tag{3.151}$$

where τ_r is the inverse of α_r.

So, the eddies drive the stratosphere away from thermal equilibrium and the radiation tries to drive it back. The largest departure from radiative equilibrium occurs in the polar winter stratosphere and summer and winter mesosphere.

In wintertime, the internal gravity waves propagate from the troposphere into the mesosphere, where they break. These breaking waves deposit energy, resulting in a strong zonal force that acts like friction, slowing down the zonal velocity. Stationary planetary waves play a similar role in the wintertime stratosphere.

Consider the equation:

$$-f_0\overline{v}^* = \frac{1}{\rho_0}\vec{\nabla}.\vec{F} + \overline{X} = \overline{G} \tag{3.152}$$

In the northern hemisphere winter, $f_o > 0$ and so wave drag force G is westward force, thus < 0. Hence, the residual meridional velocity must be > 0. By mass continuity, the residual vertical velocity must be negative, and air descends and warms.

In the southern hemisphere summer, $f_o < 0$, the zonal wind is westward, easterlies. As a result, $G > 0$ (eastward force to opposing the wind). Thus $(-|f_o|)v^* = G$ implies that $v^* > 0$, or northward drift.

3.11 Annual Mean Energetics

Tropospheric energy budget is well known, but only very few studies have been reported on the energetics of the stratosphere. Basically, the energy cycles in the troposphere and stratosphere are different (Asnani 2005). The eddy kinetic energy (KE) is generated through conversion of eddy available potential energy (AE) in the troposphere, whereas KE in the stratosphere is received from the troposphere and then gets converted into AE. The radiational differential heating in the troposphere generates zonal mean available potential energy (A_z), which gets lost by radiation in the stratosphere. Above 30 km, the mean meridional circulation is thermally driven and is maintained by the conversion of A_z to zonal mean kinetic energy (K_z) against frictional dissipation.

Julian and Labitzke (1965) had investigated the energetic of the troposphere and stratosphere up to 30 km during the months of January and February in 1963 associated with stratospheric warming events. It has been reported that the variation of energy exchange with height in the lower and middle troposphere and lower stratosphere were baroclinically active regions before and during the warming. The principal path of energy flow in these regions is from zonal to eddy available potential energy, from there to eddy kinetic energy, and hence to zonal kinetic energy. At all times eddy kinetic energy is supplied to the stratosphere by the upward flux of mechanical energy from the troposphere. The upward flux of energy to the stratosphere is large compared with the energy exchange processes occurring in the stratosphere and must be considered a significant item in the budget of the stratosphere.

The spatial and temporal variability of energy flux in the troposphere and stratosphere suggests that the sum of the meridional temperature gradient determines the principal path of energy flow between these two layers. The middle and upper stratosphere during winter are capable of maintaining a type of baroclinic circulation which is replaced by the circulation driven by the temperature during the period of major stratospheric warming. The net transfer of zonal kinetic energy to eddy kinetic energy in the vicinity of the tropopause and lower stratosphere suggests that the warming phenomenon is partly due to the result of barotropic processes. The vertical variation in the energy exchange indicates that the lower stratosphere is a region characterized by small net energy transformation separating the baroclinically active troposphere and middle stratosphere (Julian and Labitzke 1965).

Figure 3.12 depicts the annual mean energy cycle for the layers 100–10 hPa computed by Dopplick (1971) from selected radiosonde observations between altitudes 20–90°N. The eddy kinetic energy (KE) is received by the lower stratosphere from the troposphere across the tropopause through upward propagating quasi-stationary waves (Holton 1975). About half of the incoming KE is converted into AE through rising of cold air and sinking of warm air. The remaining half of the KE is converted into K_z. AE also receives equal contribution from differential heating. AE gets transformed into A_z. Thus, the eddies enhance the already existing meridional temperature gradient, known as refrigeration effect. A_z loses mostly through radiational dissipation.

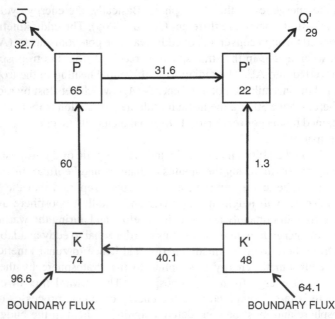

Fig. 3.12 Annual mean energy cycle for the 15–30 km region for the latitudes 20–90°N (Holton 1975, Courtesy: American Meteorological Society)

Problems and Questions

3.1. Let the pressure surface at a height of 850 hPa over Berlin decrease by 12 m southward across a distance of 240 km. Calculate the pressure gradient force at this height over Berlin.

3.2. Assume the temperature of a site 100 km north of a radar station is 4°C lower. Find the temperature change at the radar station, if the wind is blowing from northeast at a speed of 15 m s^{-1} and the air is heated at a rate of 1°C per hour.

3.3. What is the geostrophic wind speed at a station located at 30°N, if the pressure inceases by 5 hPa eastward across a distance of 200 km?

3.4. If the height of a pressure surface decreases by 200 m eastward across a distance of 700 km, calculate the geostrophic wind at 45°N latitude region.

3.5. Consider that the air density inside a cabin of 10 m length before a cyclonic storm arrives is 1 kg m^{-3}. Assume that only the western side of the cabin is fully opened and winds of 60 m s^{-1} enter from the western open side. All other sides of the cabin are fully closed, so that the air cannot leave from the other side of the cabin. In such a condition estimate the density changes within the cabin during the first 2 seconds.

3.6. Assume air flow through an open window of size 2 m width and 3 m height. Let the mass flux of air flow through the window be 2 kg m^{-2} s^{-1}. Calculate the amount of air mass flow through the window per minute and estimate the speed of the wind blowing through the window.

3.7. Let the geostrophic wind around a low pressure region located around 45°N, (where the magnitude of f is 10^{-4} s^{-1}) be 25 m s^{-1}. If the radius of curvature of the low pressure region is 40 km, estimate the speed of the cyclostrophic wind in this latitude region.

3.8. Let us assume the temperature of a parcel of air increases from 6°C to 10°C as it moves towards the east across a distance of 100 km. If the air parcel is dry and transports at 60°N latitude, calculate the vertical shear of the geostrophic wind.

3.9. Assume that the thickness of an atmospheric layer between 1,000 and 700 hPa level over a station is 3.0 km. The thickness of a neighboring station situated 700 km toward the east shows 3.2 km in the above pressure levels. Let both the stations be located in the same latitude belt, where the magnitude of the Coriolis parameter is 10^{-4} s^{-1}, calculate the components of thermal wind vector in this region.

3.10. Find the relative vorticity of a parcel of air which rotates as a solid body about the center of a low-pressure system. Assume that the tangential velocity is 12 m s^{-1}, where the radius of curvatute is 240 km.

3.11. Consider an air parcel at $30°N$ having an initial relative vorticity of 2×10^{-5} s^{-1}, moving towards north. Calculate the relative vorticity of the parcel after reaching $45°N$, $60°N$, and at the pole.

3.12. If the horizontal wind has a shear of -7 m s^{-1} across a distance of 500 km, what is the potential vorticity of air which extends upto 14 km altitude at $30°N$ latitude region.

3.13. Calculate the isentropic potential vorticity of an air parcel having the potential vorticity of 1.4×10^{-8} m^{-1} s^{-1}. Assume the air density at 12 km altitude is 0.3 kg m^{-3} and the vertical shear of the potential temperature is 2.8 K km^{-1}.

3.14. Define geostrophic wind and gradient wind. How are they related? Under what conditions does geostrophic wind become equal to the gradient wind?

3.15. What is the thermal wind? Based on thermal wind balance computations, what happens to the zonal wind in the northern hemisphere as you ascend in the atmosphere? Where would the change in zonal wind be the greatest?

3.16. Describe circulation and vorticity. What are the assumptions made to derive a quasi-geostrophic vorticity equation? Why is this equation called quasi-geostrophic?

3.17. What is potential vorticity? What happens to the relative vorticity of a parcel of air which travels across a north–south-oriented mountain barrier? Explain.

3.18. Based on the potential vorticity conservation principle, describe the relative vorticity changes of a parcel of air which moves from the north pole to the south pole.

3.19. Why is potential temperature used as a vertical coordinate? Discuss the static stability conditions of a parcel of air in the troposphere and stratosphere?

3.20. What are isentropic surfaces? How does stratospheric air move in relation to such surfaces?

3.21. Discuss the mean meridional circulation of stratospheric circulation. Describe how the stratospheric circulation is different from that of the tropospheric circulation.

3.22. What kind of temperature anomaly is associated with a transition zone between a westerly and easterly wind? How about between an easterly and westerly wind?

3.23. What phenomenon arises as a result of wave-breaking in the stratosphere? What happens to the air in the polar stratosphere immediately after a stratospheric sudden warming episode?

References

Andrews DG, Holton JR, Leovy CB (1987) Middle Atmosphere Dynamics, Academic, New York

Asnani GC (2005) Tropical Meteorology, second edition, vol. 2, Pune, India

Dopplick TG (1971) The energetics of the lower stratosphere, Quart J Royal Met Soc, 97: 209–237

Geller MA (1983) Dynamics of the middle atmosphere, Space Sci Rev, 34: 359

Gill AE (1982) Atmosphere-Ocean Dynamics, Academic, New York

Hamilton K (1998) Dynamics of the tropical middle atmosphere; A tutorial review, Atmosphere-oceans, 36: 319–354

Haynes PH (2005) Stratospheric dynamics, Annu Rev Fluid Mech, 37: 263–293

Haynes PH (2006) Stratosphere troposphere coupling, SPARC Newsletter, vol. 25

Hess SL (1959) Introduction to Theoretical Meteorology, Holt, New York

Holton JR (1975) Dynamic Meteorology of the Stratosphere and Mesosphere, Meteor Monograph, Amer Met Soc, 15(37)

Holton JR (2004) An Introduction to Dynamic Meteorology, fourth edition, Elsevier, Burlington

Holton JR, Haynes PH, McIntyre ME, Douglass AR, Rood RB, Pfister L (1995) Stratosphere troposphere exchange, Rev Geophys, 33: 403–439

IPCC (2001) Climate Change 2001, Working Group I: The Scientific basis, UNEP/WMO

Julian PR, Labitzke K (1965) A study of atmospheric energetics during January and February 1963 stratospheric warming, J Atmos Sci, 22: 597–610

Martin EJ (2006) Mid-latitude Atmospheric Dynamics, Wiley, Chichester, England

Nappo CJ, (2002) An Introduction to Atmospheric Gravity Waves, Academic, San Diego, CA

O'Neill A (1997) Observations of Dynamical Processes, The Stratosphere and its Role in the Climate System, NATO ASI Series, edited by Guy P Brasseur, vol. 54, Springer, Hiedelberg

Pedlosky J (1987) Geophysical Fluid Dynamics, second edition, Springer, New York

Pidwirny M (2006) Forces acting to create wind, Fundamentals of Physical Geography, 2nd edition, e-book (http://www.physicalgeography.net/fundamentals/7n.html)

Plumb RA (1982) The circulation of the middle atmosphere, Aust Meteorol Mag, 30: 107–121

Salby ML (1996) Fundamentals of Atmospheric Physics, Academic, San Diego, CA

Virtual Lab, The properties of potential vorticity; climatology and dynamic tropopause (http://www.virtuallab.bom.gov.au/meteofrance/cours/resource/ab03/pv_defr.gif)

Wallace JM, Hobbs PV (2006) Atmospheric Science – An Introductory Survey, second edition, Elsevier, New York

Chapter 4
Waves in the Troposphere and Stratosphere

4.1 Introduction

The atmosphere exhibits many wavelike motions with a variety of space- and timescales ranging from slow-moving planetary scale waves to much faster and smaller gravity waves, each playing important roles in the behavior of the stratosphere. It has long been known that conditions in the stratosphere are controlled by wave driving from the troposphere, but it has been assumed that the stratosphere has little effect on the troposphere. Stratospheric variations, especially variations in the strength of the polar vortex, appear to be involved in feedback processes that in turn alter weather patterns in the troposphere. Stratospheric variations are largest during the winter season, and they are influenced by changes in solar irradiance, volcanic aerosols, changes in greenhouse gases, ozone depletion, and the phase of the quasi-biennial oscillation (QBO).

Stratospheric circulation anomalies are caused mainly by wave forcing from the dense troposphere. Stochastic variations in the troposphere during northern hemispheric winter lead to high-frequency changes in the planetary wave flux upward into the stratosphere (Holton 1983). When these waves break, they deposit momentum in the stratosphere, slowing the zonal mean wind and weakening the polar vortex. The interaction of the waves with the mean flow tends to draw these zonal wind anomalies downward through the stratosphere (Andrews et al. 1987; Andrews 2000; Martin 2006).

4.2 What is a Wave?

A wave may be defined as a form or a state of disturbance advancing with a finite velocity through a medium. By means of a wave, energy is transmitted, being passed along from one part of a medium to the next by the interaction of adjoining parts.

Each wave takes along with it a certain definite quantity of energy, which remains with it. The wave may be considered as perturbation on the steady slowly changing background.

Waves in fluids result from the action of restoring forces on fluid parcels, which have been displaced from their equilibrium positions. The restoring forces may be due to compressibility, gravity, rotation, etc.

4.3 Basic Properties of Waves

When an air parcel or fluid particle is displaced from its initial position, a restoring force may cause it to return to its initial position. In such a case, because of inertia the fluid particle will generally tend to overshoot and pass its initial equilibrium position moving in the opposite direction from that in which it is initially displaced, thereby creating an oscillation around the equilibrium position. Concurrently, a wave is produced that propagates from this source or forcing region to another part of the fluid system, which itself is the physical medium of wave propagation.

A physical restoring force and a medium for propagation are the two fundamental elements of all wave motion in solids, liquids, and gases, including atmospheric waves, oceanic waves, sound (acoustic) waves, wind-induced waves, and seismic waves. The ultimate behavior of the wave is dictated by the individual properties of the restoring force responsible for wave generation and the medium through and by which the wave propagates energy and momentum.

Wave motions may be characterized by several fundamental properties, such as wave frequency, wave number, phase speed, group velocity, and dispersion relationship (which relates the wave frequency to the wave number). A typical wave motion is represented in Fig. 4.1. The period of oscillation (τ) of a wave determines the wave frequency while the horizontal and vertical spatial scales of a wave determines its horizontal and vertical wave numbers ($k = 2\pi/L_x; l = 2\pi/L_y; m = 2\pi/L_z$), where k, l, and m are wave numbers and L_x, L_y, and L_z are wavelengths in the zonal (x), meridional (y), and vertical (z) directions, respectively.

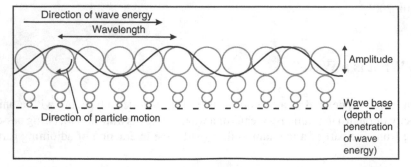

Fig. 4.1 Representation of a simple wave motion (Courtesy: University of Maryland)

A wave may be characterized by its amplitude and phase,

$$\phi(t,x,y,z) = R_e \, A \, \exp i(kx + ly + mz - vt - \alpha) \tag{4.1}$$

where ϕ represents any of the dependent flow variables, R_e the real part of the wave function ϕ, A the amplitude, $kx + ly + mz - vt - \alpha$ the phase and the phase angle. The phase angle is determined by the initial position of the wave. Lines of constant phase, such as wave crests, troughs, or any other particular part of the wave propagate through the fluid medium at a speed called the phase speed, which is given by

$$c_{px} = \frac{v}{k}; \; c_{py} = \frac{v}{l}; \; c_{pz} = \frac{v}{m} \tag{4.2}$$

In the atmosphere, waves are in a complicated form, due to the superposition of waves with different wavelengths and their nonlinear interactions, and cannot be represented by a simple, single sinusoidal wave. However, any overall observable wave shape can always be represented by a linear superposition of wave trains of different wave numbers, i.e., a Fourier series of sinusoidal components.

A wave in the x-direction may be decomposed into sinusoidal components of the form

$$\phi(x) = \sum_{n=1}^{\infty} (A_n \sin k_n x + B_n \cos k_n x) \tag{4.3}$$

where the Fourier coefficients A_n and B_n are determined by

$$A_n = \frac{2}{L} \int_0^L \pi(x) \sin \frac{2\pi n x}{L} dx \tag{4.4}$$

$$B_n = \frac{2}{L} \int_0^L \pi(x) \cos \frac{2\pi n x}{L} dx \tag{4.5}$$

The nth Fourier component or n^{th} harmonic of the wave function ϕ is defined as $A_n \sin k_n x + B_n \cos k_n x$. In deriving Eqs. (4.4), and (4.5), we have used the orthogonality relationships

$$\int_0^L \sin \frac{2\pi n x}{L} \cos \frac{2\pi m x}{L} dx = 0, \text{ for all } n, m > 0 \tag{4.6}$$

$$\int_0^L \sin \frac{2\pi n x}{L} \sin \frac{2\pi m x}{L} dx = \left\{ \begin{matrix} 0, & n \neq m \\ \frac{L}{2}, & n = m \end{matrix} \right\}, \text{ and} \tag{4.7}$$

$$\int_0^L \cos \frac{2\pi n x}{L} \cos \frac{2\pi m x}{L} dx = \left\{ \begin{matrix} 0, & n \neq m \\ \frac{L}{2}, & n = m \end{matrix} \right\} \tag{4.8}$$

If a wave is composed of a series of Fourier components of different wavelengths, then the phase speed for each individual component may also be different. If the phase speed depends on wave number, then the wave will not be able to retain its initial shape and remain coherent as it propagates in the medium, since each Fourier component is propagating at a different phase speed. In other words, the wave is *dispersive*. On the other hand, if the phase speed is independent of wave number, then the wave will retain its initial shape and remain coherent as it propagates throughout the fluid medium. This type of wave is called a *nondispersive wave*. Clearly, the relationship between wave frequency and wave number determines whether or not the wave is dispersive.

Although a dispersive wave may look like it is dissipative, dispersion and dissipation are completely different physical types of processes. In a dissipative wave, every Fourier component of the wave propagates at the same speed, while the wave amplitude decreases. Thus, individual wave groups preserve their shapes during propagation. The total potential and kinetic energy of a wave propagates with the velocity of its slow-varying modulations, which is known as the group velocity, and may be given by the relation

$$c_g = c_{gx}\hat{i} + c_{gy}\hat{j} + c_{gz}\hat{k} = \frac{\partial v}{\partial k}\hat{i} + \frac{\partial v}{\partial l}\hat{j} + \frac{\partial v}{\partial m}\hat{k} \tag{4.9}$$

If the wave is nondispersive, then the wave pattern moves throughout the medium without any change in shape of the initial waveform. This means that the (phase) velocity of the individual wave crests is equal to the (group) velocity of the slow-varying modulations or the envelope of Fourier wave components (c_g). The concept of group velocity can be illustrated by Fig. 4.2, in which the simple group of two superimposed sinusoidal waves is represented by the wave function:

$$\phi(x,t) = \exp\left\{i[(k + \Delta k)x - (v + \Delta v)t]\right\} + \exp\left\{i[(k - \Delta k)x - (v - \Delta v)t]\right\} \tag{4.10}$$

which propagates at the speed $\Delta k/\Delta v$. This speed approaches $\partial k/\partial v$, which is defined as the group velocity in the x direction as it approaches 0.

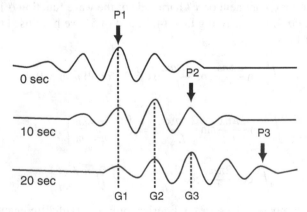

Fig. 4.2 Group velocity of two superimposed sinusoidal waves

4.4 Classification of Waves

Atmospheric waves can be classified according to their physical or geometrical properties. Firstly, they can be classified according to their restoring mechanisms; thus buoyancy or gravity waves owe their existence to stratification, while inertio-gravity waves result from a combination of stratification and Coriolis effects. Planetary or Rossby waves result from the β-effect or from the northward potential vorticity gradients.

Another type of classification is to distinguish forced waves, which must continually be maintained by an excitation mechanism of given phase speed and wave number, from free waves, which are not so maintained. Thermal tide is an example of forced waves, which is induced by the diurnal fluctuations in solar heating, while examples of free waves consist of global normal modes.

A further classification results from the fact that some waves can propagate in all directions, while others may be trapped (or *evanescent*) in some directions. Thus under some circumstances horizontally propagating planetary waves can be trapped in the vertical, while equatorial waves can propagate vertically and zonally but are evanescent with increasing distance from the equator.

Waves can also be separated into *stationary waves*, whose surfaces of constant phases are fixed with respect to the Earth, and *traveling waves*, whose phase surfaces move. Since information propagates with the group velocity and not with the phase speed, propagation can still occur in stationary waves. In the case of waves with fixed nodal surfaces, the stationary waves are called *standing waves*. The waves whose amplitudes are independent of time are known as steady waves, and those whose amplitude varies with time are called *transient waves*.

The waves that are linear, steady, frictionless, and adiabatic do not lead to any mean flow acceleration, whereas the transient and nonconservative waves lead to mean flow acceleration. Nonlinear waves can sometimes satisfy nonacceleration conditions, if they are steady and conservative.

4.5 Waves in the Atmosphere

There exist four basic modes of atmospheric wave propagation that are of interest in atmospheric science. The four types are:

> Acoustic waves – speed controlled by *temperature*
> Gravity waves – speed controlled by *stability*
> Inertial waves – speed controlled by *Coriolis parameter*
> Rossby waves – speed controlled by *latitudinal variation of Coriolis parameter*

Waves in the stratosphere and troposphere include gravity (buoyancy) waves, Rossby waves, inertio-gravity waves, forced stationary planetary waves, free traveling planetary waves, equatorial waves, and midlatitude gravity waves.

Acoustic waves (sound waves) are longitudinal waves, in which the particle oscillation is parallel to the direction of propagation. They derive their oscillations from the compression and expansion of the medium due to the compression force.

Gravity waves are those whose restoring mechanism is their buoyancy or vertical static stability. Gravity is involved in the hydrostatic balance part.

Inertio-gravity waves are those with a sufficiently long period to cause them to "feel" the rotation of Earth (i.e., the Coriolis deflection).

Rossby waves are those whose restoring mechanism is the latitudinal (north–south) gradient of potential vorticity. The more significant topography in the northern hemisphere results in air parcels that are displaced north–south. Such displacements cause changes in the planetary and relative components of vorticity in such a way that total potential vorticity is conserved. The result is north–south undulations that produce wave patterns around a circle of latitude.

Forced stationary planetary waves are Rossby waves with very long wavelengths up to 10,000 km that are generated by large-scale surface topography like the Rocky Mountains and the Himalaya–Tibet complex, or by land–sea temperature contrasts. These waves remain stationary since the topographical features or land–sea temperature contrasts force them not to move. Such waves propagate upward when westerly zonal winds are weak. They deposit their momentum in the stratosphere.

Free traveling planetary waves are planetary waves of wavelengths on the order of 10,000 km that are generated in the atmosphere at certain "natural frequencies" wherein the atmosphere is excited at a certain frequency and generates a wave. These waves propagate around a latitude band with a period of a few days.

Equatorial waves are mixed Rossby gravity waves that propagate eastward with a frequency similar to the inertio-gravity waves and westward with a frequency similar to the Rossby waves. They are driven by the change in sign of the Coriolis parameter at the equator. A type of equatorial wave is the Kelvin wave, which propagates eastward like a pure gravity wave with only zonal velocity. Kelvin waves in the ocean propagate along the thermocline or region of tightest temperature gradient. The ocean Kelvin wave is associated with the ENSO phenomenon, as these waves propagate along the thermocline, bringing warm or cold water anomalies to the eastern Pacific.

Midlatitude gravity waves are inertio-gravity waves that exist in the middle and high latitudes and can propagate into the middle atmosphere.

Now we will discuss in detail about each of the above-mentioned waves present in the atmosphere.

4.5.1 Acoustic (Sound) Waves

Consider one-dimensional, small-amplitude, adiabatic perturbations in a nonrotating, inviscid, uniform (no basic wind shear) flow. To exclude the possibility of transverse oscillations, we assume that $v = w = 0$ and $u = u(x,t)$. The momentum and pressure tendency equation takes the form

$$\frac{\partial u'}{\partial t} + U\frac{\partial u'}{\partial x} + \frac{1}{\bar{\rho}}\frac{\partial p'}{\partial x} = 0 \tag{4.11}$$

$$\frac{\partial p'}{\partial t} + U\frac{\partial p'}{\partial x} + \gamma\bar{p}\frac{\partial u'}{\partial x} = 0 \tag{4.12}$$

The above two equations may be combined into a single equation for the pressure perturbation p,

$$\left[\frac{\partial}{\partial t} + U\frac{\partial}{\partial x}\right]p' - c_s^2\frac{\partial^2 p'}{\partial x^2} = 0 \tag{4.13}$$

where $c_s^2 = \gamma RT$. Assuming a wave-like solution,

$$p' = p_0\,\exp[i(kx - ct)] \tag{4.14}$$

and substituting it into Eq. (4.13) leads to

$$v = (U \pm c_s)k \tag{4.15}$$

The above method of the assumption of wavelike solutions for the small-amplitude perturbations is also referred to as the method of normal modes. The above equation is the dispersion relation for sound waves which relates the wave frequency ω to the horizontal wave number k. From Eq. (4.15), we may obtain the horizontal phase speeds, which are given by

$$c = \frac{v}{k} = U \pm c_s \tag{4.16}$$

where $c_s = \sqrt{\gamma R \overline{T}}$. The above equation represents phase speeds $U \pm c_s$ of the upstream (leftward-moving) and downstream (rightward-moving) propagating sound waves, which are simultaneously being advected by the basic wind U. Sound waves are nondispersive since their phase speeds are independent of wave number. This nondispersive property of sound waves may also be verified by showing that the group velocities for these waves c_g are identical to their phase speeds c.

If a wave propagates in the direction of the oscillatory restoring force or parallel to the motion of the fluid particles, it is called a *longitudinal wave*. Conversely, a wave propagating in a direction perpendicular to the oscillatory restoring force or particle motions comprising the fluid is called a *transverse wave*. It can be shown that sound waves are longitudinal waves.

Consider a semi-infinite tube filled with gas whose right-hand side extends to infinity and is confined by a piston on the left-hand side. When the gas is alternatively compressed and allowed to expand by oscillating the piston in and out of the left-hand side of the semi-infinite tube, an air parcel located adjacent to the piston will be subjected to oscillate back and forth about its equilibrium position due to the oscillating horizontal pressure gradient and, concurrently, a sound wave will be excited that propagates toward the right (i.e., down the tube) at the speed $c_s = \sqrt{\gamma R \overline{T}}$. In a dry, isothermal atmosphere with a constant temperature of 300 K, a one-dimensional acoustic wave has a horizontal phase speed and group velocity of approximately 347 m s^{-1}.

4.5.2 Lamb Waves

Since sound waves do not play significant dynamic roles in affecting most atmospheric motions, they are often eliminated from the primitive governing equations, in particular those that are commonly employed in most current operational numerical weather prediction models. Although sound waves may have no particular relevance to atmospheric motions in the troposphere, which are responsible for weather, a special class of waves called Lamb waves has been observed. These waves can propagate horizontally in an isothermal atmosphere in the absence of vertical motion. In a two-dimensional, adiabatic, hydrostatic, nonrotating, inviscid, isothermal atmosphere with no vertical motion and basic wind shear, small-amplitude motions are represented as

$$\frac{\partial u'}{\partial t} + U\frac{\partial u'}{\partial x} + \frac{1}{\bar{\rho}}\frac{\partial p'}{\partial x} = 0 \qquad (4.17)$$

$$\frac{1}{\bar{\rho}}\left[\frac{\partial p'}{\partial z} + \frac{p'}{H}\right] = g\frac{\theta'}{\bar{\theta}} \qquad (4.18)$$

$$\frac{\partial p'}{\partial t} + U\frac{\partial p'}{\partial x} + \gamma\bar{p}\frac{\partial u'}{\partial x} = 0 \qquad (4.19)$$

$$\frac{\partial \theta'}{\partial t} + U\frac{\partial \theta'}{\partial x} = 0 \qquad (4.20)$$

In an isothermal atmosphere, the scale height H is constant. Equations (4.18) and (4.20) may be combined to yield an equation, which when coupled with the equation resulting from combining Eqs. (4.17) and (4.19) forms the set of equations governing the evolution of Lamb waves.

4.5.3 Shallow Water Gravity Waves

Shallow water gravity waves are horizontally propagating oscillations. Shallow water waves can exist only if the fluid has a free surface or internal density discontinuity. In shallow water gravity waves, the restoring force is in the vertical so that it is transverse to the direction of propagation. Although this type of wave is observed more often in the oceans and lakes, they are also occasionally observed in the atmosphere.

Waves are said to be shallow water when the depth is less than one twenty-fifth of their wavelength. After reaching this depth, the water particle orbits inside the wave become elliptical rather than circular as the up–down component of the motion is squeezed by the presence of the bottom, as illustrated in Fig. 4.3. The squeezing happens more quickly than the reduction of orbit size with depth so that to and fro motion extends all the way to the bottom.

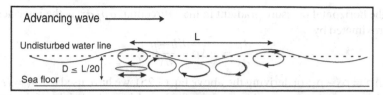

Fig. 4.3 Schematic representation of shallow-water waves (Courtesy: S Brachfeld)

Consider a nonrotating, hydrostatic, two-layer fluid system with constant densities 1 and 0 in the upper and lower layers, respectively, as shown in Fig. 4.4. Waves may propagate along the interface between the two layers of different densities. Let us assume that the density of the lower layer ρ_0 is greater than the density of the upper layer ρ_1, i.e., $\rho_1 < \rho_0$. This means the fluid system is stably stratified, i.e., heavier fluid underlies lighter fluid. The horizontal pressure gradients are independent of height in each layer if the fluid system is in hydrostatic balance, because

$$\frac{\partial}{\partial z}\left(\frac{\partial p}{\partial x}\right) = \frac{\partial}{\partial x}\left(\frac{\partial p}{\partial z}\right) = -\frac{\partial}{\partial x}(\rho g) = 0 \tag{4.21}$$

since the density is constant in each layer. We also assume that there exist no horizontal pressure gradients in the upper layer.

Considering point A in Fig. 4.4, we note that according to the hydrostatic equation,

$$\frac{p - (p + \delta p_1)}{\delta z} = -\rho_1 g \tag{4.22}$$

The above equation implies that

$$p + \delta p_1 = p + \rho_1 g \delta z = p + \rho_1 g \left(\frac{\partial (H + h')}{\partial x}\right) \delta x \tag{4.23}$$

where H is the undisturbed upstream fluid depth and h' is the perturbation or the vertical displacement from H. Similarly, we may derive the pressure at point B,

$$p + \delta p_0 = p + \rho_0 g \delta z = p + \rho_0 g \frac{\partial (H + h')}{\partial x} \delta x \tag{4.24}$$

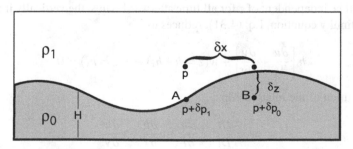

Fig. 4.4 Two layers of homogeneous incompressible fluid system

Thus, the horizontal pressure gradient in the x-direction, $\partial p/\partial x$, at the interface can be approximated by

$$\frac{\partial p}{\partial x} = g\Delta\rho\frac{\partial(h+h_s)}{\partial x} \qquad (4.25)$$

where $\Delta\rho = \rho_0 - \rho_1$. In deriving the above Eq. (4.25), we have used $h+h_s = H+h'$, where h is the instantaneous depth of the fluid, and h_s is the height of the bottom topography. Similarly, we derive the horizontal pressure gradient in the y-direction

$$\frac{\partial p}{\partial y} = g\Delta\rho\frac{\partial(h+h_s)}{\partial y} \qquad (4.26)$$

Therefore, the horizontal momentum equations become

$$\frac{\partial u}{\partial t}+u\frac{\partial u}{\partial x}+v\frac{\partial u}{\partial y}+w\frac{\partial u}{\partial z} = -g'\frac{\partial(h+h_s)}{\partial x} \qquad (4.27)$$

$$\frac{\partial v}{\partial t}+u\frac{\partial v}{\partial x}+v\frac{\partial v}{\partial y}+w\frac{\partial v}{\partial z} = -g'\frac{\partial(h+h_s)}{\partial y} \qquad (4.28)$$

where $g' = g\Delta\rho/\rho_0$ is called the reduced gravity. Assuming there is no vertical shear of the horizontal winds ($\partial u/\partial z = \partial v/\partial z = 0$) initially, it can be shown that u and v will be independent of z at any subsequent time. Under this limitation, Eqs. (4.27) and (4.28) reduce to

$$\frac{\partial u}{\partial t}+u\frac{\partial u}{\partial x}+v\frac{\partial u}{\partial y} = -g'\frac{\partial}{\partial x}(h+h_s) \qquad (4.29)$$

$$\frac{\partial v}{\partial t}+u\frac{\partial v}{\partial x}+v\frac{\partial v}{\partial y} = -g'\frac{\partial}{\partial y}(h+h_s) \qquad (4.30)$$

Integrating the continuity equation in isobaric coordinate system, from the surface of the bottom topography ($z = h_s$) to the interface ($z = H+h'$) with respect to z leads to

$$\int_{h_s}^{H+h'}\left[\frac{\partial u}{\partial x}+\frac{\partial v}{\partial y}\right]dz + \int_{h_s}^{H+h'}\frac{\partial w}{\partial z}dz = 0 \qquad (4.31)$$

Since u and v are assumed to be independent of z initially, then both $\partial u/\partial x$ and $\partial v/\partial y$ will be independent of z for all time afterward. Thus, the vertically integrated mass continuity equation, Eq. (4.31), reduces to

$$h\left[\frac{\partial u}{\partial x}+\frac{\partial v}{\partial y}\right]+w(z=h+h_s)-w(z=h_s) = 0 \qquad (4.32)$$

Substitution of the relationship

$$w = \frac{Dz}{Dt} = \frac{\partial h}{\partial t}+u\frac{\partial h}{\partial x}+v\frac{\partial h}{\partial y} \qquad (4.33)$$

into Eq. (4.32) leads to

$$h\left[\frac{\partial u}{\partial x}+\frac{\partial v}{\partial y}\right]+\left[\frac{\partial}{\partial t}(h+h_s)+u\frac{\partial}{\partial x}(h+h_s)+v\frac{\partial}{\partial y}(h+h_s)\right]-\left[u\frac{\partial h_s}{\partial x}+v\frac{\partial h_s}{\partial y}\right]=0$$

(4.34)

This can be further reduced to the expression

$$\frac{\partial h}{\partial t}+u\frac{\partial h}{\partial x}+v\frac{\partial h}{\partial y}+h\left[\frac{\partial u}{\partial x}+\frac{\partial v}{\partial y}\right]=0$$

(4.35)

since h_s, the shape of the bottom topography, is generally assumed to be independent of time. Therefore, disturbances in the two-layer shallow water system will be governed by the following reduced forms of the horizontal momentum and vertically integrated continuity equations:

$$\frac{\partial u}{\partial t}+u\frac{\partial u}{\partial x}+v\frac{\partial u}{\partial y}=-g\frac{\partial}{\partial x}(h+h_s)$$

(4.36)

$$\frac{\partial v}{\partial t}+u\frac{\partial v}{\partial x}+v\frac{\partial v}{\partial y}=-g\frac{\partial}{\partial y}(h+h_s)$$

(4.37)

$$\frac{\partial h}{\partial t}+u\frac{\partial h}{\partial x}+v\frac{\partial h}{\partial y}+h\left[\frac{\partial u}{\partial x}+\frac{\partial v}{\partial y}\right]=0$$

(4.38)

Substituting $u=U+u'$, $v=V+v'$, and $h=H+h'$, we obtain the perturbation form of Eqs. (4.36), (4.37), and (4.38) as

$$\frac{\partial u'}{\partial t}+(U+u')\frac{\partial u'}{\partial x}+(V+v')\frac{\partial u'}{\partial y}+g'\frac{\partial h'}{\partial x}=0$$

(4.39)

$$\frac{\partial v'}{\partial t}+(U+u')\frac{\partial v'}{\partial x}+(V+v')\frac{\partial v'}{\partial y}+g'\frac{\partial h'}{\partial x}=0$$

(4.40)

$$\frac{\partial h'}{\partial t}+(U+u')\frac{\partial h'}{\partial x}+(V+v')\frac{\partial h'}{\partial y}+(H+h'-h_s)\left[\frac{\partial u'}{\partial x}+\frac{\partial v'}{\partial y}\right]$$

$$=(U+u')\frac{\partial h_s}{\partial x}+(V+v')\frac{\partial h_s}{\partial y}$$

(4.41)

Here, H is, the undisturbed upstream layer depth and h' is the vertical displacement from H. Thus, h_s is no longer included in the horizontal pressure gradient forces in the momentum Eqs. (4.39) and (4.40). However, in the system of perturbation Eqs. (4.39)–(4.41), it serves as a *forcing term* for the vertical velocity $w'=Dh'/Dt$ in the vertically integrated mass continuity Eqn. (4.41).

Now let us consider the special case of small-amplitude (linear) perturbations in a single, one-layer fluid with a flat bottom and with no meridional basic state velocity component (i.e., $V = 0$). The air–water system may be approximated by this type of two-layer system since $\rho_1 = \rho_{air} << \rho_0 = \rho_{water}$. These assumptions yield $\Delta\rho = \rho_0 - \rho_1 = \rho_0$ or $\Delta\rho \approx \rho_0$.

Therefore, Eqs. (4.39)–(4.41) become

$$\frac{\partial u'}{\partial t} + U\frac{\partial u'}{\partial x} + g\frac{\partial h'}{\partial x} = 0 \tag{4.42}$$

$$\frac{\partial v'}{\partial t} + U\frac{\partial v'}{\partial x} + g\frac{\partial h'}{\partial x} = 0 \tag{4.43}$$

$$\frac{\partial h'}{\partial t} + U\frac{\partial h'}{\partial x} + H\left[\frac{\partial u'}{\partial x} + \frac{\partial v'}{\partial x}\right] = 0 \tag{4.44}$$

The above set of equations may be combined into a wave equation governing the free surface displacement perturbation h'

$$\left[\frac{\partial}{\partial t} + U\frac{\partial}{\partial x}\right]^2 h' - (gH)\left[\frac{\partial^2 h'}{\partial x^2} + \frac{\partial^2 h'}{\partial y^2}\right] = 0 \tag{4.45}$$

Applying the method of normal modes, with independent components,

$$h' = A\,\exp\left[ik(x - ct)\right] \tag{4.46}$$

and substituting it into the two-dimensional form of Eq. (4.45) (i.e., with $\partial/\partial y = 0$), we obtain the dispersion relationship for two-dimensional shallow water waves,

$$c = U \pm \sqrt{gH} \tag{4.47}$$

With no basic wind, the above equation indicates that the solution to Eq. (4.47) consists of a leftward moving and a rightward moving, propagating shallow-water waves along the free surface of the fluid with the shallow-water wave (phase) speeds of $\sqrt{gH}$, respectively. With a basic wind, these two waves are advected by the basic state flow (U). Since the phase speeds are independent of wave number, these two-dimensional shallow-water waves are nondispersive. It is straightforward to show that the group velocity is identical to the phase velocity in this two-dimensional shallow water system.

The quantity $\sqrt{gH}$ is called the shallow-water wave speed. It is the valid approximation for waves whose wavelengths are much greater than the depth of the fluid. The restriction is necessary in order that the vertical velocity be small enough so that the hydrostatic approximation is valid. For an ocean depth of 4 km, shallow-water wave speed is $\sim$200 m s^{-1}. Thus the long waves on the ocean surface travel very rapidly. It should be emphasized again that this theory applies to wave of wavelength much greater than H. Such a long wave is not generally excited by wind stresses, but may be produced by very large-scale disturbances, like underwater earthquakes, known as *Tsunami*.

Shallow water gravity waves may also occur at the interfaces within the ocean where there is sharp density gradient. In particular, the surface water is separated from the deep water by a narrow region of sharp density contrast called thermocline.

4.5.3.1 Three-dimensional Shallow-water Equation

The perturbation kinetic energy equation for a three-dimensional shallow-water system may be obtained to be (Gill 1982)

$$\left[\frac{\partial}{\partial t}+U\frac{\partial}{\partial x}\right]\left[\frac{1}{2}\rho H(u'^2+v'^2)\right]=-\rho g H\left[u'\frac{\partial u'}{\partial x}+v'\frac{\partial v'}{\partial y}\right] \tag{4.48}$$

where the quantity inside the square bracket on the left-hand side of the above equation is easily recognized to be the perturbation kinetic energy per unit area. Equation (4.48) indicates that the change in perturbation kinetic energy following the fluid motion is directly related to the advection of the perturbation height by the perturbation wind. The perturbation potential energy equation may also be derived to be

$$\left[\frac{\partial}{\partial t}+U\frac{\partial}{\partial x}\right]\left[\frac{1}{2}\rho g h'^2\right]=-\rho g H h'\left[\frac{\partial u'}{\partial x}+\frac{\partial v'}{\partial y}\right] \tag{4.49}$$

Therefore, the convergence or divergence of the perturbation velocity affects the change in perturbation potential energy per unit area following the fluid motion. The perturbation equation for the total perturbation energy per unit area for a three-dimensional, single-layer shallow-water system is given by

$$\left[\frac{\partial}{\partial t}+U\frac{\partial}{\partial x}\right]\left[\frac{1}{2}\rho H(u'^2+v'^2)+\frac{1}{2}\rho g h'^2\right]+\rho g H\left[\frac{\partial}{\partial x}(u'h')+\frac{\partial}{\partial y}(v'h')\right]=0 \tag{4.50}$$

The above equation indicates that the change in total perturbation energy per unit area is simply due to the convergence or divergence of the mass flux.

Shallow water flows may behave quite differently when upstream flow conditions are different. In other words, different flow regimes may result from different upstream flow conditions. In order to better understand this, let us consider a two-dimensional, nonrotating, small-amplitude shallow-water flow over an obstacle (see Fig. 4.5). The wave equation governing the evolution in the zonal wind perturbation is given by

$$\left[\frac{\partial}{\partial t}+U\frac{\partial}{\partial x}\right]^2 u'-(gH)\frac{\partial^2 u'}{\partial x^2}=-gU\frac{\partial^2 h_s}{\partial x^2} \tag{4.51}$$

Assuming that the flow has reached a steady state, Eq. (4.51) reduces to

$$(U^2-gH)\frac{\partial^2 u'}{\partial x^2}=-gU\frac{\partial^2 h_s}{\partial x^2} \tag{4.52}$$

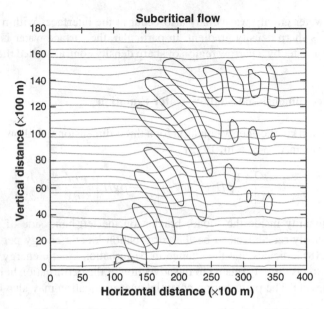

Subcritical flow

Fig. 4.5 Spatial variability of fluid due to a disturbance produced at the fluid bottom (Adapted from Holton and Durran 1993)

Integrating the above equation twice with respect to x leads to

$$u' = \frac{-gU}{(U^2 - gH)} h_s \tag{4.53}$$

Similarly, we may derive the governing equation for the free surface displacement

$$\left[\frac{\partial}{\partial t} + U\frac{\partial}{\partial x}\right]^2 h' - (gH)\frac{\partial^2 h'}{\partial x^2} = \left[\frac{\partial}{\partial t} + U\frac{\partial}{\partial x}\right]\left[U\frac{\partial h_s}{\partial x^2}\right] \tag{4.54}$$

which yields the following steady-state solution

$$h' = \frac{h_s}{(1 - F_0^{-2})} \tag{4.55}$$

Thus, we have

$$h' \to h_s, \; for \; F_0 > 1 \tag{4.56}$$

and

$$h' \leftarrow h_s, \; for \; F_0 < 1 \tag{4.57}$$

where F_0 is the shallow-water *Froude number*, which is defined as $U/\sqrt{gH}$. The Froude number represents the ratio between the kinetic and potential energies of the undisturbed, upstream basic flow. As h increases, far upstream flow will also increase. The interface or free surface will bow upward over the obstacle as shown

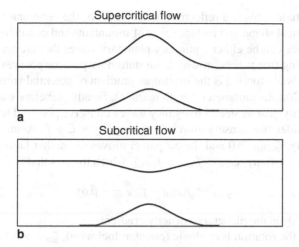

Fig. 4.6 Water flow over an obstacle produces (**a**) supercritical flow and (**b**) subcritical flow

in Fig. 4.6a. That is, physically, the upstream flow has enough kinetic energy to overcome the potential energy barrier associated with the obstacle. This flow regime is called *supercritical flow*.

On the other hand, if $F_0 < 1$, h' will decrease as $h_s(x)$ increases, as the perturbation flow converts its potential energy into enough kinetic energy to surmount the obstacle (Fig. 4.6b). Over the peak of the obstacle, the fluid reaches its maximum speed, as implied from Eq. (4.53). This flow regime is called *subcritical flow*. The Froude number F_0 is the ratio of the advection flow speed (U) to the shallow-water wave phase speed ($\sqrt{gH}$). Thus, when the flow is supercritical ($F_0 > 1$), small-amplitude disturbances cannot propagate upstream against the basic flow and any obstacle along the bottom will tend to produce a purely local disturbance. When the flow is subcritical ($F_0 < 1$), shallow-water waves are able to propagate upstream. The steady-state effect of this response is to effectively increase the layer depth upstream, which increases the potential energy of the incident flow. The potential energy is then converted into kinetic energy as the fluid surmounts the obstacle. Thus, the fluid reaches its maximum speed and the water surface dips down over the peak of the obstacle, creating a *Bernoulli or Venturi effect*.

4.5.4 Rossby Waves

Rossby waves are planetary-scale waves which are most important for stratospheric transport. Rossby waves develop where there are large-scale variations in potential vorticity, all of whose components (relative vorticity, stretching, and planetary vorticity) can be active. The restoring force of these waves is the variation in Coriolis effect with latitude. This is a reflection in dynamics of the geography of the Earth: its nearly spherical shape and its landscape of mountains and continental rises.

Rossby waves can be either stationary planetary waves that are forced by orography or traveling free waves, in which the stationary planetary waves are the most important. The wave forcing is the isentropic gradient of potential vorticity (i.e., the change of the Coriolis parameter f with latitude). Steady planetary waves conserve potential vorticity, just as steady buoyancy waves conserve potential temperature.

Let us consider the conservation of vorticity: $\eta = \zeta + f$. Assume that initial relative vorticity, $\zeta_{initial} = 0$ and the air parcel moves to another latitude by δy. By conservation of vorticity, $\zeta_{new} + f_{new} = f_{initial}$, which implies that

$$\zeta_{new} = f_{initial} - f_{new} = -\beta \delta y \tag{4.58}$$

where $\beta = df/dy$ is the planetary vorticity gradient.

For $\delta y < 0$, the rotation is cyclonic (counterclockwise), $\zeta_{new} > 0$, because

$$\left(\frac{\partial v}{\partial x} - \frac{\partial u}{\partial y} \right) = \zeta \tag{4.59}$$

For $\delta y > 0$, the rotation is anticyclonic (clockwise), $\zeta_{new} < 0$. The whole pattern moves westward.

By considering a chain of fluid parcels along a latitude circle and requiring that it is conserved, if initially a meridional displacement results in a sinusoidal displacement, it gives positive (cyclonic) vorticity for southward displacements and negative (anticyclonic) vorticity for northward displacements. It is then clear from Fig. 4.7 that this induced flow field advects the chain of fluid parcels such that the wave pattern propagates westward.

Since potential vorticity is conserved, a good starting point for looking at these waves is conservation of potential vorticity. Assume a simplified case in which the

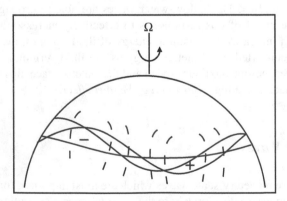

Fig. 4.7 Meridionally displaced chain of fluid parcels of vorticity perturbation field and induced velocity field (dashed arrows). Thick wavy line shows the original perturbation position, and the thin line shows the westward displacement of the pattern due to advection by the induced velocity field (Adapted from Holton 2004, Courtesy: Elsevier)

mean flow is purely zonal and that we can consider only horizontal motion on pressure surfaces (barotropic vorticity equation).

$$\left(\frac{\partial}{\partial t} + u\frac{\partial}{\partial x} + v\frac{\partial}{\partial y}\right)\zeta + \beta v = 0, \ for \ f = f_0 + \beta y \qquad (4.60)$$

Now let us take the mean and perturbed parts:

$$u = U + u'; \ v = v'; \ \zeta = \overline{\zeta} + \zeta' \qquad (4.61)$$

In addition, we can write the velocity perturbations as stream functions:

$$u' = -\frac{\partial \Psi'}{\partial y}; \ v' = \frac{\partial \Psi'}{\partial x} \qquad (4.62)$$

which yeilds to

$$\zeta' = \nabla^2 \Phi' \qquad (4.63)$$

The conservation equation becomes:

$$\left(\frac{\partial}{\partial t} + U\frac{\partial}{\partial x}\right)\nabla^2\Psi' + \beta\frac{\partial \Psi'}{\partial x} = 0 \qquad (4.64)$$

Equation (4.64) has harmonic solutions of the form:

$$\Psi' = R_e\left[\hat{\Psi}\exp(i\phi)\right] \qquad (4.65)$$

where

$$\phi = kx + ly - vt \qquad (4.66)$$

Substituting this solution back into the equation yields the dispersion relation:

$$v = Uk - \frac{\beta k}{(k^2 + l^2)^{1/2}} \qquad (4.67)$$

The zonal phase speed is $c_x = v/k$, which gives

$$c_x - U = -\frac{\beta k}{(k^2 + l^2)^{1/2}} \qquad (4.68)$$

because $\beta > 0$ in the northern hemisphere.

Recalling mean wind speed $c = v/k$ we find that the zonal phase speed

$$c_x - U < 0 \qquad (4.69)$$

Planetary waves propagate westward relative to the mean wind. Rossby waves are dispersive. The phase speeds increase rapidly with decreasing wave number (and thus increasing wavelength). Waves that are generated by orography are fixed to that location: the perturbations do not move and so the phase speed $c_x = 0$. This

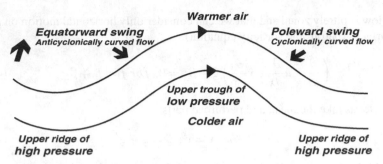

Fig. 4.8 Propagation characteristics of Rossby waves in the atmosphere (Courtesy: J. Brandon, Australian Aviation Meteorology)

implies that $U > 0$ in order for waves for propagate. Thus waves can propagate only with eastward (westerly) winds. The propagation of planetary-scale Rossby waves is shown schematically in Fig. 4.8.

Relative mean wind speed is given by

$$c_x - U = -\frac{\beta}{k^2} \qquad (4.70)$$

Therefore it is clear that phase speed is relative to mean flow and is inversely proportional to the square of the zonal wave number. Since the sign is negative in the Eq. (4.70), Rossby waves' zonal propagation is always westward in direction. Rossby waves are dispersive waves whose phase speed increases rapidly with increasing wavelength. For a typical midlatitude synoptic scale disturbance, with zonal and meridional scales and the zonal wavelength on the order of 6,000 km, the Rossby wave speed relative to the zonal flow is calculated as approximately -8 m s^{-1}, because on synoptic scale, the mean zonal wind is generally westerly and greater than 8 m s^{-1}.

Rossby waves usually move eastward, but at a phase relative to the ground that is somewhat less than the mean zonal wind speed. For longer wavelengths the westward Rossby wave speed may be large enough to balance the eastward advection by the zonal wind so that the resulting disturbance is stationary relative to the surface of the Earth. Rossby waves become stationary when

$$k^2 = \frac{\beta}{U} = k_s^2 \qquad (4.71)$$

where k_s is the stationary zonal wave number. In the upper troposphere the length scales of atmospheric circulation tend to be much greater than near the surface, as the influence of surface-related forcing must be reduced. The contrast between the warmth and thickness of the tropical atmosphere and the cold and shallowness of the polar troposphere drives a strong westerly wind at the midlatitude in this region of troposphere as shown in Fig. 4.9. The effect of potential vorticity conservation on a strong, large-scale wind is to produce large-amplitude waves as a result of any deflection from the purely zonal flow. These waves are known as Rossby waves.

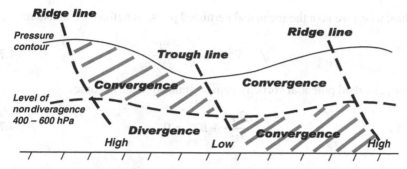

Fig. 4.9 Rossby wave pattern illustrating compensation of upper-level westerlies with convergence and divergence at lower levels (Courtesy: J. Brandon, Australian Aviation Meteorology)

In a quiescent atmosphere Rossby waves travel westward with a wave speed c, dependent on the wavelength L:

$$c = -\frac{\beta L^2}{4\pi^2} \tag{4.72}$$

where $\beta = \partial f/\partial y$ is the rate of change of Coriolis parameter, f, with latitude. However, the upper westerlies are so strong that, relative to Earth, Rossby waves travel eastward in the direction of the flow. Thus a series of snapshots at the upper atmospheric pressure fields will show large-amplitude waves moving eastward, but at a speed somewhat slower than the actual peak wind speed. Near-surface disturbances like frontal systems and depressions are embedded within the wave system and move with it, tending to be the troughs of the upper tropospheric Rossby waves (Hamilton 1998; Holton 2004).

4.5.4.1 Three-dimensional Case

Now consider a more realistic three-dimensional case. Once again we conserve potential vorticity, which we approximate with the quasi-geostrophic potential vorticity (q):

$$q = \zeta + f + \frac{1}{\rho_0}\frac{\partial}{\partial z}\left[\rho_0 \theta \frac{\partial \theta}{\partial z}\right] \tag{4.73}$$

where the terms on the right-hand side are the relative vorticity, the planetary vorticity, and the stretching term, respectively.

The solutions have the form:

$$\Psi' = Re\,[\hat{\Psi}]\,\exp(ik(x - c_x t))\sin(ly) \tag{4.74}$$

where the first part is the vertically propagating wave, the second is the longitude and time dependence, and the third is the latitudinal dependence.

Thus, when we take the mean and perturbed parts, as before, we find that:

$$q' = \nabla^2 \Psi' + \frac{f^2}{N^2} \frac{\partial^2 \Psi'}{\partial z^2} \tag{4.75}$$

and the perturbed potential vorticity conservation equation becomes

$$\frac{\partial^2 \Psi'}{\partial z^2} + B\Psi' = 0 \tag{4.76}$$

where

$$B \equiv \frac{N^2}{f_0^2} \left[\frac{D\bar{q}/Dy}{u - c_x} - (k^2 + l^2) \right] \tag{4.77}$$

B is analogous to the square of the index of refraction. The solutions are a function of $B^{1/2}$ when

$$B > 0; \ \Psi' \approx \exp(\pm iB)^{1/2} z \tag{4.78}$$

$$B < 0; \Psi' \approx \exp(-|B|^{1/2} z) \tag{4.79}$$

When $B > 0$, vertical propagation is possible. In the case of $B < 0$, the wave rapidly damps out. Note that if $U = c_x$ at some level $z = z_c$, B becomes infinite if $\partial q/\partial z$ is not equal to zero. This level is called the critical level.

Consider a constant U, so that

$$\frac{D\bar{q}}{Dy} = \beta \tag{4.80}$$

Because the equation must be solved in the vertical as well as the horizontal, the solution for the vertically propagating wave must take the density change with height into account.

$$\rho_0 = \rho_s \exp(-z/H) \tag{4.81}$$

gives a solution

$$\hat{\Psi} \propto \exp\left(\left[\frac{1}{2H} \pm \left(\frac{1}{4H^2} - B \right)^{1/2} \right] z \right) \tag{4.82}$$

if

$$\left(\frac{1}{4H^2} - B \right) > 0 \tag{4.83}$$

then

$$\hat{\Psi} \propto \exp\left(\frac{z}{2H} \right) \exp\left[\pm \left(\frac{1}{4H^2} - B \right)^{1/2} z \right] \tag{4.84}$$

It appears that the wave amplitude blows up with height, which is not supported by physics.

Remember that $u' = \partial \Psi'/\partial y$ and that the energy density of the wave is given by $\frac{1}{2}\rho u'^2$. Thus,

$$\text{Energy density} \propto \frac{1}{2}\rho_0 \exp\left(-\frac{z}{H}\right) \exp\left(\frac{2z}{2H}\right) \exp\left[\pm 2\left(\frac{1}{4H^2} - B\right)^{1/2} z\right] \quad (4.85)$$

$$\text{Energy density} \propto \frac{1}{2}\rho_0 \exp\left[\pm 2\left(\frac{1}{4H^2} - B\right)^{1/2} z\right] \quad (4.86)$$

If we do not want this energy density to become unbounded with height, we must choose the minus sign. As a result, the energy density actually vanishes with height. The planetary wave is trapped. Further, because there is no coupling in the y and z directions, the wave has no phase tilt with height, unlike gravity waves.

Now look at the vertical structure for allowed vertical propagation. If

$$\left(\frac{1}{4H^2} - B\right) < 0, \quad (4.87)$$

then

$$\Psi \propto R_e\left[\exp\left(\frac{z}{2H}\right)\exp(i(kx + mz - kct)\sin(ly)\right] \quad (4.88)$$

where

$$m = \left(B - \frac{1}{4H^2}\right)^{1/2} \quad (4.89)$$

The imaginary term indicates both vertical propagation and phase tilt with height. The wave propagates both vertically and zonally. The phase propagation $kx + mz = constant$. For $m > 0$, the constant phase line must tilt westward. Choose $m > 0$ in order to have the group velocity greater than 0 when $k > 0$.

As we found in Eq. (4.77)

$$B \equiv \frac{N^2}{f_0^2}\left[\frac{D\bar{q}/Dy}{U - c} - (k^2 + l^2)\right] \quad (4.90)$$

Putting the equation in terms of m, we arrive at the following expression:

$$U - c \equiv \beta\left[k^2 + l^2 + (f_0^2/N^2)(m^2 + 1/4H)\right]^{-1} \quad (4.91)$$

Now, $0 < m^2 < \infty$, and β, k^2, l^2, f_o^2, N^2, and H^2 are all positive. Thus, we must have

$$0 < U - c < U_c \approx \beta\left[k^2 + l^2 + (f_0^2/N^2)(1/4H^2)\right]^{-1} \quad (4.92)$$

This condition is called the *Charney-Drazin criterion*. The wind must be eastward (westerly) but less than c for Rossby waves to propagate vertically.

As k and l grow larger, U_c makes the range of wind velocities smaller for propagation. Generally, only the lowest wave number waves propagate vertically. If $c = 0$, (stationary forcing), then $0 < U < U_c$ for vertical propagation. For typical conditions, $N^2 = 5 \times 10^{-4} \, s^{-1}$ and $l = \pi/(10,000 \, \text{km})$ at $60°N$.

$U_c = 110/(s^2 + 3)$ in m s^{-1}, where $s = kacos(\phi)$

So, wave number 1 (s = 1) propagates in westerlies <28 m s^{-1}. Wave number 2 (s = 2) propagates in westerlies <16 m s^{-1}.

4.6 Atmospheric Gravity Waves

The atmosphere is basically a fluid that is being acted upon by a force due to the Earth's gravity. Under the influence of gravity, the background gas density decreases exponentially with increasing height, and so does the background gas pressure. The amplitude of the gravity wave increases exponentially with height. The physical interpretation is that in maintaining the vertical flux of wave energy constant, it offsets the decrease of background gas density.

The mechanism of the gravity waves is that when the force of Earth's gravity and the stabilizing restoring force (produced by the atmospheric density gradients) become comparable with compressibility forces, the resultant waves are gravity waves. The waves may be termed internal or surface waves according to whether the vertical wave number is pure real or pure imaginary, respectively. For internal waves, in the high-frequency limit, it behaves like simple sound waves, so it is called *acoustic gravity wave*. In the low-frequency or high period range (from several minutes to several hours), it is termed as *internal gravity wave*.

Gravity waves can be generated by many sources like jet streams, tidal waves, tropical cyclones, hurricanes, earthquakes, volcanic eruptions, nuclear explosions and thunderstorms. They propagate vertically and horizontally, dissipate, interact nonlinearly, and profoundly influence the momentum, energy, and the constituents in the atmosphere (see Fig. 4.10). They can help the mixing of chemical reactions

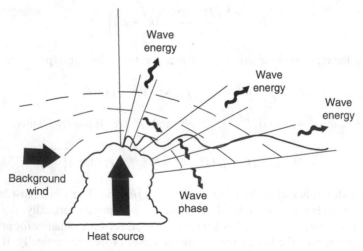

Fig. 4.10 Source and propagation of atmospheric gravity waves (Kim et al. 2003, Courtesy: Canadian Meteorological and Oceanographic Society)

in the atmosphere. Some gravity waves are observed to propagate horizontally for thousands of kilometers. Gravity waves may alter their environment and profoundly affect the circulation of the middle atmosphere on a global basis.

Basically, gravity waves act as a vehicle for energy and momentum transport into the stratosphere and mesosphere. Gravity waves can act to transfer mean horizontal momentum from the ground to levels aloft in the atmosphere or from one layer of the atmosphere to another. Flow over topography can generate stationary gravity waves that break nonlinearly in the troposphere and lower stratosphere. Such waves transfer momentum from the breaking region to the Earth's surface, and this process is thought to act as a significant drag on the eastward mean winds in the midlatitude troposphere. Other processes (such as convection, jetstream, instabilities, etc.) can produce gravity waves with nonzero horizontal phase speeds and which act to transfer mean momentum between the troposphere and the stratosphere/mesosphere.

The gravity wave–critical layer interaction is known to have several properties which are important in the dynamics of the atmosphere (Hines 1968; Hamilton 1998; Nappo 2002). These are the strong coupling between a gravity wave and the mean flow occurring at a critical layer, and the tendency for a gravity wave to develop large amplitudes and small vertical wavelengths near a critical layer. The interaction of a gravity wave with the mean flow near the critical layer results in a severe gravity wave attenuation with much of its energy and momentum being absorbed by the mean flow (Kim et al. 2003).

4.6.1 Pure Internal Gravity Waves

Internal gravity waves are transverse waves in which parcel oscillations are parallel to phase lines. Figure 4.11 shows the motion of a parcel of air moving under the

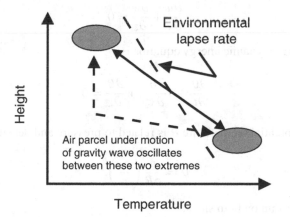

Fig. 4.11 Gravity wave effect on the oscillation of air parcel between two extreme limits (Courtesy: WK Hocking 2001)

influence of buoyancy wave, plotted in terms of height and temperature. The parcel moves adiabatically, which means it heats and cools according to the adiabatic lapse rate. At the top of its oscillation, it is coolest, and therefore has the lowest density. The environment lapse rate is also shown in the same figure. At the top of its oscillation, the parcel of air is cooler than its surroundings. This is consistent with the stable atmospheric condition. If the environmental lapse is unstable, gravity waves cannot exist, since a parcel of air which is displaced vertically upward will continue to rise, rather than oscillate. At the bottom of its motion, the parcel of air is warmer and less dense than its surroundings, so it is forced to rise again.

Let us consider the linearized equations for two-dimensional internal gravity waves. Assume Boussinesq approximation, in which density is treated as constant except where it is coupled with gravity in the buoyancy term of the vertical momentum equation. In this approximation the atmosphere is considered to be incompressible and local density variations are assumed to be small perturbation of the constant basic state density field. Since the vertical variation of the basic state density is neglected except where coupled with gravity, the validity of Boussinesq approximation is confined to the motions in which the vertical scale is less than the atmospheric scale height (H ~ 8 km).

After neglecting the effects of rotation, the basic equations for two-dimensional motion of an incompressible atmosphere can be written as:

The momentum equation in x, z plane,

$$\frac{\partial u}{\partial t} + u\frac{\partial u}{\partial x} + w\frac{\partial u}{\partial z} + \frac{1}{\rho}\frac{\partial p}{\partial x} = 0 \qquad (4.93)$$

$$\frac{\partial w}{\partial t} + u\frac{\partial w}{\partial x} + w\frac{\partial w}{\partial z} + \frac{1}{\rho}\frac{\partial p}{\partial z} + g = 0 \qquad (4.94)$$

The continuity equation

$$\frac{\partial u}{\partial x} + \frac{\partial w}{\partial z} = 0 \qquad (4.95)$$

And the thermodynamic energy equation is

$$\frac{\partial \theta}{\partial t} + u\frac{\partial \theta}{\partial x} + w\frac{\partial \theta}{\partial z} = 0 \qquad (4.96)$$

where the potential temperature θ is related to pressure and density by

$$\theta = \frac{p}{\rho R}\left(\frac{p_0}{p}\right)^k \qquad (4.97)$$

Taking logarithm on both sides:

$$\ln \theta = \ln p - (\ln \rho + \ln R) + \frac{R}{C_p}(\ln p_0 - \ln p) \qquad (4.98)$$

$$\ln \theta = \frac{\ln p}{\gamma} - \ln p + \text{const} \tag{4.99}$$

applying perturbation

$$\rho = \rho_0 + \rho'; \; p = \overline{p}(z) + p'; \; \theta = \overline{\theta}(z) + \theta'; \; u = U + u'; \; w = w' \tag{4.100}$$

where the basic state zonal flow U and the density ρ_0 are both assumed to be constant. The basic-state pressure field must satisfy the hydrostatic equation.

$$\frac{D\overline{p}}{Dz} = -\rho_0 g \tag{4.101}$$

While the basic-state potential temperature must satisfy, the linearized equations are obtained by substituting Eq. (4.100) in Eqs. (4.93)–(4.98) and neglecting all terms that are products of perturbation variables. The last two terms in Eq. (4.94) are approximated as

$$\frac{1}{\rho}\frac{\partial p}{\partial z} + g = \frac{1}{\rho_0 + \rho'}\left(\frac{D\overline{p}}{Dz} + \frac{\partial p'}{\partial z}\right) + g \tag{4.102}$$

$$\frac{1}{\rho}\frac{\partial p}{\partial z} + g \approx \frac{1}{\rho_0}\frac{D\overline{p}}{Dz}\left[1 - \frac{\rho'}{\rho_0}\right] + \frac{1}{\rho_0}\frac{\partial p'}{\partial z} + g = \frac{1}{\rho_0}\frac{\partial p'}{\partial z} + \frac{\rho'}{\rho_0}g \tag{4.103}$$

where Eq. (4.99) has been used to eliminate $\overline{p}$. The perturbation form of Eq. (4.99) is obtained by noting that

$$\ln(\overline{\theta} + \theta') = \gamma^{-1}\ln(\overline{p} + p') - \ln(\rho_0 + \rho') + \text{constant} \tag{4.104}$$

$$\ln\left[\overline{\theta}\left(1 + \frac{\theta'}{\overline{\theta}}\right)\right] = \gamma^{-1}\ln\left[\overline{p}\left(1 + \frac{p'}{\overline{p}}\right)\right] - \ln\left[\rho_0\left(1 + \frac{\rho'}{\rho_0}\right)\right] + \text{constant} \tag{4.105}$$

Now let us consider that $\ln(1 + \theta) \approx \theta$ for any $\theta \ll 1$. We find with the aid of Eq. (4.104) that Eq. (4.105) may be approximated by

$$\frac{\theta'}{\overline{\theta}} \simeq \frac{1}{\gamma}\frac{p'}{\overline{p}} - \frac{\rho'}{\rho_0} \tag{4.106}$$

solving for ρ' yields

$$\rho' = -\rho_0\frac{\theta'}{\overline{\theta}} - \frac{p'}{c_s^2} \tag{4.107}$$

where $c_s^2 \equiv \overline{p}\gamma/\rho_0$ is the square of the speed of sound.

For buoyancy wave motion,

$$\left| \frac{\rho_0 \theta'}{\overline{\theta}} \right| \gg \left| \frac{p'}{c_s^2} \right| \tag{4.108}$$

that is the density fluctuations due to pressure changes are small compared to those due to temperature changes. Therefore, to the first approximation,

$$\frac{\theta'}{\overline{\theta}} = \frac{\rho'}{\rho_0} \tag{4.109}$$

Using Eqs. (4.103) and (4.109), the linearized set of equations can be written as follows:

$$\left[\frac{\partial}{\partial t} + U \frac{\partial}{\partial x} \right] u' + \frac{1}{\rho_0} \frac{\partial p'}{\partial x} = 0 \tag{4.110}$$

$$\left[\frac{\partial}{\partial t} + U \frac{\partial}{\partial x} \right] w' + \frac{1}{\rho_0} \frac{\partial p'}{\partial z} - \frac{\theta'}{\overline{\theta}} g = 0 \tag{4.111}$$

$$\frac{\partial u'}{\partial x} + \frac{\partial w'}{\partial z} = 0 \tag{4.112}$$

$$\left[\frac{\partial}{\partial t} + U \frac{\partial}{\partial x} \right] \theta' + w' \frac{d\overline{\theta}}{dt} = 0 \tag{4.113}$$

We can eliminate p' by differentiating Eq. (4.110) with regard to z and Eq. (4.111) with regard to x, and substrating them will obtain

$$\left[\frac{\partial}{\partial t} + U \frac{\partial}{\partial x} \right] \left[\frac{\partial w'}{\partial x} - \frac{\partial u'}{\partial z} \right] - \frac{g}{\overline{\theta}} \frac{\partial \theta'}{\partial x} = 0 \tag{4.114}$$

which is the y component of the vorticity equation. With the help of Eqs. (4.112) and (4.113), u' and θ' can be eliminated from Eq. (4.114) to yield a single equation for w.

$$\left[\frac{\partial}{\partial t} + U \frac{\partial}{\partial x} \right]^2 \left[\frac{\partial^2 w'}{\partial x^2} + \frac{\partial^2 w'}{\partial z^2} \right] + N^2 \frac{\partial^2 w'}{\partial x^2} = 0 \tag{4.115}$$

where $N^2 = gDln\overline{\theta}/Dz$ is the square of the buoyancy frequency, which is assumed to be a constant.

Equation (4.115) has harmonic wave solutions of the form

$$w' = \hat{w} \exp\{i\phi\} \tag{4.116}$$

where $\phi = kx + mz - vt$. Substituting the solutions in Eq. (4.115) and solving for it gives

$$(v - Uk)^2 (k^2 + m^2) - N^2 k^2 = 0 \tag{4.117}$$

so that

$$\hat{v} = v - Uk = \pm \frac{Nk}{(k^2 + m^2)^{1/2}} = \pm \frac{Nk}{|k|} \tag{4.118}$$

In this equation, $\hat{v}$ is the *intrinsic frequency*, the frequency relative to mean wind speed. Here the plus sign is taken for eastward phase propagation and the minus sign for westward phase propagation, relative to mean wind speed.

If we let $k > 0$ and $m < 0$, then the lines of constant phase will tilt eastward with increasing height. The positive root in Eq. (4.118) corresponds to eastward and downward propagation relative to mean flow with the horizontal and vertical phase speeds (relative to mean flow) given by $c_x = \hat{v}/k; c_z = \hat{v}/m$, respectively.

The component of group velocity is given by

$$c_{gx} = \frac{\partial v}{\partial k} = U \pm \frac{Nm^2}{(k^2 + m^2)^{3/2}} \tag{4.119}$$

$$c_{gz} = \frac{\partial v}{\partial m} = \pm \frac{-Nkm}{(k^2 + m^2)^{3/2}} \tag{4.120}$$

The vertical component of group velocity has a sign opposite to that of the vertical phase speed relative to mean flow. It can be seen that the group velocity vector is parallel to the lines of constant vector. Internal gravity has thus a remarkable property that the group velocity is perpendicular to the direction of propagation (Holton 2004).

4.6.1.1 Source of Gravity Waves

The atmosphere can be excited at frequencies only up to the Brunt-Vaisala frequency, N. If the frequency is less than N, then we will get waves that are excited on the slant paths.

Assume now that there is a mean zonal wind.

$$v_{\text{total}} = v + Uk = Uk \pm (N\cos\alpha) = Uk \pm \frac{Nk}{(m^2 + k^2)^{1/2}} \tag{4.121}$$

Now the group velocities are:

$$c_{gx} = \frac{\partial}{\partial k} v_{\text{total}} = U \pm \frac{Nm^2}{(m^2 + k^2)^{3/2}} \quad \text{and} \tag{4.122}$$

$$c_{gz} = \pm \frac{Nkm}{m(m^2 + k^2)^{3/2}} \tag{4.123}$$

The group velocity is increased to the east by the addition of the mean zonal wind.

It should be pointed out that the actual air molecules are not moving very far from the position that they would have with the mean zonal wind. Therefore, the actual motion of the air molecules in the stratosphere is confined within 100 m or so.

4.6.2 Inertia-gravity Waves

In the case of inertial and gravitationally stable fluid flow, the parcel displacements are resisted by both buoyancy and rotation. The waves generated from the resulting oscillation are known as inertia gravity waves. The dispersion relation for such waves can be evaluated using an alternate method (Holton 2004).

Let us consider the parcel oscillations along a tilted path in the y–z plane as shown in Fig. 4.12. For the vertical displacement δs, the buoyancy force parallel to the slope of the parcel oscillation is $-N^2\delta z \cos\alpha$. For meridional displacement δy, the Coriolis (inertial) force component parallel to the slope of the parcel path is $-f^2\delta y \sin\alpha$, where we have assumed that the geostrophic basic flow is constant with latitude. Thus the equation for the parcel oscillation is represented as

$$\frac{D^2}{Dt^2}\delta s = -(f\sin\alpha)^2\delta s - (N\cos\alpha)^2\delta s \tag{4.124}$$

where δs is the the perturbation of parcel displacement.

The frequency of the dispersion relationship can be represented as

$$v^2 = N^2\cos^2\alpha + f^2\sin^2\alpha \tag{4.125}$$

Normally $N^2 > f^2$, which shows that the inertia-gravity frequencies lie in a range between $f \leq |v| \leq N$. The frequency approaches N as the trajectory slope comes close to the vertical, and approaches f as the trajectory slopes toward the horizontal. For typical midlatitude tropospheric conditions, inertia-gravity wave periods are in the range of 12 min to 15 hr. Rotational effects dominate only when $f^2 \sin^2\alpha$ in Eq. (4.125) is similar to the magnitude of $N^2\cos^2\alpha$. This requires that

$$\tan^2\alpha \sim \frac{N^2}{f^2} = 10^4 \tag{4.126}$$

in which case $v \ll N$ in Eq. (4.125). This shows that, only low-frequency waves are modified significantly by the rotation of the Earth.

The parcel derivation can be further verified by using the linearized dynamical equations including rotation. The small parcel trajectory slopes of relatively long period waves are altered significantly by rotation. This necessitates that the horizontal scales are much greater than the vertical scales for these waves.

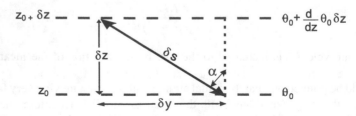

Fig. 4.12 Parcel o scillation for inertia gravity waves in a meridional plane (Adapted from Holton 2004, Courtesy: Elsevier)

Now let us assume a motionless basic state, the linearized equations are replaced by the perturbation equations as:

$$\frac{\partial u'}{\partial t} - fv' + \frac{1}{\rho_0}\frac{\partial p'}{\partial x} = 0 \tag{4.127}$$

$$\frac{\partial v'}{\partial t} + fu' + \frac{1}{\rho_0}\frac{\partial p'}{\partial y} = 0 \tag{4.128}$$

$$\frac{1}{\rho_0}\frac{\partial p'}{\partial z} - \frac{\theta'}{\theta}g = 0 \tag{4.129}$$

$$\frac{\partial u'}{\partial x} + \frac{\partial v'}{\partial y} + \frac{\partial w'}{\partial z} = 0 \tag{4.130}$$

$$\frac{\partial \theta'}{\partial t} + w'\frac{D\bar\theta}{Dz} = 0 \tag{4.131}$$

In Eq. (4.131), θ' can be eliminated by applying the hydrostatic relationship from Eq. (4.129) as

$$\frac{\partial}{\partial t}\left(\frac{1}{\rho_0}\frac{\partial p'}{\partial z}\right) + N^2 w' = 0 \tag{4.132}$$

Now letting

$$(u',v',w',p'/\rho_0) \quad R_e \ [(\hat u,\hat v,\hat w,\hat p)\,\mathrm{exp}\,i(kx + ly + mz - vt)] \tag{4.133}$$

Applying Eq. (4.133) in Eqs. (4.127), (4.128), and (4.132) we get

$$\hat u = \frac{(vk + ilf)}{(v^2 - f^2)}\hat p \tag{4.134}$$

$$\hat v = \frac{(vk + ikf)}{(v^2 - f^2)}\hat p \tag{4.135}$$

$$\hat w = -\frac{vm}{N^2}\hat p \tag{4.136}$$

With the support of Eq. (4.130) we produce the dispersion relation for hydrostatic waves.

$$v^2 = f^2 + N^2(k^2 + l^2)m^{-2} \tag{4.137}$$

In Eq. (4.137), when $(k^2 + l^2)/m^2 \ll 1$, vertical propagation is possible (m real) the frequency must satisfy the inequality

$$f \le |v| \le N \tag{4.138}$$

when we let

$$\sin^2\alpha \to 1, \quad \cos^2\alpha = (k^2 + l^2)/m^2 \tag{4.139}$$

which is consistent with hydrostatic approximation.

If the axes are chosen to make $l = 0$, it may be shown that the ratio of vertical and horizontal components of group velocity is given by

$$\left| \frac{c_{gz}}{c_{gx}} \right| = \left| \frac{k}{m} \right| = \frac{(v^2 - f^2)^{1/2})}{N} \tag{4.140}$$

Now we can see that for fixed v, inertia-gravity waves propagate more closely to the horizontal than that of pure internal gravity waves. However, in the latter case the group velocity vector is again parallel to the lines of constant phase.

Eliminating $\hat{p}$ between Eqs. (4.134) and (4.135) for the case $l = 0$ yields the relationship $\hat{v} = if\hat{u}/v$. It can be verified that if it is real, the perturbation horizontal motions satisfy the relations

$$u' = \hat{u}\cos(kx + mz - vt) \tag{4.141}$$

$$v' = \hat{u}(f/v)\sin(kx + mz - vt) \tag{4.142}$$

Thus, the horizontal velocity vector rotates anticyclonically with time. As a result, the parcels follow an elliptical trajectory in a plane of orthogonal wave number vector. Equations (4.141) and (4.142) also show that the horizontal velocity vector turns anticyclonically with height for waves with upward energy propagation.

4.6.2.1 Interactions of Gravity Waves with the Mean Flow

To understand how gravity waves interact with the mean flow, let us recall the dispersion relation:

$$\omega = Uk \pm (N\cos\alpha) = Uk \pm \frac{Nk}{(m^2 + k^2)^{1/2}} \tag{4.143}$$

$$c_x - U = \pm \frac{N}{(m^2 + k^2)^{1/2}} \tag{4.144}$$

$$c_x = \pm \frac{N}{m(m^2 + k^2)^{1/2}} \tag{4.145}$$

$$c_{gx} = U \pm \frac{Nm^2}{(m^2 + k^2)^{3/2}} \tag{4.146}$$

$$c_{gx} = \pm \frac{Nkm}{(m^2 + k^2)^{3/2}} \tag{4.147}$$

Choose the positive sign, so that when $m < 0$ (downward phase speed), $k > 0$ so that we get upward propagation. Thus, $c_x > U$.

Choose the negative sign, so that when $m < 0$ (downward phase speed), $k < 0$ so that we get upward propagation. Thus, $c_x < U$.

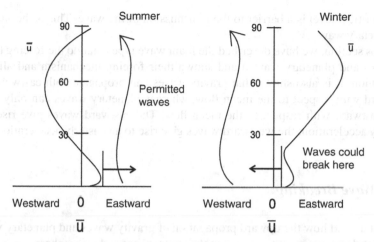

Fig. 4.13 Conditions for the vertical propagation of gravity waves (Adapted from Andrews et al. 1987, Courtesy: Elsevier)

The conditions favorable for vertical propagation of gravity waves are schematically shown in Fig. 4.13.

The mean zonal wind thus determines which gravity waves are allowed to propagate upward. Flow across mountains and heating creates a broad spectrum of gravity waves, both eastward and westward propagation, but the mean zonal flow filters these gravity waves.

We see that the permitted waves have $c_x > U$ in the summer, but $c_x < U$ in the winter. Recall that for planetary waves, $c_x < U$, which occurs in the winter.

Since $L_x \gg L_z$, $k \ll m$. Thus, we can approximate the equations above as the following:

$$c_x = \frac{\omega}{k} = \overline{U} \pm \frac{N}{(m^2 + k^2)^{1/2}} \tag{4.148}$$

which implies

$$c_x - U \approx \frac{N}{m} \tag{4.149}$$

since typically, $k \ll m$ ($L_x \gg L_z$). Therefore, we can approximate:

$$m \approx \frac{N}{c_x - \overline{U}}; \ as \ c_x \to \overline{U}; \ m = \frac{2\pi}{L_z} \to \infty \tag{4.150}$$

The level at which this happens is called the critical level, z_c, just as for planetary waves. Note that the wavelength of the wave goes to zero.

The group velocity also goes to zero as c_x approaches U:

$$c_{gz} = \frac{\partial \omega}{\partial m} \approx -\frac{Nk}{m^2} \ because \ m \to \infty \tag{4.151}$$

The critical level is a barrier to the transmission of the wave. This is the same as for planetary waves.

In this section, we have described the main wave types that do the forcing (gravity waves and planetary waves) and shown their forcing mechanism and allowed propagation. It is also shown that gravity waves can propagate both eastward and westward with respect to the mean flow, whereas planetary waves can only propagate eastward with respect to the mean flow. The eastward waves give rise to a westerly acceleration; the westward waves give rise to an easterly acceleration.

4.6.3 Wave Breaking

Now, let us find how the upward propagation of gravity waves and planetary waves create accelerations on zonal winds in the stratosphere and mesosphere.

Considering the vertical change in density in the gravity wave generation, we get perturbation solutions of the following form:

$$(u', v', w', \Phi') = \exp\left(\frac{z}{2H}\right) \mathrm{Re}\left[(\hat{u}, \hat{v}, \hat{w}, \hat{\Phi}) \exp\{i(kx + ly + mz - vt)\}\right] \quad (4.152)$$

$$\hat{u} = \frac{k}{v} \hat{\Phi}; \quad \hat{v} = \frac{l}{v} \hat{\Phi}; \quad \hat{w} = -\frac{\omega}{N^2}\left[m - \frac{i}{2H}\right]\Phi \quad (4.153)$$

dispersion relation,

$$v^2 = \frac{N^2 k^2}{m^2 + 1/(4H^2)} \quad (4.154)$$

At isentropic surfaces, the gravity waves accelerate when the phase propagates downward and moves in the eastward direction. The wave amplitude grows because $|u'| \sim \exp(z/2H) \sim |c - \bar{u}|$, and that $\frac{1}{2}\rho_0 u'^2 \sim$ constant. The isentropic surfaces get progressively more distorted until the isentropic surface becomes quasi-vertical ($\partial\theta/\partial z = 0$) get convective instability – and the wave breaks (see Fig. 4.14). Convective overturning generates local turbulence. Most important, the wave can no longer increase in amplitude and perturbations must decrease above this altitude. The level at which $\partial\theta/\partial z = 0$ is called the saturation level, z_s. When the zonal phase speed, c_x, approaches the zonal wind speed, U. This is the critical level, z_c, which is above z_s.

Thus, for $z < z_s$ (no energy deposition)

$$\frac{\partial}{\partial z}(\rho\overline{u'w'}) = 0 \quad (4.155)$$

for $z_s < z < z_c$ (wave breaking; energy deposition)

$$\frac{\partial}{\partial z}(\rho\overline{u'w'}) \neq 0 \quad (4.156)$$

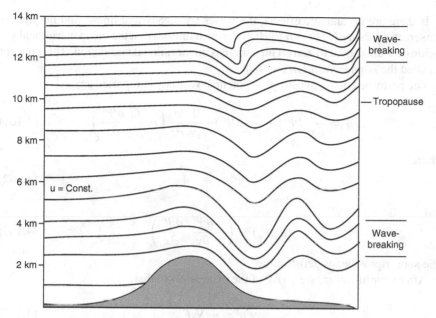

Fig. 4.14 Acceleration of gravity waves in isentropic surfaces (Adapted from Andrews et al. 1987, Courtesy: Elsevier)

Thus, the wave drag and the gravity wave contribution to the forcing G occur between the saturation level and the critical level.

4.7 Planetary Wave Forcing

For planetary waves, the zonal mean momentum equation and the thermodynamic energy equations have the form:

$$\frac{\partial \bar{u}}{\partial t} - f_0 \bar{v}^* = \rho_0^{-1} \vec{\nabla} \cdot \vec{F} + \overline{X} \equiv \overline{G} \tag{4.157}$$

$$\frac{\partial \overline{T}}{\partial t} + N^2 \frac{H}{R} \bar{w}^* = \frac{\overline{J}}{c_p} \tag{4.158}$$

where $\vec{F}$ is the *Eliassen-Palm (EP) flux* (see section 4.10.1), which arises from large-scale eddies, and X is the forcing from the small-scale eddies, like gravity wave drag. G is the total zonal drag force. $\vec{F} = \hat{j}F_y + \hat{k}F_z$, where the two components are given by:

$$F_y = -\rho_0 \overline{u'v'} \tag{4.159}$$

and

$$F_z = \rho_0 f_0 R \overline{v'T'} / (N^2 H) \tag{4.160}$$

If there are no diabatic effects, the waves are steady and linear, and the flow is conservative, then $\partial U/\partial t = \partial T/\partial t = 0$, which implies that the mean meridional velocities are also equal to zero. Thus, there is no meridional circulation. This concept is called the *nonacceleration theorem*.

The perturbed quasi-geostrophic potential vorticity is reported as

$$q' = \frac{\partial^2}{\partial x^2}(\Psi') + \frac{\partial^2}{\partial y^2}(\Psi') + \frac{1}{\rho_0}\frac{\partial}{\partial z}\left(\rho_0 \frac{f_0^2}{N^2}\frac{\partial \Psi'}{\partial z}\right) \tag{4.161}$$

where

$$\Psi' = \frac{\Phi'}{f_0} \tag{4.162}$$

and

$$(u'_g, v'_g) = \left(-\frac{\partial \Psi'}{\partial y}, \frac{\partial \Psi'}{\partial x}\right) \tag{4.163}$$

The subscript g means geostrophic.

After simplification, we arrive at the expression so that:

$$\overline{v'q'} = \frac{1}{\rho_0}\vec{\nabla}.\vec{F} \tag{4.164}$$

$$\frac{\partial U}{\partial t} - f_0\bar{v}^* = \overline{v'q'} + \overline{X} \equiv \overline{G} \tag{4.165}$$

Where the left-hand side of the equation represents time-rate-of-change of the zonal mean wind residual Coriolis force, $\overline{v'q'}$ indicates the northward eddy flux of the potential vorticity, $\overline{X}$ is the small-scale forcing, and $\overline{G}$ is the total forcing.

Note that the northward eddy flux of potential vorticity is nonzero only if the planetary wave is either growing or shrinking in time.

$$\left(\frac{\partial}{\partial t} + U\frac{\partial}{\partial x}\right)q' + v'\frac{\partial \bar{q}}{\partial x} = Z' \tag{4.166}$$

where Z' is all transient forcing.

We multiply by

$$\rho_0\left(\frac{q'}{\partial \bar{q}/\partial y}\right) \tag{4.167}$$

and take the zonal average to get the equation:

$$\frac{\partial}{\partial t}\left(\frac{\rho_0}{2}\frac{\overline{q'^2}}{\partial \bar{q}/\partial y}\right) + \vec{\nabla}\cdot\vec{F} = \rho_0(\overline{Z'q'})\left(\frac{\overline{Z'q'}}{\partial \bar{q}/\partial y}\right) \equiv D \tag{4.168}$$

where the first term is the time-rate-of-change of the wave activity density:

$$A = \left(\frac{\rho_0}{2}\frac{\overline{q'^2}}{\partial \bar{q}/\partial y}\right) \tag{4.169}$$

If the waves are steady, then $\partial A/\partial t = 0$. If the waves are adiabatic and inviscid, then $D = 0$. Thus the divergence of the Eliassen-Palm flux is zero.

Conditions favorable for planetary wave forcing are: (i) planetary waves forcing exists only when $\vec{\nabla} \cdot \vec{F} \neq 0$; (ii) the Eliassen-Palm (EP) flux F is usually directed upward and toward the equator; (iii) F is parallel to the group velocity of coherent waves. When

$$\frac{\partial A}{\partial t} > 0; \ \vec{\nabla} \cdot \vec{F} = \rho_0(\overline{v'q'}) < 0 \qquad (4.170)$$

Thus, if $q' > 0$, then $v' < 0$. Thus growing waves transport potential vorticity equatorward. In this sense, the growing planetary waves tend to smooth out the mean potential vorticity distribution.

4.8 Equatorial Waves

Near the equator, atmospheric waves acquire a rather different character and they are created by distinct mechanisms. The amplitude of the equatorial wave is maximum in the neighborhood of the geographical equator and decreases rapidly, away from the equator. These waves are considered responsible for the semiannual oscillations in the upper stratosphere and lower mesosphere and quasi-biennial oscillation in the middle and lower stratosphere. In the troposphere, the equatorial waves control the Walker circulations, Madden and Julian oscillations, and even the El Niño-southern oscillations (Asnani 2005).

Equatorial waves are an important class of eastward and westward propagating disturbances in both the atmosphere and the ocean that are trapped about the equator. Diabatic heating by organized tropical convection can excite atmospheric equatorial waves, which control the spatial and temporal distribution of convective heating (Dunkerton and Delisi 1985; Holton 2004).

Equatorial waves play a significant role in stratosphere–troposphere interactions, which are not confined to the tropical region, but extend into the extratropics. The two classes of well-established equatorial waves, Kelvin waves and mixed Rossby gravity waves, are discussed in this section.

Figure 4.15 illustrates the pressure and wind distribution of Kelvin waves and mixed Rossby gravity waves. Kelvin waves on either side of the equator show symmetric nature, whereas MRG waves are unsymmetric in their pressure and velocity distribution. These two equatorially trapped wave modes are believed to play a crucial role in forcing the quasi-biennial oscillation (QBO) of the lower tropical stratosphere. The characteristic features of Kelvin waves and MRG waves in the equatorial lower stratosphere are shown in Table 4.1.

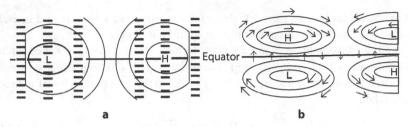

Fig. 4.15 Pressure and wind distribution of (**a**) Kelvin waves and (**b**) mixed Rossby gravity waves (Adapted from Matsuno 1966, Courtesy: American Meteorological Society)

Table 4.1 Characteric features of MRG waves and Kelvin waves in the atmosphere (Adapted from Andrews et al. 1987; Asnani 2005)

Feature	Mixed Rossby gravity wave	Kelvin wave
Period	~4.5 days	~15 days
Horizontal wavelength	~10,000 km	~30,000 km
Vertical wavelength	~6 km	~ 8 km
Phase speed relative to ground	~23 m s^{-1} moves westward	~25 m s^{-1} moves eastward
Amplitude of wave perturbation		
(a) zonal wind u	~3 m s^{-1}	~8 m s^{-1}
(b) meridional wind v	~3 m s^{-1}	0
(c) temperature T	~ 1°C	~ 3°C
(d) geopotential height Z	~30 m s^{-1}	~4 m s^{-1}
(e) vertical velocity w	~0.15 cm s^{-1}	~0.15 cm s^{-1}
Tilt	Westward as they go up	Eastward as they go up
Vertical flux momentum	Westerly momentum upward but the coupled meridional circulation carries easterly momentum upward	Carries westerly momentum upward
Poleward flux of sensible heat in stratosphere	Transport heat poleward in both hemispheres	No heat transport either poleward or equatorward
Group velocity in vertical	Individual phase velocity travels downward but the group velocity is upward	Individual phase velocity travels downward but group travels upward
Region of absorption	Penetrates through the zone of westerlies but gets absorbed in the zone of easterlies	Penetrates through the zone of easterlies but gets absorbed near the zone of transition between easterlies and westerlies

4.8.1 Kelvin Waves

A Kelvin wave is a wave in the ocean or atmosphere that balances the Earth's Coriolis force against a topographic boundary such as a coastline. A feature of a Kelvin wave is that it is nondispersive, i.e., the phase speed of the wave crest is equal to

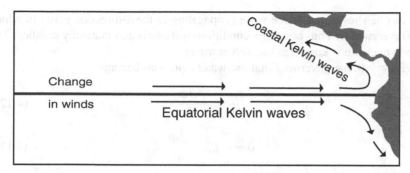

Fig. 4.16 Propagation of equatorial ocean Kelvin waves (Courtesy: Naval Postgraduate School, USA)

the group speed of the wave energy for all frequencies. This means that it retains its shape in the alongshore direction over time. Kelvin waves in the ocean always propagate with the shoreline on the right in the northern hemisphere, and with the shoreline on the left in the southern hemisphere, as shown in Fig. 4.16.

4.8.1.1 Equatorial Kelvin Wave

An equatorial Kelvin wave is a special type of Kelvin wave in which the equator acts analogously to a topographic boundary for both the northern and southern hemispheres. This wave always propagates eastward and exists only on the equator. Equatorial Kelvin waves are often associated with anomalies in surface wind stress.

Equatorial waves propagate towards the east in the northern hemisphere, using the equator as a wave guide. Coastal Kelvin waves propagate around the northern hemisphere oceans in a counterclockwise direction using the coastline as a wave guide. These waves, especially the surface waves are very fast-moving, typically with speeds of ~ 2.8 m s^{-1}, or about 250 km day^{-1}.

Both atmospheric and oceanic equatorial Kelvin waves play an important role in the dynamics of El Niño-southern oscillation, by transmitting changes in conditions in the western Pacific to the eastern Pacific.

4.8.1.2 Derivation of Kelvin Wave Equation

In the free modes of oscillation of a shallow water model of infinite horizontal extent, let us consider an infinite half plane, $y > 0$. We take

$$f = f_0; \ \Phi_s = 0; \ U = 0 \qquad (4.171)$$

and consider waves of same amplitude. We assume the boundary conditions as (i) $v = 0$ at $y = 0$, i.e., no flow through the wall; and (ii) (u, v, Φ) finite at $y = \infty$.

It can be shown that a free wave propagating in the x-direction exists in which $y = 0$ everywhere. Thus boundary condition (i) above is automatically satisfied. The wave in question is known as the *Kelvin wave*.

With $v = 0$, the governing shallow-water equations become

$$\frac{Du}{Dt} = -\frac{\partial \Phi}{\partial x} \tag{4.172}$$

$$0 = -f_0 u - \frac{\partial \Phi}{\partial y} \tag{4.173}$$

$$\frac{D\Phi}{Dt} = -\Phi \frac{\partial u}{\partial x} \tag{4.174}$$

The linear perturbation equations are therefore

$$\frac{\partial u'}{\partial t} = -\frac{\partial \Phi'}{\partial x} \tag{4.175}$$

$$0 = -f_0 u' - \frac{\partial \Phi'}{\partial y} \tag{4.176}$$

$$\frac{\partial \Phi'}{\partial t} = -\overline{\Phi} \frac{\partial u'}{\partial x} \tag{4.177}$$

We assume the solution in the form

$$(u', \Phi') = \left[\hat{u}(y), \hat{\Phi}(y)\right] \exp\left[i(kx - vt)\right] \tag{4.178}$$

Here the amplitudes $\hat{u}$ and $\hat{\Phi}$ are functions of y. We shall see that this is necessary to satisfy the equations and the boundary conditions.

Substituting Eq. (4.178) into Eqs. (4.175), (4.176), and (4.177), we have

$$-iv\hat{u} = -ik\hat{\Phi} \tag{4.179}$$

$$0 = f_0 \hat{u} + \frac{\partial \hat{\Phi}}{\partial y} \tag{4.180}$$

$$-iv\hat{\Phi} = -\overline{\Phi} ik\hat{u} \tag{4.181}$$

From Eq. (4.179), we can write

$$\hat{u} = \frac{k}{v}\hat{\Phi} \tag{4.182}$$

Equation (4.181) gives

$$\hat{u} = \frac{v}{k}\frac{1}{\overline{\Phi}}\hat{\Phi} \tag{4.183}$$

Equating the above two expressions for it gives

$$\frac{v}{k}\frac{\hat{\Phi}}{\bar{\Phi}} = \frac{k}{v}\hat{\Phi} \tag{4.184}$$

$$v^2 = k^2\bar{\Phi} \tag{4.185}$$

$$v = \pm k\sqrt{\bar{\Phi}} \tag{4.186}$$

$$c = \frac{v}{k} = \pm\sqrt{\bar{\Phi}} = \pm\sqrt{gH} \tag{4.187}$$

That is, the phase speed is independent of rotation.
Substituting for $\hat{u}$ from Eq. (4.179) into Eq. (4.180) gives

$$\frac{\partial\hat{\Phi}}{\partial y} = -f_0\left[\frac{k}{v}\right]\hat{\Phi} \tag{4.188}$$

The solution to this is

$$\hat{\Phi}(y) = \hat{\Phi}(0)\,\exp\left[-f_0\left(\frac{k}{v}\right)\hat{\Phi}\right] \tag{4.189}$$

In order to satisfy the boundary condition that it be finite at $y = \infty$, we see that only the positive root given by Eq. (4.186) is allowable. Choosing this, we see that

$$\hat{\Phi}(y) = \hat{\Phi}(0)\,\exp(-y/R) \tag{4.190}$$

where $R = \sqrt{\bar{\Phi}}/f_0 = Rossby\ radius$
i.e., the amplitude of the wave decreases exponentially away from the boundary and $c = +c_0$.
Assuming $\bar{\Phi}$ is real, we see that

$$\Phi' = \hat{\Phi}(0)\,\exp[-y/R]Cosh(x - c_0t) \tag{4.191}$$

$$u' = \frac{1}{c_0}\hat{\Phi}(0)\,\exp[-y/R]Cosh(x - c_0t) \tag{4.192}$$

Thus, the Kelvin wave has the following properties: (i) the motion is parallel to the wall and is in geostrophic balance, i.e., $v' = 0$; $u' = -1/f_0(\partial\Phi'/\partial f)$; (ii) the direction of propagation of the wave is such that, for an observer facing in the direction of the propagation, the boundary is to the right; (iii) the phase speed is the same as for a pure gravity wave in a nonrotating system, i.e., the Kelvin wave is nondispersive; and (iv) the amplitude of the motion falls off exponentially away from the boundary, the characteristic scale being the Rossby radius of deformation.

4.8.2 Mixed Rossby Gravity Waves

Mixed Rossby gravity (MRG) wave (also called *Yanai wave*) is an equatorial wave with a dispersion relation asymptotic to that for equatorial Kelvin waves for large positive (eastward) zonal wave numbers and asymptotic to that for equatorial Rossby waves for large negative (westward) zonal wave numbers.

Pure inertia-gravity waves are obtained by choosing a free surface shallow-water model on an f- plane. Rossby waves are obtained by choosing a shallow water model with a rigid lid (i.e., a nondivergent barotropic model) on a β plane (Lindzen 1967).

4.8.2.1 Derivation of Mixed Rossby Gravity Wave Equation

Consider the case of mixed inertia-gravity and Rossby waves by choosing a free surface shallow-water model on a β plane. Thus $f = f_0 + \beta y$. We assume a basic state of rest ($U = 0$) and no bottom tropography ($\Phi_s = 0$).

We use the shallow-water equations in vorticity- divergence form. As seen earlier, this form of the equation is

$$\frac{D}{Dt}(\zeta + f) = -(\zeta + f)\left[\frac{\partial u}{\partial x} + \frac{\partial v}{\partial y}\right] \tag{4.193}$$

$$\frac{D}{Dt}(D) + (D)^2 - 2J(u,v) = -\nabla^2 \Phi_T + (f\rho - \beta u) \tag{4.194}$$

$$\frac{D\Phi}{Dt} = -\Phi D \tag{4.195}$$

Where

$$\zeta = \frac{\partial v}{\partial x} - \frac{\partial u}{\partial y} \tag{4.196}$$

$$D = \frac{\partial u}{\partial x} + \frac{\partial v}{\partial y} \tag{4.197}$$

J is the Jacobian and Φ_T represents the upper topography. The linearized form of the above equations for a resting base state with $\Phi_s = 0$ (in which case $\Phi_T = \Phi$) is

$$\frac{\partial \zeta'}{\partial t} + \beta v' = -fD' \tag{4.198}$$

$$\frac{\partial D'}{\partial t} = -\nabla^2 \Phi' + (f\zeta' - \beta u') \tag{4.199}$$

$$\frac{\partial \Phi'}{\partial t} = -\overline{\Phi}D' \tag{4.200}$$

We now make the approximation of taking f as a constant f_0 except where it occurs in differentiated form (this is equivalent to assuming that the motions do not extend over too great a distance in latitude).We also assume that the perturbation quantities are independent of y. Equations (4.198), (4.199), and (4.200) then become

$$\frac{\partial}{\partial t}\left(\frac{\partial v'}{\partial x}\right) + \beta v' = -f_0\frac{\partial u'}{\partial x} \tag{4.201}$$

$$\frac{\partial}{\partial t}\left(\frac{\partial u'}{\partial x}\right)\beta v' = -\frac{\partial^2\Phi'}{\partial x^2} + f_0\frac{\partial v'}{\partial x} - \beta u' \tag{4.202}$$

$$\frac{\partial\Phi'}{\partial t} = -\overline{\Phi}\frac{\partial u'}{\partial x} \tag{4.203}$$

Seeking solutions of the form

$$(u',v',\Phi') = (\hat{u},\hat{v},\hat{\Phi})\exp\left[i(kx - vt)\right] \tag{4.204}$$

we have

$$(-iv)(ik)\hat{v} + \beta\hat{v} = -f_0ik\hat{u} \tag{4.205}$$

$$(-iv)(ik)\hat{u} = k^2\hat{\Phi} + f_0ik\hat{u} - \beta\hat{u} \tag{4.206}$$

$$(-iv)\hat{\Phi} = -\overline{\Phi}ik\hat{u} \tag{4.207}$$

i.e.,

$$\left(v + \frac{\beta}{k}\right)\hat{v} = if_0\hat{u} \tag{4.208}$$

$$\left(v + \frac{\beta}{k}\right)\hat{u} = k\hat{\Phi} - if_0\hat{v} \tag{4.209}$$

$$v\hat{\Phi} = k\overline{\Phi}\hat{u} \tag{4.210}$$

We use Eqs. (4.208) and (4.210) to eliminate $\hat{u}$ and $\hat{v}$ from Eq. (4.209)

$$\hat{u} = \left[\frac{v}{k\overline{\Phi}}\right]\hat{\Phi} \tag{4.211}$$

Hence Eq. (4.208) becomes

$$\hat{v} = -\left[\frac{if_0}{\left(v + \frac{\beta}{k}\right)}\right]\hat{u} = -\left[\frac{if_0}{\left(v + \frac{\beta}{k}\right)}\right]\left(\frac{v}{k\overline{\Phi}}\right)\hat{\Phi} \tag{4.212}$$

Substituting in Eq. (4.209), we have

$$\left(v + \frac{\beta}{k}\right)\left(\frac{v}{k\overline{\Phi}}\right)\hat{\Phi} = k\hat{\Phi} + \left[\frac{f_0^2}{\left(v + \frac{\beta}{k}\right)}\right]\left(\frac{v}{k\overline{\Phi}}\right)\hat{\Phi} \tag{4.213}$$

$$\left(v + \frac{\beta}{k}\right)\left[v\left(v + \frac{\beta}{k}\right) - k^2\overline{\Phi}\right] = f_0^2 v \tag{4.214}$$

This is a cubic equation for v. We shall see that two of the roots are close approximations to pure inertia-gravity waves while the third root is a close approximation to a pure Rossby wave.

Case I: Suppose $|v| > \beta/k$ (i.e., the frequency is much greater than the Rossby wave frequency).

Then Eq. (4.213) can be approximated by

$$v\left[v^2 - k^2\overline{\Phi}\right] = f_0^2 v \qquad (4.215)$$

Hence we have

$$v = \pm\sqrt{k^2\overline{\Phi} + f_0^2} \qquad (4.216)$$

$$c = \pm\sqrt{\overline{\Phi} + \frac{f_0^2}{k^2}} \qquad (4.217)$$

i.e., we have the inertia-gravity wave solution.

Note: The assumption $|v| \gg \beta/k$ is justified a posteriori, if $v^2 \gg [\beta/k]^2$ using Eq. (4.217) becomes

$$f_0^2 + k^2\overline{\Phi} \gg \left(\frac{\beta}{k^2}\right)^2 \qquad (4.218)$$

$$1 + 4\pi^2\left(\frac{R}{L}\right)^2 \gg \left(\frac{\beta}{f_0 k^2}\right)^2 \qquad (4.219)$$

where $R = \sqrt{\overline{\Phi}/f_0}$ is known as Rossby radius of deformation.
But

$$\frac{\beta}{f_0} = \left(\frac{2\Omega\cos\theta/a}{2\Omega\sin\theta}\right) = \frac{1}{a\tan\theta} \approx \frac{1}{a}; \; if \; \theta = 45° \qquad (4.220)$$

Therefore the above inequality holds if

$$1 + 4\pi^2\left(\frac{R}{L}\right)^2 \gg \frac{1}{4\pi^2}\left[\frac{1}{a^2}\right] \qquad (4.221)$$

This obviously holds if $L/a \ll 1$, which is a reasonable assumption in the context of inertia-gravity waves.

Case II: Suppose $v^2 \ll \overline{\Phi}k^2$ (i.e., the frequency is much less than the inertia-gravity wave frequency) and suppose in addition that $\beta/k^2 \ll \sqrt{\overline{\Phi}}$ (i.e., the Rossby wave phase speed is much less than the gravity wave phase speed).

Equation (4.214) can be approximated, making use of the first of these approximations, by

$$\left(v + \frac{\beta}{k}\right)\left(v\frac{\beta}{k} - k^2\overline{\Phi}\right) = f_0^2 v \qquad (4.222)$$

and this can be approximated using the first and second approximation, by

$$\left(v+\frac{\beta}{k}\right)\left(-k^2\overline{\Phi}\right)=f_0^2 v \tag{4.223}$$

Hence we have

$$v=-\frac{\beta}{k}\left(\frac{k^2\overline{\Phi}}{f_0^2+k^2\overline{\Phi}}\right) \tag{4.224}$$

which gives

$$c=-\frac{\beta}{k^2}\left[\frac{1}{1+(L^2/4\pi^2 R^2)}\right] \tag{4.225}$$

Thus, we have the Rossby wave solution slightly modified by gravitational effects through the term $L^2/4\pi^2 R^2$. For this term to be significant we must have $L > R$. Its effect can slow down the very long waves.

Note: Using Eq. (4.224) our assumption $v^2 \ll \overline{\Phi}k^2$ is justified a posteriori if

$$\left[\frac{\beta}{k}\right]^2\left[\frac{1}{1+(L^2/4\pi^2 R^2)}\right] \ll \overline{\Phi}k^2 \tag{4.226}$$

i.e., if

$$\frac{\beta}{k^2} \ll \sqrt{\overline{\Phi}}\left[1+\frac{L^2}{4\pi^2 R^2}\right] \tag{4.227}$$

This is satisfied a priori if the second assumption $\beta/k^2 \ll \sqrt{\overline{\Phi}}$ is satisfied. In the real atmosphere, the Rossby wave phase speed β/k^2 is small, amounting to only tens of m s^{-1}. For even the longest waves, the inertia-gravity wave speed $\sqrt{\overline{\Phi}}$ is large, having the value 313 m s^{-1} if we take $H = 10$ km (the approximate depth of the troposphere). Thus Eq. (4.227) holds good and the approximations are justified.

4.9 Vertical Propagating Atmospheric Waves

The equatorial waves (both gravity and Rossby types) may propagate vertically under certain conditions and the shallow-water model must be replaced by a continuously stratified atmosphere in order to examine the vertical structure. It turns out that vertically propagating equatorial waves share a number of physical properties with ordinary gravity modes.

Vertically propagating gravity waves in the presence of rotation were considered for the simple situation in which the Coriolis parameter was assumed to be constant and the waves were assumed to be sinusoidal in both x and y directions. Inertia-gravity waves can propagate vertically only when the wave frequency satisfies inequality $f < v < N$. Thus in middle latitudes, waves with periods in the range of several days are generally vertically trapped, so that they are not able to propagate significantly into the stratosphere. As the equator is approached, however, the decreasing Coriolis frequency should allow vertical propagation to occur for lower frequency waves. Thus in the equatorial region there is a possibility for existence of long periods of vertically propagating internal gravity waves.

Let us consider linearized perturbation on an equatorial β plane. The linearized equations of motion, continuity equation, and first law of thermodynamics are expressed in log-pressure coordinates as follows

$$\frac{\partial u'}{\partial t} - \beta y v' = -\frac{\partial \Phi'}{\partial x} \tag{4.228}$$

$$\frac{\partial v'}{\partial t} - \beta y u' = \frac{\partial \Phi'}{\partial y} \tag{4.229}$$

$$\frac{\partial u'}{\partial x} + \frac{\partial v'}{\partial y} + \frac{1}{\rho_0} \frac{\partial}{\partial z}(\rho_0 w') = 0 \tag{4.230}$$

$$\frac{\partial^2 \Phi'}{\partial t \partial z} + \omega' N^2 = 0 \tag{4.231}$$

Assume that the perturbations are zonally propagating waves, and propagate vertically with vertical wave number m. Due to the basic density stratification, there will also be an amplitude growth in height proportional to $\rho_0^{-1/2}$. The x, y, and z dependencies can thus be separated as

$$\begin{bmatrix} u' \\ v' \\ \omega' \\ \Phi' \end{bmatrix} = \exp\left[\frac{z}{2H}\right] \begin{bmatrix} \hat{u}(y) \\ \hat{v}(y) \\ \hat{\omega}(y) \\ \hat{\Phi}(y) \end{bmatrix} \exp\left[i(kx + mz - vt)\right] \tag{4.232}$$

Substituting from Eq. (4.232) into Eqs. (4.228)–(4.231) yields a set of ordinary differential equation for meridional structure

$$-iv\hat{u} - \beta y \hat{v} = -ik\hat{\Phi} \tag{4.233}$$

$$-iv\hat{v} - \beta y \hat{u} = -\frac{\partial \hat{\Phi}}{\partial y} \tag{4.234}$$

$$-ik\hat{u} + \frac{\partial \hat{v}}{\partial y} + i\left(m + \frac{i}{2H}\right)\hat{\omega} = 0 \tag{4.235}$$

$$v\left(m - \frac{i}{2H}\right)\hat{\Phi} + \hat{\omega}N^2 = 0 \tag{4.236}$$

4.9.1 Vertically Propagating Kelvin Waves

Matsuno (1966) and Lindzen (1967, 1971) illustrated that in a baroclinic atmosphere, there can be forced modes that would propagate vertically, transferring energy and momentum in the vertical. It has been shown that for Kelvin wave, the

upward component of the group velocity is exactly equal to the downward component of the phase velocity. In the case of Kelvin waves, the perturbation equation
can be simplified further. Setting $\hat{v} = 0$ in Eqs. (4.233) and (4.234) and eliminating
$\hat{\omega}$ between Eqs. (4.235) and (4.236), we obtain

$$-iv\hat{u} = -ik\hat{\Phi} \tag{4.237}$$

$$-\beta y\hat{u} = -\frac{\partial \hat{\Phi}}{\partial y} \tag{4.238}$$

$$-v\left(m^2 + \frac{1}{4H^2}\right)\hat{\Phi} + \hat{u}kN^2 = 0 \tag{4.239}$$

Equation (4.237) can be used to eliminate $\hat{\Phi}$ in Eqs. (4.238) and (4.239). This
yields two independent equations that the field of $\hat{u}$ must satisfy. The first of these
determines the meridional distribution of $\hat{u}$. The second is simply the dispersion
equation.

$$c^2\left(m^2 + \frac{1}{4H^2}\right) - N^2 = 0 \tag{4.240}$$

where $c^2 = (v^2/k^2)$.

If we assume that $m^2 \gg 1/4H$, as it is true for most observed stratospheric Kelvin
waves, (4.240) reduces to the dispersion relationship for internal gravity waves
(Eq. (4.137)) in the hydrostatic limit. For waves in stratosphere that are forced disturbance in the troposphere, energy propagation (i.e., the group velocity) must have
upward component. Therefore the phase velocity must have a downward component. Kelvin waves must propagate eastward ($c > 0$) if they are to be trapped equatorially. However, eastward phase propagation requires $m < 0$ for downward phase
propagation. Thus, the vertically propagating Kelvin wave has phase lines that tilt
eastward with height (Holton 2004).

4.9.2 Vertically Propagating Rossby Gravity Waves

Similar to the Kelvin wave mode, the other most significant forced vertically propagating equatorial waves is the mixed Rossby gravity wave. These westward propagating waves are thermally damped in the stratosphere where they interact with the
mean flow producing westward acceleration (Holton and Tan 1980). The equatorial modes shown in Eqs. (4.233)–(4.236) can be combined and if we assume that
$m^2 \gg 1/4H^2$, the resulting meridional structure equation is

$$\hat{\Phi} = \frac{N^2}{m^2} \tag{4.241}$$

For $n = 0$ mode, the dispersion is then

$$|m| = \frac{N}{v^2}(\beta + vk) \tag{4.242}$$

When $\beta = 0$ we get the dispersion relationship for hydrostatic internal gravity waves. The role of β effect in Eq. (4.247) is to break the symmetry between eastward ($v > 0$) and westward ($v < 0$) propagating waves. Eastward propagating modes have shorter vertical wavelengths than westward propagating modes. Vertically propagating $n = 0$ modes can exists only for $c = v/k > -\beta/k^2$. Because $k = s/a$, where s is the number of wavelengths around a latitude circle, this condition implies that for $v < 0$ solutions exist only for frequencies satisfying the inequality

$$|v| < \frac{2\Omega}{s} \tag{4.243}$$

For frequencies that do not satisfy Eq. (4.243), the wave amplitude will not decay away from the equator and it is not possible to satisfy boundary conditions at the pole.

After simplification, the meridional structure of the horizontal velocity and geopotential perturbation for $n = 0$ mode can be expressed as

$$\begin{bmatrix} \hat{u} \\ \hat{v} \\ \hat{\Phi} \end{bmatrix} = v_0 \begin{bmatrix} i|m|N^{-1}vy \\ 1 \\ ivy \end{bmatrix} \exp\left[-\frac{\beta|m|y^2}{2N}\right] \tag{4.244}$$

The westward propagating $n = 0$ mode is generally referred to as Rossby gravity mode. For upward energy propagation this mode must have downward phase propagation ($m < 0$) just like an ordinary westward propagating internal gravity wave. The resulting wave structure in the $x - z$ plane is situated at a latitude north of the equator. Of particular interest is the fact that poleward moving air has positive temperature perturbation and vice versa so that the eddy heat flux contribution to the vertical EP flux is positive.

4.10 Energetics of Vertical Propagating Waves

Atmospheric waves of various scales are generated in the troposphere. In this section, we will concentrate mainly on the vertical propagating waves and their involvement in transferring energy between troposphere and stratosphere. This gives more insight into the dynamical processes involved in the stratosphere–troposphere interactions.

Let us confine our attention to the large-scale planetary waves. The planetary waves are generated in the troposphere by mechanical obstruction to air flow by mountains and land-sea contrasts and also due to differential heating. Many of the large-scale waves propagate into the stratosphere and even extend to the mesosphere. These waves interact with, and hence modify, the mean flow at different levels in the middle atmosphere. It is now generally believed that the large-scale general circulation of the stratosphere is driven by the energy received from the lower troposphere through vertical propagation of wave energy across the tropopause.

The vertical propagation of waves in the atmosphere is treated theoretically in the following three ways: (i) Eliassen-Palm formulation, (ii) Charney-Drazin theory, and (iii) Lindzen's theory.

4.10.1 Eliassen-Palm Approach

Consider the case of stationary waves induced by mountains under adiabatic, frictionless, quasi-static, and quasi-geostrophic conditions (Eliassen and Palm 1961). The basic zonal mean flow (U) can be represented as

$$fU = -\frac{\partial \Phi}{\partial y} \tag{4.245}$$

$$\alpha_0 = -\frac{\partial \Phi}{\partial p} \tag{4.246}$$

Differentiate Eq. (4.245) with respect to p and Eq. (4.246) with regard to y and subtract. We get

$$f\frac{\partial U}{\partial p} = \frac{\partial \alpha_0}{\partial y} \tag{4.247}$$

where α_0 is the specific volume and Φ is the geopotntial in basic state.

Linearized components of the momentum equation and continuity equation in isobaric coordinates can be written as

$$U\frac{\partial u'}{\partial x} + v'\frac{\partial U}{\partial y} + \omega'\frac{\partial U}{\partial p} - fv' = -\frac{\partial \Phi'}{\partial x} \tag{4.248}$$

$$U\frac{\partial v'}{\partial x} + fu' = -\frac{\partial \Phi'}{\partial y} \tag{4.249}$$

$$U\frac{\partial}{\partial x}\left[\frac{\partial \Phi'}{\partial p}\right] - fv' + \frac{fU}{\partial p} + \sigma \omega' = 0 \tag{4.250}$$

$$\frac{\partial u'}{\partial x} + \frac{\partial v'}{\partial y} + \frac{\partial \omega'}{\partial p} = 0 \tag{4.251}$$

where $\sigma = -\frac{\alpha}{\theta}\frac{\partial \theta}{\partial p} = \left[\frac{N}{g}\frac{RT}{p}\right]^2$ is the static stability parameter $\tag{4.252}$

and $N = \sqrt{g/\theta(\partial\theta/\partial p)}$ is the Brunt-Vaisala frequency, u', v', Φ', and ω' are the perturbation quantities.

Now multiply Eq. (4.248) by u, Eq. (4.249) by v, Eq. (4.250) by $\sigma^{-1}(\partial\Phi'/\partial p)$, and Eq. (4.251) by Φ and then add together. We get

$$\frac{\partial}{\partial x}(EU + \Phi'u') + \frac{\partial}{\partial y}(\Phi'v') + \frac{\partial}{\partial p}(\Phi'\omega') = -u'v'\frac{\partial U}{\partial y} - u'\omega'\frac{\partial U}{\partial p} + f\left[\frac{v'}{\sigma}\frac{\partial U}{\partial p}\frac{\partial \Phi'}{\partial p}\right]$$

(4.253)

where

$$E = \left[\frac{u'^2}{2} + \frac{v'^2}{2} + \frac{\alpha'^2}{2\sigma}\right]$$

(4.254)

is the wave energy, which is the sum of kinetic energy and available potential energy.

The left-hand side of Eq. (4.253) represents the divergence of wave energy flux and the right-hand side shows the source terms. The first two terms on the right-hand side correspond to the conversion of kinetic energy in the basic current into kinetic energy of the wave motion. The last term on the right-hand side stands for the conversion of available potential energy (APE) of the basic current into APE of the wave motion.

Taking the zonal mean (indicated by angle brackets) of Eq. (4.253), we get

$$\frac{\partial}{\partial y}<\Phi'v'> + \frac{\partial}{\partial p}<\Phi'\omega'> = -<u'v'>\frac{\partial U}{\partial y} - <u'\omega'>\frac{\partial U}{\partial p} + \frac{f}{\sigma}\frac{\partial U}{\partial p}\left\langle\overline{\frac{v'\partial\Phi'}{\partial p}}\right\rangle$$

(4.255)

After simplification of Eq. (4.255), we get the meridional fluxes of geopotential wave energy as

$$<\Phi'v'> = U\left[\frac{1}{\sigma}\frac{\partial U}{\partial p}\left\langle v'\frac{\partial\Phi'}{\partial p}\right\rangle - <u'v'>\right]$$

(4.256)

The first term on the right-hand side shows that the zonal mean current U decreases with pressure or decreases with height. The last term indicates that the meridional momentum flux is opposite to the direction of the meridional flux of geopotential energy.

In a similar manner, the vertical flux of geopotential energy is obtained from Eq. (4.255) as

$$<\Phi'\omega'> = U\left[\frac{1}{\sigma}\left(f - \frac{\partial U}{\partial y}\right)\left\langle v'\frac{\partial\Phi'}{\partial p}\right\rangle - <u'\omega'>\right]$$

(4.257)

The first term on the right-hand side gives the information that for westerly current, vertical flux of geopotential wave energy will be upward when flux of sensible heat is northward, corresponding to westward tilt of wave troughs and ridges with height. The last term, $-<u'\omega'>$, is the vertical momentum flux, which is negative, and is opposite to the direction of vertical flux of geopotential energy.

We can represent the flux of geopotential wave energy in the (y, p) plane by a pseudo-stream function ψ given by

$$<\Phi'v'> = U\frac{\partial\psi}{\partial p}$$

(4.258)

$$< \Phi' \omega' > = -U \frac{\partial \psi}{\partial y} \qquad (4.259)$$

The curves where ψ = constant resemble streamlines for the flow of energy. The flux between two adjacent streamlines is not constant, but varies with U. According to Eliassen-Palm formulation, the wave energy flux tends toward zero as one approaches a singular line $U = 0$, so that it seems that the wave energy cannot be transported across such a line. This theory is intended for stationary wave motions, where horizontal phase speed is zero. Hence this conclusion is a particular case of a more general theorem that a wave traveling vertically gets trapped and absorbed at that level in the atmosphere where the basic air flow has the same horizontal velocity as the horizontal velocity of the wave.

4.10.1.1 Eliassen-Palm (EP) Flux Diagram

Meridional fluxes of westerly momentum and sensible heat can be easily calculated from constant pressure charts at various levels in the atmosphere. As per the Eliassen-Palm theory, these parameters give fluxes of geopotential wave energy in the meridional-vertical (y, p) plane. Streamlines can then be drawn in (y, p) plane, giving the direction and intensity of flow of energy in this plane.

Figure 4.17 is a schematic diagram representing the flux of wave energy in (y, p) plane. The curves of ψ = constant streamlines for the flux of wave energy. It may be noted that the flux of wave energy in a channel between two adjacent streamlines is not a constant, but varies along the channel in proportion to U as indicated by the length of the arrows in the diagram. Thus when energy flows toward increasing

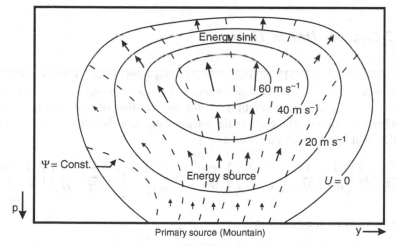

Fig. 4.17 Schematic representation of meridional flow of wave energy. Solid lines represent zonal wind isotachs, dashed lines shows wave energy streamlines, and arrows indicate wave energy flux (Eliassen and Palm 1961; Asnani 2005)

value of U, the wave energy flux increases in the direction of the streamline, carrying energy from the basic zonal flow. The opposite phenomenon occurs in the direction of decreasing U. In any way, wave energy cannot cross the singular line $U = 0$.

4.10.1.2 EP Flux Vector

As per the assumption of Eliassen-Palm theorem, define a vector $\vec{F}$ in the vertical meridional plane as

$$F_y = - <u'v'> \tag{4.260}$$

$$F_p = -f \left[\frac{<v'\theta'>}{\partial\theta/\partial p} \right] \tag{4.261}$$

It can be seen that $\vec{\nabla} \cdot \vec{F} = <v'q'>$, where q' is the eddy potential vorticity, given as

$$q' = \frac{\partial v'}{\partial x} - \frac{\partial u'}{\partial y} + f\frac{\partial}{\partial p}\left[\frac{<\theta>}{\partial <\theta>/\partial p} \right] \tag{4.262}$$

The divergence of the vector $\vec{F}$ can change the mean zonal wind field or induce circulations in the meridional plane (Eliassen and Palm 1961).

The eddies can be split up into standing and transient components. Transient and standing eddy fluxes are presented for winter and summer seasons, in the troposphere and lower stratosphere.

Observational and theoretical study of Eliassen Palm fluxes and their divergence is getting importance due to the insight which they provide for energetics of the atmosphere.

4.10.2 Charney-Drazin Theory

Charney and Drazin (1961) studied vertical propagation of wave energy in the case of large-scale waves for which quasi-static and quasi-geostrophic assumptions are justified. They derived the equation for the vertical flux of wave energy for the vertical propagation of planetary scale waves connecting the eddy momentum flux and eddy heat flux in the y direction as

$$\left\{ \frac{\partial^2}{\partial y^2} + f_0^2 \left(\frac{\partial}{\partial z} - \frac{1}{H} \right)\left(\frac{1}{N^2}\frac{\partial}{\partial z} \right) \right\}\frac{\partial\psi_0}{\partial t} = -\frac{\partial^2 M}{\partial y^2} - f_0^2 \left(\frac{\partial}{\partial z} - \frac{1}{H} \right)\left(\frac{1}{N^2}\frac{\partial B}{\partial y} \right) \tag{4.263}$$

where the eddy momentum flux

$$M = \overline{\frac{\partial\psi'}{\partial x}\frac{\partial\psi'}{\partial y}} = \overline{u'v'} \tag{4.264}$$

and eddy heat flux

$$B = \frac{\overline{\partial \phi'}}{\partial x} \frac{\partial \psi'}{\partial y} = \frac{\overline{v' \partial \psi'}}{\partial z} \tag{4.265}$$

In Eqs. (4.263)–(4.265), H is the scale height, N is the frequency of Brunt-Vaisala oscillations, ψ is the stream function, bar indicates horizontal average, prime represents perturbation, and 0 indicates the undisturbed state.

By assuming perturbations, the differential equation can be derived of the form

$$\frac{D^2 B}{Dz^2} + n^2 B = 0 \tag{4.266}$$

In this equation

$$n^2 = -\left\{ \frac{(k^2 + l^2)N^2}{f_0^2} + \left[\frac{N^2}{\overline{\rho}} \right]^{1/2} \frac{D^2}{Dz^2} \left[\frac{\overline{\rho}}{N^2} \right]^{1/2} \right\} + \frac{N^2}{U - c} \left\{ \frac{\beta}{f_0^2} - \frac{1}{\overline{\rho}} \frac{D}{Dz} \left[\frac{\overline{\rho}}{N^2} \frac{Du_0}{Dz} \right] \right\} \tag{4.267}$$

This equation resembles one-dimensional wave propagation in a medium of variable refractive index n. If n^2 is positive, the wave is internal and propagates in the vertical direction. If n^2 is negative, then the wave is trapped.

For an isothermal atmosphere with constant zonal wind velocity U, then

$$n^2 = \frac{1}{4H^2} \frac{N^2}{f_0^2} \left\{ (k^2 + l^2) - \frac{\beta}{U - c} \right\} \tag{4.268}$$

In this case, if n^2 is positive

$$0 < (U - c) < \left[\frac{\beta}{(k^2 + l^2) + (f_0^2 / 4H^2 N^2)} \right] \equiv U_c \tag{4.269}$$

Where U_c is the critical mean wind speed. Energy gets trapped when $(U - c)$ is negative or $(U - c) > U_c$.

Charney and Drazin (1961) applied their theory for planetary scale waves in the troposphere, stratosphere, and mesosphere. They noted that large amounts of planetary wave energy are produced in the troposphere. In the midlatitude region, this energy cannot penetrate far above the middle stratosphere throughout the year because there are either weak easterly or strong westerly zonal winds above the tropopause. Wave energy gets trapped in these regions.

During spring or for a brief period in autumn, the wave energy of the troposphere can penetrate into the upper stratosphere and mesosphere, but only a small amount of energy can penetrate in these levels. In general, synoptic scale and mesoscale waves are unable to transmit their energy upward into the middle atmosphere. The lower stratosphere acts as a filter preventing the shorter-wave energy from escaping into the upper stratosphere and mesosphere. Thus, Rossby waves, shorter than planetary-scale waves, are not seen in the upper stratosphere.

Small-scale waves in the lower troposphere gradually disappear in the upper troposphere. Therefore, the standing planetary-scale motions in the upper stratosphere and mesosphere are not much influenced by the small-scale motions in the troposphere. In the case of planetary waves, vertical propagation of wave energy is possible only when the prevailing zonal winds are westerly and also less than the crtical wind speed of 38 m s^{-1}. Upward propagation of planetary-scale wave energy is generally not possible when the zonal winds are easterly or when the zonal winds are westerly with more than the critical wind speed.

The limitation of the Charney-Drazin theory is that it is based on quasigeostrophic approximation with β plane centered at 45°N. This theory could extend to tropical regions by substituting appropriate values for β in the near equatorial regions.

4.10.3 Lindzen's Theory

Lindzen (1971) has formulated the problem of vertical wave propagation in terms of equivalent depth that depends on horizontal wavelength, direction of horizontal wave propagation, and on the period of wave and given theoretical treatment of vertical propagation of wave energy in the near-equatorial region. In this approach, two β planes, one centered at the equator and the other centered at 45°N, were considered. It assumed the basic state as one of rest and indicated how the results would be modified for basic state to be one of uniform zonal motion. By considering small perturbations on the basic state, the nonlinear perturbations in the primitive equation models were ignored. After assuming separability of variables, the second-order differential equation in the vertical involving equivalent depth thus takes the form as

$$\frac{D^2V_n}{Dz^2} + \left[\frac{\kappa}{Hh_n} - \frac{1}{4H^2} \right] V_n = -\frac{\kappa}{H}S_n \qquad (4.270)$$

where the subscript n stands for the latitudinal wave number in y direction, $\kappa = R/c_p$, S_n is the component of diabatic heating, h_n is the equivalent depth, V_n is the amplitude of perturbation component velocity in y direction. For forced oscillations, diabatic heating is nonzero; while for free oscillations, diabatic heating is zero.

In the above equation, if the wave solution of this vertical structure is oscillatory type, the wave propagates vertically upward. If the solution is of exponential type, the wave cannot propagate upward and gets trapped.

Lindzen's theory explains the vertical propagation of planetary-scale waves in the equatorial and nonequatorial latitudes. Critical latitudes separating equatorial and nonequatorial latitudes are those where Coriolis parameter f equals Doppler shifted frequency of the wave.

In the equatorial region, all equatorial modes can propagate vertically, usually with very short vertical wavelengths. Whereas, in the nonequatorial latitudes, all eastward traveling waves are vertically trapped. The westward traveling waves with periods less than 5 days are also vertically trapped. Westward traveling waves with periods longer than 5 days are not trapped vertically and they can propagate vertically.

In the nonequatorial latitudes Lindzen's conclusion, that eastward propagating waves will not propagate vertically, is similar to that suggested by Charney and Drazin (1961). Lindzen's outcome that both the eastward and westward moving planetary waves can propagate vertically in the near-equatorial latitudes is consistent with the observational evidences and became useful in constructing the theory of equatorial QBO, in which eastward moving Kelvin waves and westward moving mixed Rossby gravity waves propagate vertically in the near-equatorial region.

4.11 Mechanism of Quasi-Biennial Oscillation

Most part of the middle atmosphere is dominated by the annual cycle, whereas in the tropical lower stratosphere the major signal in the zonal velocity is the quasi-biennial oscillation (QBO). The QBO wind extremes are about 30 m s^{-1} easterly and 20 m s^{-1} westerly with peak amplitudes near 20 hPa. It is a quasi-regular vacillation with a mean period of 28 months with alternating easterly and westerly and it dominates the signal of the tropical wind field between 70 and 10 hPa (see section 1.7.8 for details). The latitude height cross section of the zonal wind amplitude is shown in Fig. 4.18. The QBO is equatorially trapped, with a half-width near 15° latitude (Hamilton 1984, 1998; Baldwin et al. 2001).

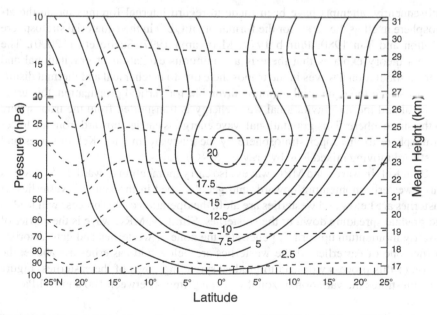

Fig. 4.18 Latitude–height cross section of the amplitude of QBO (Adapted from US Navy; Asnani 2005)

Even though the equatorial phenomenon is reasonably well understood, the problem of how the equatorially confined QBO forcing can induce a signal in the troposphere and also to the extratropics of comparable or even larger amplitudes remains unsolved.

4.11.1 Solar QBO Connection

Earlier studies hypothesized that stratospheric QBO is driven by the Sun. It was speculated that there is some quasi-biennial oscillations in solar energy reaching the Earth's surface. These periodic fluctuations originating from the Sun arrive as thermal input into the atmosphere and would induce a QBO in the equatorial lower stratosphere. Under certain conditions, such a forcing could cause downward propagating oscillations in zonal wind and temperature (Lindzen and Holton 1968; Holton and Lindzen 1972; Lindzen 1987). The major limitation of this theory is to explain the observed increase in westerly winds over the equator through meridional advection of absolute angular momentum. Still some of the observational evidences (e.g., Salby and Callaghan 2000; Labitzke 2006) indicate a connection between solar activity and QBO.

4.11.2 Planetary Wave Forcings

Subsequently, attempts have been made to regard internal forcings inside the atmosphere itself as the cause for the formation of QBO in the equatorial stratosphere (Holton and Tan 1980; Plumb 1977; Maruyama 1997; Scaife et al. 2000). The current theory explains that there is a continuous excitation of gravitational and inertia-gravitational waves in the troposphere due to mechanical and thermal disturbances. Since the equatorial region is favorable for vertical propagation, this gravitational and inertia-gravitational wave energy is transported from the troposphere to the stratosphere. This gravitational wave energy is carried upward from the lower stratosphere to the upper stratosphere by the eastward moving Kelvin waves and westward moving mixed Rossby gravity (MRG) waves.

Both Kelvin waves and MRG waves also carry momentum upward. Kelvin waves carry westerly momentum upward, and get absorbed in the zone of westerlies. If easterlies are below and westerlies are aloft, the operation of this process will lead to the gradual spreading down of the westerlies. Similary, MRG wave is the carrier of easterly momentum upward. It penetrates the zone of westerlies and gets absorbed in the zone of easterlies. If the westerlies are below and easterlies are above, the action of this process will lead to gradual spreading down of the easterlies. Figure 4.18 illustrates the variation of zonal wind amplitudes between 100 and 10 hPa.

There is a strong semiannual oscillation existing at 35 km level which provides alternating easterly and westerly zonal currents. Westerlies at this level help the absorption of Kelvin waves and spread of westerlies below (Hamilton 1984; 2002). Easterlies at 35 km level facilitate the absorption of MRG waves and spread of easterlies below.

When the descending westerly wind regime reaches the tropopause level through the absorption of Kelvin waves, the MRG waves, which were previously absorbed by the easterlies in the lower stratosphere, become free to carry easterly momentum upward until they encounter the upper level easterly wind zone near 35 km. Then they start organizing the easterly regime downward (Haynes 1998).

At the time of descending, easterly wind regime reaches the tropopause level through the absorption of MRG waves, then the Kelvin waves, which were previously absorbed by the lower-level stratospheric westerlies, become free to carry the westerly momentum upward until they meet the upper-level westerly wind zone near 35 km.

If the semiannual oscillation at 35 km is in such a phase that it cannot absorb the wave coming from below in the form of Kelvin wave or MRG wave, then the wave simply continues to propagate upward above the 35–40 km region. In such cases the trapping of wave and momentum at 35 km level is delayed till the appropriate zonal wind reappears at that level. Since the easterlies and westerlies follow each other after 3 months, this trapping may be delayed for a maximum period of 3 months. The period of this particular cycle QBO gets extended by that much time.

Modeling studies could successfully simulate quantitatively the observed QBO, based on the present theory (Holton and Austin 1991). Still this theory fails to explain the complete phenomena of the formation and existence of QBO in the atmosphere.

4.11.3 Limitations of the Planetary Wave Theory

The present theory assumes that the basic state of the stratosphere is one in which wave energy is being continuously injected from below across the tropopause in the form of Kelvin waves and MRG waves. In the real case, the stratosphere has got a seasonal zonal motion of its own. The interaction between the zonal motion and these waves has not been considered adequately.

During the Asian summer monsoon season, tropical easterly jet streams appear in the upper troposphere over the monsoon region. The theory cannot explain how the wave momentum from the lower troposphere crosses these strong easterlies and penetrates into the lower stratosphere as MRG waves (Asnani 2005).

The current theory is applicable only to the observed features of stratospheric QBO; no attempt has been made to explain the presence of the observed QBO signal in other meteorological and oceanographic elements of the Earth's geosphere.

4.12 Sudden Stratospheric Warming

Sudden stratospheric warming (SSW) is one of the dramatic events in the atmosphere. Occasionally, very large planetary wave amplitude pulses are observed in the stratosphere, often associated with the development of a blocking situation in the troposphere. Within 7–10 days following the sudden increase in wave amplitude, the temperature in the stratosphere increases dramatically over the poles and even the zonal wind may change from westerly to easterly.

4.12.1 Evolution of Warming

During the prewarming stage an intense polar night jet stream with strong vertical wind shear will appear. At this time a blocking situation develops in the troposphere. There is amplification of wave numbers 1 and 2 occuring in the upper troposphere and lower stratosphere. There are two types (Type A and Type B) of wave amplification which takes place during the prewarming period. In Type A, intensification of wave number 2 occurs about 2 weeks prior to the onset of the major warming. This is closely followed by an increase in the amplitude of wave number 1 with a subsequent weakening of wave number 2. During this time, wave number 1 reaches a peak about 1 week prior to the warming. In Type B warming, wave number 1 maintains a large amplitude for a long period. Type A is the most frequently occurring type of stratospheric warming.

An important character of sudden warming is its downward propagation from 45 km into the lower stratosphere. This feature is not zonal in character, but has nonzonal structure, with a planetary wave tilting westward with height. In the middle latitudes, warming in the stratosphere (40 km) is followed by cooling in the mesosphere (65 km). Simultaneously, the opposite occurs in low latitudes, with stratospheric cooling and mesospheric warming, but the magnitude of the changes in the lower latitudes is one order less compared to that noted in higher latitudes.

4.12.2 Theory of Stratospheric Warming

Sudden stratospheric warming evolves in the darkness of the polar night without any apparent external source of heating. However, in a major warming event, the mean temperature of the stratosphere increases at the rate of $10°C$ per day and within a week, the normal north–south temperature gradient gets reversed. It is believed that this occurs mainly due to the convergence of heat and momentum fluxes associated with the amplified transient planetary-scale waves. This again produces an induced meridional circulation which decelerates the westerly jet.

Earlier researchers were deceived by the apparent downward propagation of the warming and easterlies and looked for an upper energy source. Further studies, like that of Matsuno (1971), have shown that the source of energy lies in the troposphere. In this theory it is explained that a sudden increase in stationary planetary wave amplitude in the troposphere is accompanied by vertical propagation of transient wave energy into the stratosphere.

The transient wave energy passes through the stratosphere without any absorption, but gets trapped in the mesosphere causing convergence of heat and easterly momentum. Due to lower density in the upper levels, the effect on temperature and wind becomes substantial. The prewarming temperature gradient becomes weak and gets reversed. The westerly jet weakens and may be replaced by easterly winds.

The appearance of the easterly flow above the westerly flow produces a critical level with zero wind velocity in the mean current, where the horizontal speed of the mean current is equal to the horizontal phase speed of stationary waves in the troposphere. The easterly regime descends as the westerly wind gets destroyed. The critical level of wave energy thus comes down. The processes of energy absorption and sequence of events may be so fast and mixed that the appearance of easterlies may look simultaneous throughout the stratosphere (Holton and Austin 1991).

The vertically rising wave energy from the troposphere is spread over a band of synoptic and planetary-scale waves. Since the stratosphere is opaque to synoptic-scale waves, the transmissivity of the planetary-scale wave increases. As a result, wave numbers 1 and 2 are dominant in the upward propagation of wave energy from the troposphere into the stratosphere. It is believed that wave number 1 prepares conditions favorable for stratospheric sudden warming, whereas wave number 2 acts as an effective source of energy for the stratospheric warming.

Prior to the incidence of SSW, the stratosphere is so tuned that relatively small perturbation gets enormously amplified within a short period, a phenomenon called resonance. Association between the tropospheric blocking situation, amplification of standing planetary waves in the troposphere and lower stratosphere, and the subsequent appearance of sudden warming in the stratosphere is viewed as manifestation of the same phenomenon in the troposphere, stratosphere, and to a certain extent in the mesosphere. The difference among various situations lies in the differences in the depths of tuned conditions, favorable for resonance.

Since the energy source lies in the troposphere, tuning of the troposphere leads to resonance in the troposphere and the appearance of blocking situation. If the middle atmosphere also gets tuned at or around the time of resonance and blocking situation in the troposphere, then there is resonant response of the middle atmosphere also (Schoberl 1978). This may be the reason why all stratospheric warmings are accompanied or preceded by the tropospheric blocking phenomenon, but all tropospheric blockings are not accompanied by SSW.

It is hypothesized that sudden increase in planetary wave amplitude in the troposphere during winter triggers vertically propagating transient wave energy. It can be considered that troposphere and middle atmosphere constitute one system tuned for the occurrence of resonance; the upward flux of wave energy from the upper troposphere into the lower stratosphere cannot be taken as a lower boundary condition for the stratospheric warming. Thus upward flux will have to be taken as an internal phenomenon for the troposphere–stratosphere–mesosphere system.

Problems and Questions

4.1. Discuss the basic properties of waves in the atmosphere. How the atmospheric waves are classified? Distinguish the fundamental characteristics of the waves in the equatorial region and mid-latitudes.

4.2. Calculate the horizontal phase speed and group velocity of one-dimensional acoustic wave in a dry and isothermal atmosphere with a constant temperature of $30°$ What would be the speed of this acoustic wave in the tropopause level, where the temperature is $-40°C$?

4.3. What are shallow water gravity waves? Why it is called so? Discuss the three-dimensional structure of shallow water equation.

4.4. Show that an upward propagating wave with constant kinetic energy, the velocity in the wave will vary with an amplitude of $\exp(z/2H)$, where H is the scale height. If a gravity wave at $100\,km$ has an amplitude of $100\,ms^{-1}$, what would be its amplitude at the surface where it originates? (Assume mean temperature is $260\,K$)

4.5. Average depth of Indian Ocean is about $5\,km$, calculate the speed of shallow water waves in Indian ocean region, where the average depth is maintained.

4.6. Assume two Tsunamis of same intensity have generated in the central Pacific and Atlantic Oceans. Find the tsunami of which ocean moves faster? Why?

4.7. Explain the salient feature of gravity waves in the atmosphere. Distinguish the properties of pure internal gravity waves and inertia gravity waves.

4.8. Find the period of an internal gravity wave of horizontal wavelength $100\,km$ and vertical wavelength $5\,km$ in the upper stratosphere, where $N^2 = 3 \times 10^{-4}\,s^{-2}$. How long this wave take to propagate through a vertical distance of $20\,km$?

4.9. What are the characteristics of Rossby waves? Why Rossby waves are considered as important for meteorological predictions?

4.10. What is a standing planetary wave? How do such waves move? What happens when these standing planetary waves reach the stratosphere?

4.11. Calculate the propagation speed of wavenumber 1 of a stationary forcing Rossby wave a $60°N$ where the critical zonal wind speed is $22.5\,ms^{-1}$ (Given $N^2 = 5 \times 10^{-5}\,s^{-1}$, $1 = 3 \times 10^{-7}\,m$)

References

Andrews DG (2000). An Introduction to Atmospheric Physics, Cambridge University Press
Andrews DJ, Holton JR, Leovy CB (1987) Middle Atmosphere Dynamics, Academic, New York
Asnani GC (2005) Tropical Meteorology, second edition, vol. 2, Pune, India

Baldwin MP, Gray LJ, Dunkerton TJ, Hamilton K, Haynes PH, Randel WJ, Holton JR, Alexander MJ, Hirota I, Horinouchi T, Jones DBA, Kinnersley JS, Marquardt C, Sato K, Takahashi M (2001) The quasi-biennial oscillation, Rev Geophys, 39 (2): 179–229

Brachfeld S, Montclair State University, Montclair, New Jersey

Brandon J : Recreational Aviational Australia, Lecture Notes on Aviation Meteorology, Met Images (*http://www.auf.asn.au/metimages/rossby1.gif*) (*http://www.auf.asn.au/metimages/rossby2.gif*)

Charney JG, Drazin PG (1961) Propagation of planetary scale waves from the lower into the upper atmosphere, J Geophys Res, 66: 83–109

Dunkerton TJ, Delisi DP (1985) Climatology of the equatorial lower stratosphere, J Atmos Sci, 42: 376–396

Eliassen A, Palm E (1961) On the transfer of energy in stationary mountain waves, Geophys Publ, 22: 1–23

Gill AE (1982) Atmosphere-Ocean Dynamics, Academic, New York

Hamilton K (1984) Mean wind evolution through the quasi-biennial cycle in the tropical lower stratosphere, J Atmos Sci, 41: 2113–2125

Hamilton K (1998) Dynamics of the tropical middle atmosphere; A tutorial review, Atmosphere-Oceans, 36: 319–354

Hamilton K (2002) On the quasi-decadal modulations of the stratospheric QBO period, J Climate, 15: 2562–2565

Haynes PH (1998) The latitudinal structure of the quasi-biennial oscillation, Quart J Royal Met Soc, 124: 2645–2670

Hines CO (1968) Gravity waves in the presence of wind shear and dissipative processes, North-Holland, Amsterdam

Hocking WK (2001) Buoyancy (gravity) waves in the atmosphere (*http://www.physics.uwo.ca/~whocking/p103/parcel.gif*

Holton JR (1983) The stratosphere and its links to the troposphere, Large Scale Dynamical Processes in the Atmosphere, edited by B Hoskins and R. Pierce, Academic, New York

Holton JR (2004) An Introduction to Dynamic Meteorology, fourth edition, Elsevier Academic, Burlington

Holton JR, Austin J (1991) The influence of the QBO on sudden stratospheric warming, J Atmos Sci, 48: 607–618

Holton JR, Durran D (1993) Convectively generated stratospheric gravity waves: The role of mean wind shear in Coupling Processes in the Lower and Middle Atmosphere edited by E. V. Thrane, T. A. Blix, D. C. Fritts), Kluwer, Netherlands

Holton JR, Lindzen RS (1972) An updated theory for the quasi-biennial cycle of the tropical stratosphere, J Atmos Sci, 29: 1076–1080

Holton JR, Tan HC (1980) The influence of the equatorial quasi-biennial oscillation on the global atmospheric circulation at 50 mb, J Atmos Sci, 37: 2200–2208

Kim YJ, Eckermann SD, Chun HY (2003) An overview of the past, present and future of gravity wave drag parametrization for numerical climate and weather prediction models, Atmosphere-Ocean, 41: 65–98

Labitzke K (2006) Solar variations and stratospheric response, Space Sci Rev, 125: 247–260

Lindzen RS (1967) Planetary waves on β-planes, Mon Wea Rev, 95: 441–451

Lindzen RS (1971) Equatorial planetary waves in shear, Part I, J Atmos Sci, 28: 609–622

Lindzen RS (1987) On the development of the theory of the QBO, Bull Amer Meteor Soc , 68: 329–337

Lindzen RS, Holton JR (1968) A theory of the quasi biennial oscillation, J Atmos Sci, 25: 1095–1107

Martin EJ (2006) Mid-latitude atmospheric dynamics, Wiley, Chichester, England

Maruyama T. (1997) The quasi-biennial oscillation (QBO) and equatorial waves – A historical review, Meteorol Geophys, 48: 1–17

Matsuno T (1966) Quasi-geostrophic motions in the equatorial area, J Met Soc Japan, 45:25–43

Matsuno T (1971) A dynamical model of the sudden stratospheric warming, J Atmos Sci, 28: 1479–1494

Naval Postgraduate School (2003) Department of Oceanography, Lecture Notes, US Government
    Website : *http://www.oc.nps.edu/webmodules?ENSO/images/kelvin.gif*
Nappo CJ, (2002) An Introduction to Atmospheric Gravity Waves, Academic, San Diego, CA
Plumb RA (1977) The Interaction of two internal waves with the mean flow: implications for the
    theory of the quasi-biennial oscillation, J Atmos Sci, 34: 1847–1858
Salby M, Callaghan P (2000) Connection between the solar cycle and QBO The missing Link,
    J Climate, 13: 2652–2662
Scaife AA, Butchart N, Warner CD, Stainforth D, Norton WA, Austin J (2000) Realistic quasi-
    biennial oscillations in a simulation of the global climate, Geophys Res Let., 27: 3481–3484
Schoeberl MR (1978) Stratospheric warming: Observations and theory, Rev Geophys Space Phys,
    16: 521–538
University of Maryland (2004) Department of Geology, Shoreline processes *http://www.
    geol.umd.edu/~jmerck/geo1100/images/31/wave.schematic.gif*

Chapter 5
Chemical Processes in the Stratosphere and Troposphere

5.1 Introduction

Chemistry of the lower and middle atmosphere plays a critical role in stratosphere–troposphere coupling processes by controlling the abundances and distribution of natural and anthropogenic agents, such as greenhouse gases, aerosols, and clouds, which regulate the incoming and outgoing radiation, temperature, and tropospheric weather events. The atmospheric constituents are closely coupled through chemical processes as well as dynamics and radiation.

Stratospheric chemistry involves hundreds of different gases, and the interaction of those gases with one another. The ultraviolet and visible parts of the solar radiation are necessary for initiating many of the reactions in the atmosphere. Since ozone screens biologically harmful ultraviolet (UV) light from the Sun, it is a highly significant component of the atmosphere. Most of the stratospheric chemistry is centered around ozone.

Tropospheric chemistry is mostly about gases released into the atmosphere from the biosphere below. They are generally more complex compounds than those involved in stratospheric chemistry. In addition to the UV and visible radiation, infrared radiation also plays an important role in the chemistry of the troposphere. Tropospheric ozone abundances vary widely both in time and space, from the short lifetime of a few days to a few months. This large regional and temporal variability makes it difficult to use available observations to validate the global budget of tropospheric ozone and quantify the changes since the preindustrial era (UNEP 2002; Saltzman et al. 2004).

Changes in tropospheric weather and climate can also affect atmospheric chemistry significantly. For example, a change in water vapor abundance can alter the ability of the atmosphere to oxidize trace gases. A change in temperature or water vapor abundance can modify the chemical and physical properties of aerosols and can alter the rates of chemical transformations in the atmosphere. Temperature and precipitation changes can also affect emissions from the surface. Biotic emissions will modify as ecosystems shift, and atmospheric mineral dust loading may vary with increased desertification or with changes to the meteorological systems that loft the dust. These interactions and feedback processes are complex and poorly understood.

Extensive information on the chemistry of the atmosphere is available in the literature (e.g., Wayne 1991; Salby 1996; Dessler 2000; Pitts and Pitts 2000; Newman and Morris; Brasseur and Solomon 2005).

In this chapter, some of the basic concepts of the chemistry involved in stratosphere–troposphere interactions are discussed.

5.2 The Absorption Cross Section

The absorption cross section denotes the ability of a particular gas to absorb radiation at a given wavelength. The fraction of incident radiation to the absorbed part is given by the relation

$$\frac{dI}{I} = n\sigma \tag{5.1}$$

where I is the incident radiation and dI is the absorbed part, n is the number of molecules, and σ is the absorption coefficient which is represented in units of area.

The absorption coefficient of a given gas depends on the wavelength of the incident radiation, and is highly variable in nature. In the case of ozone, the absorption cross section is small for wavelengths above about 325 nm. As we move to higher energies (shorter wavelengths), the cross section increases, with a peak value of about 10^{-17} cm^2 at a wavelength of 255 nm. The cross section decreases as we move to still higher energies (shorter wavelengths).

The energy of a photon is related to its wavelength, with longer wavelengths having less energy. Photons with too little or too much energy cannot be absorbed by the atom or molecule. Thus, the absorption cross sections tend to have maxima and minima at particular wavelengths determined by the structure of the atom or molecule.

5.3 Chemical Kinetics

For understanding the chemical processes in the atmosphere, knowledge of the rate of a process is important. The activation energy of a chemical reaction is the energy required to be added to a system before a reaction can take place, and effectively represents a barrier toward the reaction. In the following section, we explain, the factors that determine the rates of chemical reactions in the atmosphere (for details see Dessler 2000).

5.3.1 First-Order Reactions

A first-order reaction is one in which a reactant spontaneously transforms itself into one or more products, like radioactive decay and isomerization. The rate of a first-order reaction is equal to the product of a rate coefficient and the abundance of the reactant X

$$\text{Rate} = k[X] \tag{5.2}$$

where k is the first-order rate coefficient for the reaction, and $[X]$ is the abundance of X, expressed either in number density or volume mixing ratio. In the absence of any other sources or sinks of X, the time derivative of $[X]$ is the same as the instantaneous rate of reaction

$$\frac{d[X]}{dt} = -k[X] \tag{5.3}$$

Since the rate of reaction is always positive (Solomon 1977a), the negative sign indicates that X is diminishing.

Integrating the above equation, we get the expression for X as a function of time t as

$$[X](t) = [X]_0 exp(-kt) \tag{5.4}$$

where $[X]_0$ is the concentration of X at time $t = 0$. Over any time interval $1/k$, $[X]$ is reduced by a factor of $1/e$ (0.368). Therefore, $1/k$ is often referred to as *e-folding time* or lifetime of X.

5.3.2 Second-Order Reaction

In second-order reaction, two reactants are involved. Consider the reaction between chlorine monoxide (ClO) and atomic oxygen (O).

$$ClO + O \rightarrow Cl + O_2 \tag{5.5}$$

The rate of the above reaction can be written as

$$\text{Rate} = k[ClO][O] \tag{5.6}$$

which is the product of a rate coefficient (k) and the concentration of the two reactants: ClO and O.

Second-order reaction rate coefficients are generally functions of temperature and have units of cm^3 molecules^{-1} s^{-1}. The abundance of the reactants of Eq. (5.5) must be expressed in number density units.

In the absence of any other sources or sinks of the reactants or products, the second-order rate can be written, based on the conservation of mass, as

$$\text{Rate} = -\frac{d[ClO]}{dt} - \frac{d[O]}{dt} - \frac{d[Cl]}{dt} = -\frac{d[O_2]}{dt} \tag{5.7}$$

5.3.3 *Three-Body Reaction*

A three-body reaction (sometimes known as association reaction) is one in which two reactants combine to form a single product molecule, like the formation of chlorine nitrate ($ClONO_2$).

$$ClO + NO_2 \xrightarrow{M} ClONO_2 \tag{5.8}$$

The above reaction takes place in two steps. In the first step, the reactants collide to form an excited intermediate molecule:

$$ClO + NO_2 \xrightarrow{M} ClONO_2^* \tag{5.9}$$

The sign $*$ indicates excited state. The $ClONO_2^*$ molecule contains considerable amount of internal energy, so that it can collide with a third body, M (generally N_2 or O_2), which carries away some of the excess energy of the excited molecule, thereby stabilizing it.

$$ClONO_2^* + M \rightarrow ClONO_2 + M \tag{5.10}$$

It can also decompose back into ClO and NO_2, in which case there is no net reaction:

$$ClONO_2^* \rightarrow ClO + NO_2 \tag{5.11}$$

If the pressure is sufficiently high (large [M]), then the excited intermediate molecule formed will collide with a third body M and form the stable molecule. In this case, the formation rate of $ClONO_2$ is equal to the rate of formation of the excited intermediate $ClONO_2^*$. This is known as the high-pressure limit of reaction.

At low pressure, [M] is small and the excited intermediate $ClONO_2^*$ usually decomposes back to ClO and NO_2. In this case, the formation rate of $ClONO_2$ is set by the collision rate between $ClONO_2^*$ and M, which is known as the low-pressure limit. At pressures between these limits, the rate of reaction is set by a combination of the two processes.

$$\text{Rate} = k^*[X][Y] \tag{5.12}$$

where k^* is the effective second-order rate coefficient for the three-body reaction, and [X] and [Y] are the abundances of the reactants. For these reactions, the rate coefficient k^* depends on both temperature and pressure.

5.4 Thermal Decomposition Reaction

Thermal decomposition occurs when a molecule splits into fragments following nonreactive collisions with other molecules. Consider the decomposition of the chlorine monoxide dimer (ClOOCl):

$$ClOOCl \xrightarrow{M} ClO + ClO \tag{5.13}$$

Here, the collision between ClOOCl and M transfers energy to ClOOCl and creates the excited intermediate ClOOCl*. This excited intermediate can either collide with another M, leading to the restabilization of the intermediate, or the excited intermediate can defragment:

$$ClOOCl^* \rightarrow ClO + ClO \tag{5.14}$$

As a first-order process, the rate of reaction of Eq. (5.13) can be written as:

$$\text{Rate} = k^T [ClOOCl] \tag{5.15}$$

Where k^T is the thermal decomposition rate coefficient (s^{-1}). k^T can be calculated from the rate of association reaction and the equilibrium constant between the reactant and products.

5.5 Equation of Continuity

The continuity equation for a constituent X at a given point in the stratosphere can be represented as

$$\frac{\partial [X]}{\partial t} = P - L[X] - \nabla.V[X] \tag{5.16}$$

The terms on the right-hand side of the continuity equation represent the sources and sinks of X in unit volume. The first term on the right-hand side, P, represents the production which is equal to the amount of X produced per unit volume per second. This term always leads to a positive rate of change of time. The second term stands for the loss, which is the product of loss frequency L and the abundance of X and always leads to a negative rate of change of time. The last term is the divergence of X representing net transport of X in or out of the unit volume by atmospheric motions, which can be either positive or negative. If the sources and sinks balance, then the net change with the abundance of X vanishes, i.e., $\partial [X]/\partial t$ becomes zero.

In the stratosphere, the production and loss terms (P and $L[X]$) of the continuity equation are much larger than the transport term ($\nabla.V[X]$); therefore the transport term can be generally neglected to a first approximation in Eq. (5.16) as

$$\frac{\partial [X]}{\partial t} = P - L[X] \tag{5.17}$$

The source and sink terms for a particular species vary depending upon the season and location.

5.6 Ozone Photochemistry

Basics of ozone photochemistry began with Chapman (1930), who hypothesized that UV radiation was responsible for ozone production. Ozone photochemistry is

Stratospheric ozone production

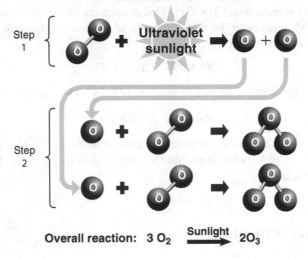

Overall reaction: $3 O_2 \xrightarrow{\text{Sunlight}} 2O_3$

Fig. 5.1 The process of ozone photochemical production. Ozone is formed when an energetic ultraviolet photon splits an oxygen molecule (O_2) to give rise to oxygen atoms. These oxygen atoms quickly react with other oxygen molecules to form ozone (WMO 2007)

driven by the interaction of the Sun's radiation with various gases in the atmosphere, particularly oxygen. A schematic representation of the photochemical reaction for the production and destruction of ozone in the atmosphere is shown in Fig. 5.1. The atomic oxygen is formed by the splitting (dissociation) of O_2 by high-energy ultraviolet photons ($\lambda < 242$ nm) by the reaction

$$O_2 + h\nu \rightarrow O + O \tag{5.18}$$

where h is the Planck's constant, $\nu = c/\lambda$, c is the speed of light, and λ is the wavelength of the photon.

The photolysis of oxygen molecules by solar radiation is relatively slow in the lower and middle stratosphere because the photons of sufficient energy have already been absorbed by molecular oxygen in the upper stratosphere. Few such photons are able to penetrate deeply into the atmosphere.

Oxygen atoms are highly reactive, and quickly react with oxygen molecules to form ozone as in the following reaction:

$$O_2 + O + M \rightarrow O_3 + M \tag{5.19}$$

where M represents any other molecule (most probably N_2 or O_2). The third neutral body (M) is needed for the energy balance of the reaction. This particular reaction proceeds at a very fast rate. The lifetime (defined as the time required for the abundance of oxygen atoms (O) to decrease by about 63%, known as the *e-folding* timescale) is very short in the stratosphere, typically less than 1 second. Hence, oxygen atoms almost immediately form ozone after they are dissociated.

Atomic oxygen can also be lost by the recombination

$$O + O + O \rightarrow O_3 \tag{5.20}$$

But this chemical reaction is important only at higher altitudes where atomic oxygen atoms are more abundant. In the lower and middle stratosphere reaction Eq. (5.19) is the more dominant loss mechanism for O.

Ozone strongly absorbs UV radiation. The ozone molecule is dissociated by UV photons into O and O_2 by the reaction

$$O_3 + h\nu \rightarrow O_2 + O \tag{5.21}$$

Because the O atoms have such short lifetimes, they quickly recombine to form ozone, converting the energy of the photons at these wavelengths into thermal energy. Most of the ozone production in the atmosphere occurs in the tropical upper stratosphere and mesosphere.

Figure 5.2 shows the vertical profiles of Chapman's photochemical ozone production based on density, flux, and absorption rate. The total mass of ozone produced per day over the globe is about 400 million metric tons. The global mass of ozone is relatively constant at about 3 billion metric tons, meaning the Sun produces about 12% of the ozone layer each day.

Chapman also hypothesized that ozone is lost by a reaction with the free oxygen atoms, which balances ozone production. The basic ozone loss reaction is

$$O_3 + O \rightarrow O_2 + O_2 \tag{5.22}$$

This reaction should be relatively slow in most of the atmospheric regions since ozone concentrations as a total share of the atmosphere are quite small. Equations (5.18)–(5.21) represent the so-called Chapman cycle for ozone chemistry.

The Chapman profile

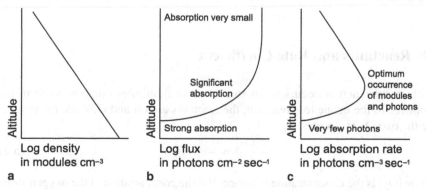

Fig. 5.2 Chapman's photochemical ozone production based on density, flux, and absorption rate with height in the stratosphere (Adapted from Science and Society, Columbia University)

5.7 Limitations of Chapman Cycle

Even though the Chapman cycle, as mentioned in Eqs (5.18)–(5.21), provides the basic mechanisms for ozone production and destruction in the stratosphere, they fail to explain the observed ozone distribution in the atmosphere. Since the Sun is overhead in the tropics, the total amount of ozone in the tropics is quite low. The Chapman chemical reactions lead to predicted ozone amounts twice as high as actual tropical observations. In contrast, middle to high latitude prediction of ozone by the Chapman cycle are too low to explain the observations. Not only the global distribution (latitudinal variation) but also the seasonal variations are not explained by the Chapman cycle.

The problems with Chapman chemistry result from two processes. First, there are other ozone loss reactions with gases containing chlorine, bromine, nitrogen, and hydrogen that contribute to the overall ozone loss process.These reactions create additional pathways from O_x back to O_2. Second, there exists an equator-to-pole stratospheric circulation known as the Brewer-Dobson circulation (for details see section 7.2) that transports ozone from the photochemical production region in the tropics to the middle and high latitudes. The Brewer-Dobson circulation decreases tropical ozone amounts, while increasing extratropical amounts.

In general, O_x is formed by the photolysis of O_2. These odd oxygen compounds are destroyed by the reaction of O_3 and O on timescales of months to years in the lower stratosphere and days in the upper stratosphere. On timescales of minutes to hours (i.e., timescales that are short compared to O_x lifetimes), the sum of O_3 and O is nearly constant. However, these species rapidly cycle back and forth and in every minute photochemically interconvert all the O_3 destroyed by UV photolysis. This leads to the formation of free O atoms, and all of the O atoms are immediately consumed in reactions with O_2 to reform O_3 in a fraction of a second. Since lifetime lasts for minutes to hours in the tropical middle to lower stratosphere, and oxygen atoms last for less than a second, most of the odd oxygen exists in the form of ozone. Thus, the odd oxygen exists mostly in the form of ozone in the middle and lower stratosphere.

5.8 Reactants and Rate Coefficient

A chemical reaction occurs when two or more molecules combine to form new products. In the ozone loss reaction, the reactants are O_3 and O, while the products are the two O_2 molecules.

$$\frac{d[O_3]}{dt} = -k[O_3][O] \tag{5.23}$$

where $[O_3]$ is the concentration of ozone, $[O]$ the concentration of the oxygen atoms, and k is the reaction rate coefficient.

Reactants not only have kinetic energy and potential energy, but also internal energy associated with the strength of the bonds holding the molecules together. Reactions that absorb heat or energy are known as endothermic reactions, while those that give off energy are exothermic reactions. The photolysis of oxygen molecules to produce atomic oxygen is an endothermic reaction, since this reaction requires the energy of UV photons. The reactions of $O_3 + O$ to form the two oxygen molecules are exothermic since it gives off energy.

The loss rate is proportional to the concentrations of O_3, O, and the reaction rate constant. The $O_3 + O$ reaction rate constant is about three times slower than the $O_2 + O$ reaction rate constant, and there are typically more than 100,000 oxygen molecules for each ozone molecule. Hence, the rate at which an oxygen atom will react with ozone is quite small. Figure 5.3 gives the latitudinal distribution of ozone climatology in the entire stratosphere during the month of January.

In Chapman's model, the loss process balances the photochemical production from the photolysis of oxygen molecules. The oxygen and ozone molecules are rapidly interconverted. The source of odd oxygen (O_x) is the photolysis of oxygen molecules, which is a relatively slow process with a timescale of many weeks at 30 km over the equator. The loss of odd oxygen is the reaction of ozone and oxygen atoms, which also is a slow process with a comparable timescale. The time between creation and destruction (lifetime) of O_x is much longer than the lifetime of either O_3 or O individually because of this rapid interconversion. All the ozone in a given air parcel is destroyed many times over during the course of a single day when the

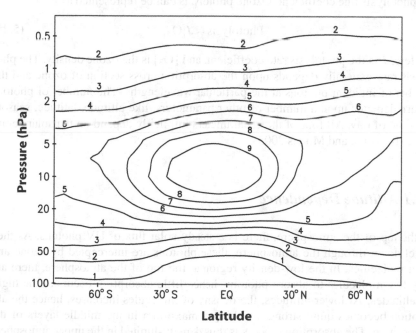

Fig. 5.3 Latitudinal variation of the ozone climatology in the stratosphere during the month of January (Source: SPARC Data Center)

parcel is in sunlight. Indeed, at an altitude of 30 km above the equator, the lifetime of an ozone molecule due only to UV photolysis is less than 1 hour. However, ozone is reformed in the parcel at almost exactly the same rate through the reaction between O and O_2. Hence, ozone concentrations in the middle stratosphere change only very slowly over the long timescales on the order of weeks to months (Pitts and Pitts 2000).

5.9 Ozone Photolysis

Solar energy is capable of breaking chemical bonds to produce highly reactive radical species. When ozone absorbs photons with wavelengths below 325 nm, an oxygen atom breaks away from the molecule. The reaction can be represented as

$$O_3 + h\nu \leftrightarrow O_2 + O \tag{5.24}$$

The dissociation of ozone molecules by UV photons is known as ozone photolysis. The heat released by this reaction explains why the stratosphere is warmer than the upper troposphere. The very existence of the stratosphere is due to this heating. The absorption of UV photons by ozone both shields the surface from UV and heats the local atmosphere.

The photolysis rate is proportional to the density of the gas (e.g., O_3 or O_2) and the photolysis rate coefficient. Ozone photolysis can be represented as

$$\text{Photolysis} = J[O_3] \tag{5.25}$$

Here, J is the photolysis rate coefficient, and $[O_3]$ is the ozone density. The photolysis rate generally depends upon the absorption cross section of ozone and the number of incident photons at the particular wavelengths. The number of photons in turn depends upon a number of other parameters, like altitude, latitude, season, and time of day. All four of these parameters implicitly depend on the solar zenith angle (Newman and Morris 2003).

5.9.1 Altitude Dependence

At the top of the atmosphere, there is a steady solar flux of UV photons. As they travel down through the atmosphere, these photons are intercepted by ozone and other molecules. In the low-density region at the top of the atmosphere, there are very few molecules to absorb photons, hence little absorption occurs at the highest altitudes. At lower altitudes, the density of molecules increases, hence the absorption becomes quite strong, reaching a maximum in the middle layers of the atmosphere. The absorption process is thus density-limited in the upper atmosphere and photon-limited in the lower atmosphere.

5.9.2 Latitude Dependence

During noontime, the Sun appears much higher in the sky over the tropics than at middle latitudes. The path of the solar rays through the Earth's atmosphere is much longer in the middle latitudes than in the tropics. The longer the path that light must travel through the atmosphere, the more molecules the light will encounter, and the more photons that get absorbed. So at a given altitude, the photolysis rate decreases from tropics poleward.

5.9.3 Seasonal Dependence

Since the photolysis rate depends on the angle of the Sun, it also depend upon the seasons. The Sun is much higher in the sky overhead during the summer than during the winter. The path length of light from the Sun through Earth's atmosphere will therefore be shorter in summer than in winter. The shorter the path length at a given location, the lesser the absorption of UV photons, and at a given latitude more photons are available and greater is the photolysis rate. Therefore, photolysis rates will have maximum values during summer and minimum values during winter. A seasonal cycle is thus observed in photolysis rates.

5.9.4 Diurnal Dependence

At night when there is no sunlight, photolysis rates drop to zero. In the morning and late afternoon, the Sun is lower in the sky than near noontime. The path lengths are therefore relatively long at sunrise and sunset and short at noontime. Photolysis rates for a given altitude and latitude are faster at noontime than at sunrise and sunset.

5.10 Heterogeneous Reactions

Heterogeneous reaction is a chemical process that involves solid, liquid, and gaseous phases, and hence it is a multiphase process. Such a reaction occurs on or in condensed particles that are in contact with gaseous molecules. Such reactions are the physico-chemical mechanism behind the Antarctic ozone hole. Heterogeneous reactions are extremely significant, since they release chlorine and bromine from reservoir species into reactive forms. The importance of these reactions results from the freeing of chlorine from relatively benign chlorine forms into highly reactive forms, and the removal of reactive nitrogen species (NO_x) into more stable forms such as nitric acid (HNO_3).

There are five basic heterogeneous reactions which are important in ozone chemistry. The five reactions are given below.

$$ClONO_2 + HCl \rightarrow Cl_2 + HNO_3 \tag{5.26}$$

$$ClONO_2 + H_2O \rightarrow HOCl + HNO_3 \tag{5.27}$$

$$N_2O_5 + HCl \rightarrow ClNO_2 + HNO_3 \tag{5.28}$$

$$N_2O_5 + H_2O \rightarrow 2HNO_3 \tag{5.29}$$

$$HOCl + HCl \rightarrow Cl_2 + H_2O \tag{5.30}$$

Rates for these reactions depend on a number of factors, including particle type, surface area, and temperature. If H_2O is in the liquid phase, N_2O_5 simply dissolves in water to give HNO_3. If it is in the vapor phase, it needs a surface or a particle for Eq. (5.29) to occur. This reaction does not have a temperature-dependent sticking coefficient, and is therefore important wherever and whenever we find particles in the stratosphere (e.g., after volcanic eruptions). In fact, the nitrogen chemistry of the middle latitude stratosphere cannot be explained without including Eq. (5.29). Because the other reaction rates increase dramatically in low temperatures, they are extremely important to an understanding of the ozone budget of the polar stratosphere.

In Eq. (5.26) the HCl (hydrochloric acid) and $ClONO_2$ (chlorine nitrate) molecules are mostly nonreactive in their gaseous state. However, when these molecules are dissolved in liquids such as sulfuric acid–water solutions, they become highly reactive with one another. The absorption of HCl and $ClONO_2$ on sulfate aerosol particle in the stratosphere results in the release of Cl_2 (molecular chlorine) in gaseous form (desorption) and the retention of a nitric acid (HNO_3) molecule with the aerosol particle.

Latitudinal ditributions of HCl and HNO_3 in the stratosphere during the month of January are depicted in Fig. 5.4 and Fig. 5.5, respectively. The concentration of HCl is found to increase with height in the stratosphere. Near the stratopause level, the concentration of HCl becomes ten times higher than near the tropopause. There is a strong upwelling of HCl concentration noted in the tropical region, which extends upto 40 km. In the upper stratosphere, the latitudinal variation of HCl is found to be the minimum. The concentration of HNO_3 is mainly confined in the lower stratosphere. Higher amount of HNO_3 is seen in higher latitudes with peak values in the winter polar region (see Fig. 5.5).

The nonreactive HNO_3 formed remains in a solid (frozen) state on the surfaces of the PSCs. As the PSCs undergo sedimentation, the HNO_3 is carried out of the stratosphere. This process leads to the removal of nitrogen from the stratosphere in a process called denoxification. Since HNO_3 photolysis results in the formation of reactive NO_2, which in turn reacts with ClO to form the reservoir $ClONO_2$ species, the removal of nitrogen from the stratosphere means that there is more reactive, ozone-destroying ClO (Brasseur and Solomon 2005).

The chlorine species Cl_2 and HOCl created in Eqs. (5.26), (5.27), and (5.30) are short-lived. They are quickly photolyzed by sunlight even in visible wavelengths.

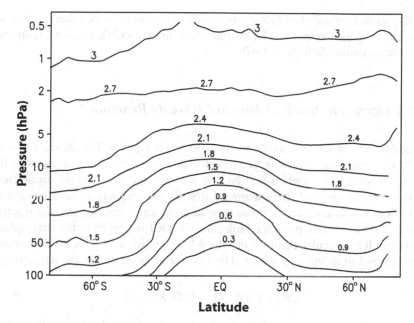

Fig. 5.4 January mean values of HCl in the stratosphere from UARS MLS as a function of height with latitude (Source: SPARC Data Center)

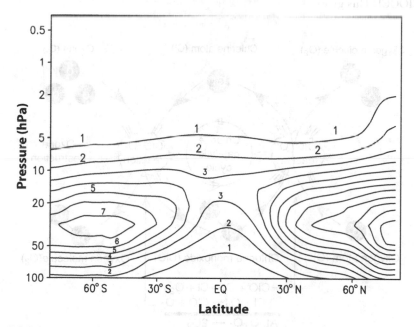

Fig. 5.5 January mean values of HNO$_3$ in the stratosphere from UARS MLS as a function of height with latitude (Source: SPARC Data Center)

The Cl_2 molecules are photolyzed into two chlorine atoms, which then participate in the Cl_x catalytic cycles. The HOCl photolysis liberates ClO, another Cl_x species that destroys ozone (Solomon 1997b).

5.10.1 Ozone Destruction Due to ClO–ClO Reaction

Ozone destruction reaction is schematically shown in Fig. 5.6. The amount of ozone loss by ClO (chlorine monoxide) can be calculated based on two assumptions. In the first assumption, it is implicit that NO_x is not involved in removing ozone from the stratosphere. Second, it is assumed that the stratosphere is cold enough to form PSCs, which leads to the conversion of nonreactive chlorine into the reactive form ClO. The chlorine peroxide molecule, ClOOCl is formed in the stratosphere when two ClO molecules combine in a three-body collision. This molecule is sometimes referred to as the *ClO dimer*. The *ClO dimer* is created in the termolecular reaction.

$$ClO + ClO + M \rightarrow ClOOCl + M \tag{5.31}$$

Based on these two assumptions, we can write that the production of ClOOCl by the ClO–ClO termolecular reaction is exactly balanced by the photolytic destruction of ClOOCl. This gives us

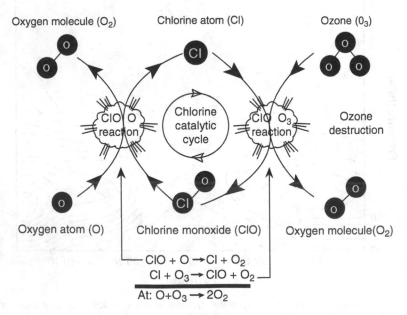

Fig. 5.6 Schematic representation of ozone destruction reaction (WMO 2007)

$$k[\text{ClO}]^2[\text{M}] = J[\text{ClOOCl}] \tag{5.32}$$

The photolysis rate constant is denoted by J, while the ClO–ClO reaction rate coefficient is k, and $[\text{M}]$ is the density of the other molecule (M) involved.

The direct loss of ozone is through the bimolecular reaction $\text{Cl} + \text{O}_3 \rightarrow \text{ClO} + \text{O}_2$. Using this reaction, we can balance the production and loss of ClO as

$$2k[\text{ClO}]^2 = k_2[\text{Cl}][\text{O}_3] \tag{5.33}$$

Here, k_2 is the $\text{Cl} + \text{O}_3$ reaction rate coefficient. The most dominant Cl_x species are ClO and Cl_2O_2; hence we write

$$[\text{Cl}_x] = [\text{ClO}] + 2[\text{Cl}_2\text{O}_2] = [\text{ClO}] + k[\text{ClO}]^2[\text{M}]/J(\text{C}) \tag{5.34}$$

Equation (5.32) is used to estimate ClOOCl in the above expression. We can now solve Eq. (5.34) and substitute into Eq. (5.33) to calculate the ozone loss rate in terms of Cl_x. Solving the equations, we find that

$$[\text{ClO}] = \frac{J}{4k}\left[\sqrt{1 + \frac{8k}{J}[\text{Cl}_x]} - 1\right] \tag{5.35}$$

For small values of $[\text{Cl}_x]$, the concentration of $[\text{ClO}]$ is almost equal to $[\text{Cl}_x]$. Using this formula, we can calculate the ozone loss rate in the ozone hole using estimates of Cl_x.

Figure 5.7 shows the monthly mean (January) chlorine monoxide (ClO) distribution using data from microwave limb sounder (MLS) instrument onboard the upper atmospheric research satellite (UARS) as a function of height and potential vorticity. ClO concentration does not show much latitudinal variation in the lower stratosphere. In the upper stratosphere ClO concentration is quite high over the northern hemispheric middle latitude region. Over the tropics, its concentration does not change much with height.

Preindustrial values of Cl_x were smaller than 3 ppbv (three reactive chlorine molecules per billion molecules of air). Solving for these preindustrial conditions, it is found that the timescale for ozone loss was greater than an entire season. Hence, the chlorine activated by PSCs did not cause much ozone loss. Current Cl_x stratospheric values are greater than 3 ppbv, giving an ozone loss timescale of about 2.5 weeks. This is why the Antarctic ozone hole now develops seasonally, but did not do so in the past (Newman and Morris 2003).

The assumptions invoked here are reasonable for Antarctic conditions. In the Arctic spring, however, temperatures are not cold and persistent enough to keep all of the chlorine in the form of Cl_x via heterogeneous processes on PSCs. In particular, once ClO is formed from HCl and ClONO_2 via a heterogeneous reaction, the ambient NO_2 will react to reform ClONO_2. This is because denoxification of the Arctic stratosphere does not occur. The paucity of PSCs means that HNO_3 is not carried out of the stratosphere via PSC sedimentation, and when sunlight returns, HNO_3 is

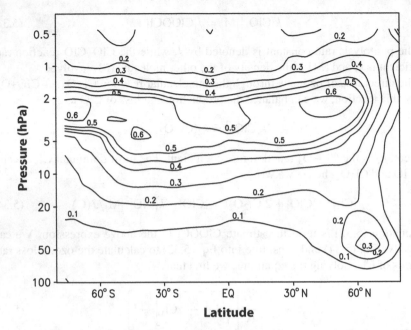

Fig. 5.7 Monthly averages of ClO observations from UARS MLS as a function of height and potential vorticity expressed in terms of equivalent latitude. Contour interval is 0.1 ppmv (Source: SPARC Data Center)

photolyzed by UV light to recreate NO_x species. Hence, it cannot be assumed that all of the chlorine is in the reactive Cl_x form.

5.10.2 Chlorine and Nitrogen Activation/Deactivation

The heterogeneous reactions (particularly Eq. (5.26) are crucial to ozone loss for two reasons. First, the chlorine is liberated from the HCl and $ClONO_2$ reservoir species (enabling the catalytic reactions); and second, the NO_2 is locked up as nonreactive HNO_3 and cannot deactivate chlorine. Hence, the chlorine is free to destroy ozone at an extremely efficient rate, via the Cl_x catalytic cycles.

5.11 Catalytic Loss

To understand the observed distribution of stratospheric ozone, a sound knowledge in catalytic loss processes is essential. A catalyst is a substance, usually present in small amounts, that causes chemical reactions without itself being consumed by those reactions. Several catalytic cycles are important in ozone chemistry.

5.11.1 Hydrogen Catalytic Loss

Hydrogen is transported into the stratosphere in the form of methane (CH_4) and water vapor (H_2O) molecules. Both of these molecules have large sources in the troposphere. Methane, a long-lived tracer, is efficiently carried from its surface source region into the stratosphere by Brewer-Dobson tropical lifting motion. Methane is able to pass through the tropical tropopause into the lower stratosphere. Depletion of methane instead occurs in the upper stratosphere, where there are enough of the excited free oxygen atoms to trigger the oxidation reactions that convert methane into other species, including, water vapor.

Figure 5.8 shows the monthly mean (January) water vapor distribution using data from UARS, Halogen Occulation Experiment (HALOE), and MLS as a function of height and potential vorticity. Maximum water vapor concentration is found in the tropical lower stratosphere. Due to lifting action from the troposphere, the water vapor mixing ratio contours are bowed upward in the middle and upper stratosphere. Higher values of water vapor concentration are noted in the summer stratopause region. Water vapor actually begins to increase above the lower stratosphere, with elevated values in the lower mesosphere.

On rainy days, water vapor can reach very high amounts in the troposphere, but is not efficiently transported into the stratosphere. Most of the water is frozen out as

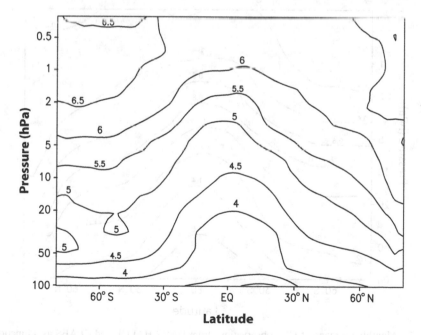

Fig. 5.8 Monthly averages of H_2O observations from UARS HALOE and MLS as a function of height and potential vorticity expressed in terms of equivalent latitude. Contour interval is 0.5 ppmv (Source: SPARC Data Center)

air moves through the very cold tropopause region in the tropical upper troposphere through the stratospheric circulation. As moist air is lofted upward through the tropical tropopause, water vapor freezes into ice crystals, which do not readily enter the stratosphere. Water is carried into the stratosphere at only a few parts per million. Hence, the lower stratosphere tends to be very dry. Water vapor mixing ratio, however, increases in the middle to upper stratosphere and into the mesosphere due to the oxidation reactions of methane.

5.11.2 Methane Photodissociation Reactions

Latitudinal variation of January mean methane distribution is shown in Fig. 5.9 with data from UARS, HALOE, and cryogenic limb array etalon spectrometer (CLAES). The figure indicates that the methane originates in the troposphere and decrease steadily through the stratosphere. The upward bowing in the tropics of methane mixing ratio contours is due to the lifting of the Brewer-Dobson circulation. Tropical region shows the maximum concentration and the peak values are centered over the equator in the lower middle stratosphere. In the upper stratosphere a tilt toward the summer hemisphere can be seen.

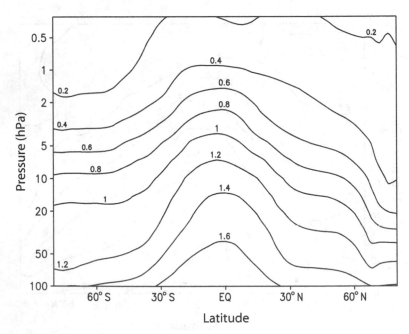

Fig. 5.9 Monthly averages of CH_4 observations from UARS HALOE and CLAES as a function of height and potential vorticity expressed in terms of equivalent latitude. Contour interval is 0.2 ppmv (Source: SPARC Data Center)

The source of methane in Earth's atmosphere can be traced to its release at the surface through a variety of sources: wood combustion, coal mining, oil and gas drilling and refining, landfills, wetland rice cultivation, crop residue burning, industrial activity, and digestive action by grazing animals (i.e., cow flatulence).

Since the tropical tropopause is the very cold boundary between the troposphere and the stratosphere, water vapor is frozen out when moist air is lofted upward through the tropopause. Air entering the stratosphere, consequently, is rather dry. On the other hand, methane remains unaffected by the cold temperatures as it passes through this boundary. Only when methane reaches the upper stratosphere is it depleted via oxidation reactions with OH (Rasmussen and Khalil 1981). These reactions lead to the production of water vapor molecules. Indeed, each methane molecule eventually is converted into two molecules of water vapor in the middle to upper stratosphere via the following two oxidation reactions.

In the first reaction, methane is converted into water vapor by a reaction with the hydroxyl radical OH.

$$CH_4 + OH \rightarrow CH_3 + H_2O \tag{5.36}$$

The second reaction involves a series of steps that begins with methane reacting with the singlet D oxygen atom, denoted $O(^1D)$. This is a free oxygen atom that is in an excited state. The reaction is

$$CH_4 + O(^1D) \rightarrow CH_3 + OH \tag{5.37}$$

The result is a hydroxyl radical and a leftover methyl radical (CH_3), which quickly reacts via

$$CH_3 + O_2 + M \rightarrow CH_3O_2 + M \tag{5.38}$$

$$CH_3O_2 + NO \rightarrow CH_3O + NO_2 \tag{5.39}$$

$$CH_3O + O_2 \rightarrow HCHO + HO_2 \tag{5.40}$$

Reactions of HCHO (formaldehyde) with the hydroxyl radical result in the production of another water vapor molecule in the region between 35 and 45 km

$$HCHO + OH \rightarrow CHO + H_2O \tag{5.41}$$

Above about 65 km, photodissociation of methane becomes the dominant mechanism for methane loss.

So the effect of the freeze-drying of air at the cold tropical tropopause is to leave water vapor relatively scarce in the lower stratosphere, while oxidation of methane increases the relative amount of water vapor in the upper stratosphere. Methane, on the other hand, is relatively plentiful in the lower stratosphere (near its tropospheric source region), while it becomes relatively depleted in the upper stratosphere due to the above oxidation reactions. These processes help explain the appearance of water vapor and methane profiles through the stratosphere (Whiticar 1993).

5.11.3 HO_x Catalytic Cycles

In the ozone photochemistry, the important species, methane and water vapor, radicals transport and release hydrogen radicals into the stratosphere. These activated hydrogen radicals that are released can then participate in the destruction of odd oxygen, i.e., ozone, through a variety of catalytic cycles.

Each water vapor molecule can be transformed into two molecules of HO_x (reactive hydrogen) through reaction with O atoms via a reaction of water vapor with the singlet D oxygen atom.

$$H_2O + O(^1D) \rightarrow 2OH \tag{5.42}$$

Recall that $HO_x = OH + HO_2$. In this case, the reactive hydrogen exists in the form of two liberated OH (hydroxyl radical) molecules which become the catalyst in a pair of reactions with odd oxygen (O_x) that result in a net loss of O_x, by which we mean a net loss of both ozone molecules and free oxygen atoms.

$$OH + O_3 \rightarrow HO_2 + O_2$$
$$HO_2 + O \rightarrow OH + O_2$$

$$\boxed{\text{Net: } O_3 + O \rightarrow 2\,O_2} \tag{5.43}$$

Notice that the net effect of the reactions is simply a conversion of two odd oxygen molecules into two molecules of O_2. OH is recovered at the end of the cycle.

This catalytic cycle involving HO_x will only be disrupted if HO_x is lost through another mechanism. Several reactions can remove HO_x from this cycle

$$OH + HO_2 \rightarrow H_2O + O_2 \tag{5.44}$$

$$OH + NO_2 + M \rightarrow HNO_3 + M \tag{5.45}$$

$$HO_2 + NO_2 + M \rightarrow HNO_4 + M \tag{5.46}$$

The hydrogen species on the right-hand side of these equations (i.e., H_2O, HNO_3, and HNO_4) are known as *reservoir species*, which are chemical compounds that store like a reservoir in a nonreactive form. These species act as stores of hydrogen, locking up HO_x and preventing its participation in the catalytic cycle outlined above. H_2O, HNO_3, and HNO_4 react very slowly with odd oxygen.

$$OH + O_3 \rightarrow HO_2 + O_2$$
$$HO_2 + O_3 \rightarrow OH + O_2 + O_2$$

$$\boxed{\text{Net: } 2O_3 \rightarrow 3O_2} \tag{5.47}$$

Again, notice that HO_x ($OH + HO_2$) is neither produced nor destroyed by this cycle, but merely acts as a catalyst for converting two molecules of ozone into three molecules of oxygen. The importance of this reaction is that it does not require free oxygen atoms for the ozone loss.

Two other reactions are important for O_x loss in the upper stratosphere. The first one involves a free hydrogen atom as an intermediate compound. Two oxygen atoms are converted into a single diatomic oxygen molecule.

$$H + O_2 + M \rightarrow HO_2 + M$$
$$HO_2 + O \rightarrow OH + O_2$$
$$\boxed{\text{Net: } 2O \rightarrow O_2} \tag{5.48}$$

The second reaction involves HO_x and the loss of two odd oxygens (an ozone molecule and an oxygen atom). The two odd oxygens are converted into two diatomic oxygen molecules, as shown in Eq. (5.43).

Another pair of catalytic cycles is also important in the lower stratosphere. These cycles involve interaction with the chlorine or bromine cycles. These cycles are the most complicated of the catalytic cycles, but yield the same result: catalytic destruction of odd oxygen. In these equations, Z can be either chlorine (Cl) or bromine (Br):

$$ZO + HO_2 \rightarrow HOZ + O_2$$
$$HOZ + h\nu \rightarrow OH + Z$$
$$OH + O_3 \rightarrow HO_2 + O_2$$
$$Z + O_3 \rightarrow ZO + O_2$$
$$\boxed{\text{Net: } 2O_3 \rightarrow 3O_2} \tag{5.49}$$

5.11.4 NO_x Catalytic Cycles

The loss of odd oxygen can also occur through catalytic cycles involving nitrogen species in the form of reactive nitrogen. Reactive nitrogen, denoted NO_x, includes NO (nitric oxide) and NO_2 (nitrogen dioxide). Like reactive hydrogen, reactive nitrogen species have their origins in the troposphere. Approximately 90% of stratospheric NO_x comes from tropospheric N_2O (nitrous oxide, also known as laughing gas). Like H_2O and CH_4, N_2O is transported upward into the stratosphere mainly through the tropical tropopause by the lifting of the Brewer-Dobson circulation.

Figure 5.10 shows mean January distribution of N_2O in the stratosphere as seen by the cryogenic limb array etalon spectrometer (CLAES) instrument aboard the upper atmospheric research satellite (UARS). The highest values of N_2O are observed near its source region in the lower tropical troposphere. The concentration of N_2O is quite high in northern hemispheric latitudes during winter.

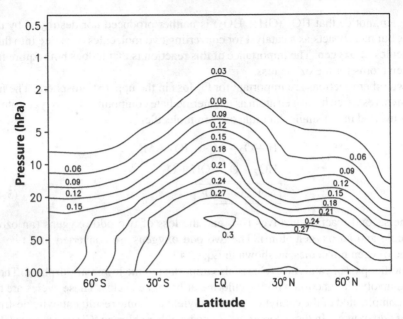

Fig. 5.10 Monthly averages of N_2O observations from UARS CLAES as a function of height and potential vorticity expressed in terms of equivalent latitude. Contour interval is 0.03 ppmv (Source: SPARC Data Center)

Reactive nitrogen species are formed from N_2O via the reaction of nitrogen dioxide with the singlet D oxygen atom to form two molecules of nitric acid.

$$N_2O + O(^1D) \rightarrow 2NO \tag{5.50}$$

This reaction transfers nitrogen from the inert species, N_2O, into the reactive species, NO.

Another way that N_2O is lost is via photolysis. An energetic UV photon is able to dissociate N_2O into molecular nitrogen, N_2, and the singlet D oxygen atom.

$$N_2O + h\nu \rightarrow N_2 + O(^1D) \tag{5.51}$$

N_2 is a very long-lived, nonreactive species. Even though N_2 is the most abundant gas in the atmosphere, it does not contribute to photochemical processes.

Sources of N_2O at the ground level include oceans, forest soils, combustion, biomass burning, and fertilizers. The amounts of nitrogen generated annually by these processes are estimated to be in the range of 4.4–10.5 trillion (Tg).

Stratosphere contains many of the nitrogen species, but they are not involved in stratospheric nitrogen photochemistry, or in N_2O. For example, $BrONO_2$ (bromine nitrate) can be photolyzed by energetic photons to form Br (bromine) and NO_3 (nitrogen trioxide). The reaction is given by

$$BrONO_2 + h\nu \rightarrow Br + NO_3 \tag{5.52}$$

Bromine is not represented, because it is not a nitrogen species. The reactive forms of nitrogen drive their own catalytic cycle, analogous to the HO_x cycles.

In the first step, NO reacts with O_3 to form NO_2 and O_2. The NO_2 then reacts with O to reform NO. In the process, both an ozone and oxygen molecule are destroyed (i.e., two O_x molecules), while the NO is reformed

$$NO + O_3 \rightarrow NO_2 + O_2$$
$$NO_2 + O \rightarrow NO + O_2$$
$$\boxed{Net: O_3 + O \rightarrow 2\,O_2} \tag{5.53}$$

The catalytic cycle results in the loss of two odd oxygen molecules without loss of the catalytic NO_x species.

5.11.5 Temperature Dependence of NO_x Catalytic Reactions

Reactions between chemical species are temperature-dependent. The temperature dependence for NO_x species is particularly strong.

Figure 5.11(a) shows the reaction rate of $NO + O_3 \rightarrow NO_2 + O_2$ as a function of temperature over a range of temperatures characteristic of the stratosphere. This

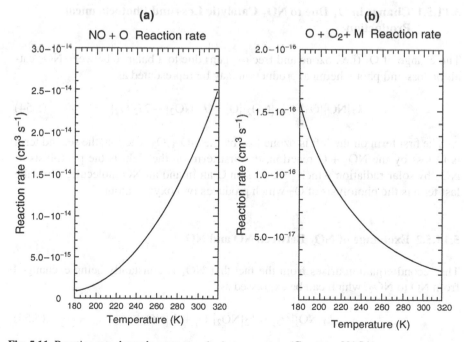

Fig. 5.11 Reaction rate depends on stratospheric temperature (Courtesy: NASA)

type of reaction is referred to as a *bimolecular reaction*, as it involves two molecules. It can be seen that the $NO + O_3$ rate increases very steeply as the temperature increases. This steep increase is associated with the faster speeds and energies that molecules acquire at higher temperatures. The statistical behavior of large collections of molecules shows that the average energy of a molecule is related to its temperature. As the temperature increases, the molecules move faster, and the probability for the collision and reaction of the two molecules becomes higher. Hence, the reaction rate increases with temperature. These reaction rates are also dependent on other factors, such as the energy associated with the molecular bonds.

The termolecular (three molecule) reaction rate, on the contrary, tends to decrease with increasing temperature. Figure 5.11(b) shows the reaction rate of $O + O_2 + M \rightarrow O_3 + M$ as a function of temperature over a range of temperatures characteristic at about 40 km in altitude. In this reaction, the rate decreases as temperature increases. The reason for this decrease is that the probability of three molecules colliding simultaneously to react falls sharply as the molecules move about faster with increase in temperature.

The temperature dependence of these NO_x reaction rates can be used to show why ozone and temperatures are anticorrelated in the upper stratosphere. Recall that the change in concentration of a molecule is due to both photolysis and chemical reactions. The following three chemical equations allow us to solve for ozone concentration as a function of temperature and NO_x.

5.11.5.1 Change in O_x Due to NO_x Catalytic Loss and Photochemical Production

The change of O_x (i.e., ozone and free oxygen) due to a balance between NO_x catalytic loss and photochemical production may be represented as

$$k_1[NO][O_3] + k_2[NO_2][O] = J_1[NO_2] + 2J_2[O_2] \tag{5.54}$$

The first term on the left is ozone loss by the $NO + O_3$ reaction, the second term is O loss by the $NO_2 + O$ reaction, the first term on the right is the photolysis of NO_2 by solar radiation which produces an O atom and an NO molecule, while the last term is the photolysis of O_2 which produces two oxygen atoms.

5.11.5.2 Exchange of NO_x Between NO and NO_2

The second equation arises from the fact that NO_x is constantly being exchanged from NO to NO_2, which can be expressed as

$$k_1[NO][O_3] = k_2[NO_2][O] + J_1[NO_2] \tag{5.55}$$

This indicates that the production of NO_2 is equal to the loss of NO_2. The term on the left-hand side of the equation is NO_2 production by the $NO + O_3$ reaction. The first term on the right-hand side is the loss of NO_2 by the $NO_2 + O$ reaction, and the second term is the loss of NO_2 by photolysis.

5.11.5.3 Exchange of O_x Between O and O_3

The third equation arises because odd oxygen, O_x, like reactive nitrogen, NO_x, is constantly being altered from free oxygen, O, to ozone, O_3. It can be expressed as

$$k_3[O][O_2][M] = J_3[O_3] \tag{5.56}$$

The first term is the ozone production by the $O + O_2 + M$ termolecular reaction, while the term on the right-hand side is the photolysis of ozone.

We can combine the three equations above to solve for the ozone concentration as a function of temperature and NO_x, which forms a relatively complicated expression.

Figure 5.12 shows that ozone amounts decrease with increasing temperature. The decrease is a result of the temperature dependence of the $NO + O_3$ reaction which is strongly temperature-dependent. As temperatures increase, this reaction speeds up, destroying ozone faster, and reducing the ambient amount of ozone. This reaction rate temperature dependence explains the observed ozone–temperature relationship in the upper stratosphere, and illustrates the impact of temperature-dependent reaction rates on the stratosphere.

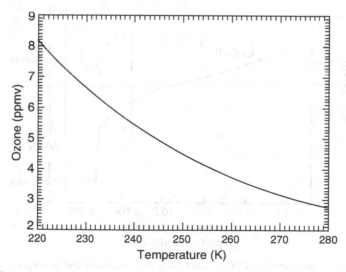

Fig. 5.12 Ozone versus temperature at 40 km altitude during equinox (Courtesy: NASA)

5.11.6 Sources of Chlorine

The sources of chlorine are from the troposphere. Chlorine atoms are bound up in the various man-made chlorofluorocarbon (CFC) and hydrochlorofluorocarbon (HCFC) molecules, including F-11, f12, f113, f114, CCl_4, CH_3CCl_3, CH_3Cl, HCFC-22. Because of their long lifetime these molecules can be transported across the tropopause into the stratosphere. Chlorofluorocarbons (CFCs) were developed in the 1930s as a safe, nontoxic refrigerant alternative to ammonia. CFCs are cheap to manufacture, nonflammable, and insoluble. They gained enormous usage worldwide from the 1940s to the 1980s.

Because CFCs have very long lifetimes and are not water-soluble, they can be transported upward into the stratosphere, where they are subject to photodissociation and liberate chlorine. Indeed, the current stratospheric chlorine content arises mostly from CFCs. This represents a dramatic change to the atmosphere by human activity. While natural sources of chlorine exist, most of these are restricted to lower part of the atmosphere and it is the release of chlorine atoms from the photolysis of CFC molecules that provide most of the observed chlorine in the stratosphere at the present time. In the upper stratosphere, the high-energy UV radiation on the ozone layer can break the CFC bonds, releasing chlorine to participate in its own catalytic cycle destroying odd oxygen.

Figure 5.13 shows the abundance of carbon tetrafluoride (CF_4) and CFC-11 in the troposphere and stratosphere. CF_4 is completely unreactive at altitudes up to at

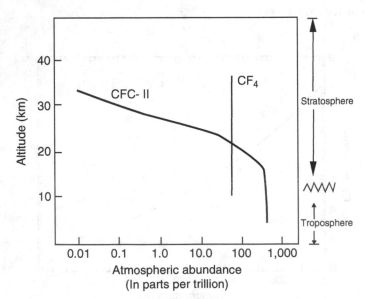

Fig. 5.13 Vertical distribution of CFC-11 and CF4 in the troposphere and stratosphere (Courtesy: WMO/UNEP 1995)

least 50 km in the atmosphere. Measurements show it to be nearly uniformly distributed throughout the atmosphere. CFC-11 is unreactive in the lower atmosphere (below about 15 km) and is similarly uniformly mixed there, as shown in Fig. 5.13. However, the abundance of CFC-11 decreases as the gas reaches higher altitudes, because it is broken down by high-energy solar ultraviolet radiation. Chlorine released from this breakdown of CFC-11 and other CFCs remains in the stratosphere for several years, where every chlorine atom destroys many thousands of molecules of ozone.

5.11.7 Chlorine Catalytic Reactions

Cl can react with O_3 to form ClO and O_2. The reaction can be written as

$$Cl + O_3 \rightarrow ClO + O_2 \tag{5.57}$$

A principal loss of ozone in the upper stratosphere is the Cl/ClO reaction, and is represented as

$$Cl + O_3 \rightarrow ClO + O_2$$
$$ClO + O \rightarrow Cl + O_2$$
$$\boxed{\text{Net: } O_3 + O \rightarrow 2\,O_2} \tag{5.58}$$

Here the reactive chlorine compounds are Cl and ClO, denoted Cl_x. The net reaction is to convert two odd oxygens, an ozone molecule and an oxygen atom, into two molecules of diatomic oxygen. Since there is not much free oxygen atom available in the lower stratosphere, this reaction may not be the major loss mechanism for polar lower stratospheric ozone.

As in the NO_x case, several reactions exists that transform reactive chlorine into reservoir species.

$$Cl + CH_4 \rightarrow HCl + CH_3 \tag{5.59}$$

$$ClO + HO_2 \rightarrow HOCl + O_2 \tag{5.60}$$

$$ClO + NO_2 + M \rightarrow ClONO_2 + M \tag{5.61}$$

The chlorine reservoir species HCl (hydrochloric acid), HOCl (hypochlorous acid), and $ClONO_2$ (chlorine nitrate) are characterized by a variety of lifetimes, determined by their photolysis rates. HCl is the longest, with a lifetime on the order of weeks. HOCl is the shortest, with a lifetime on the order of hours, as HOCl quickly photolyzes in sunlight.

In the lower stratosphere, several other catalytic cycles involving chlorine have important effects on the ozone balance. Yet another cycle involves the photolysis of NO_3 and $ClONO_2$.

The reactions are

$$ClONO_2 + hv \rightarrow Cl + NO_3$$
$$NO_3 + hv \rightarrow NO + O_2$$
$$NO + O_3 \rightarrow NO_2 + O_2$$
$$Cl + O_3 \rightarrow ClO + O_2$$
$$ClO + NO_2 + M \rightarrow ClONO_2 + M$$

$$\boxed{Net: 2O_3 \rightarrow 3O_2}$$ (5.62)

The three-body reaction of ClO with NO_2 and some other molecule M deactivates the reactive chlorine species ClO into nonreactive chlorine species $ClONO_2$. The net effect of this set of reactions involving both reactive and nonreactive forms of chlorine and nitrogen is to convert two molecules of ozone into three molecules of diatomic oxygen.

5.11.8 Cl_x Catalytic Reactions and the Antarctic Ozone Hole

The chemical loss of polar ozone during winter and spring occurs primarly by two gas-phase catalytic cycles that involve ClO_x radicals (Molina and Molina 1987) and bromine (McElroy et al. 1986). These two critical Cl_x catalytic cycles are responsible for destroying stratospheric ozone in the Antarctic, leading to the formation of the ozone hole (Antarctic ozone hole is described in detail in Chapter 6). In both cases, two ozone molecules are converted into three diatomic oxygen molecules.

First catalytic cycle

$$ClO + ClO + M \rightarrow Cl_2O_2 + M$$
$$Cl_2O_2 + hv \rightarrow Cl + ClO_2$$
$$ClOO + M \rightarrow Cl + O_2 + M$$
$$2(Cl + O_3 \rightarrow ClO + O_2)$$

$$\boxed{Net: \ 2O_3 \rightarrow 3O_2}$$ (5.63)

Second catalytic cycle

$$BrO + ClO \rightarrow Br + ClOO$$
$$ClOO + M \rightarrow Cl + O_2 + M$$
$$Cl + O_3 \rightarrow ClO + O_2$$
$$Br + O_3 \rightarrow BrO + O_2$$

$$\boxed{Net: \ 2O_3 \rightarrow 3O_2}$$ (5.64)

There are no oxygen atoms involved in either of these two catalytic cycles. Yet reactive chlorine is still created when the ClO dimer (ClOOCl) is photolyzed by UV light, liberating chlorine atoms, which then destroy ozone. Hence, these reactions can destroy ozone in the lower stratosphere, where there are very few O atoms. However, it turns out that these reactions are relatively unimportant throughout the stratosphere most of the time, except in the presence of polar stratospheric clouds (PSCs). In addition, it is only in the presence of PSCs that reactive nitrogen species, NO_x, are locked up as nitric acid, which stops the reaction of NO_2 and the reactive chlorine species ClO. This allows ClO levels to increase dramatically. The result is the NO_x and Cl_x catalytic destruction of ozone.

Interference cycles can reduce ozone loss rates. These interference reactions of Cl_x with NO_x can cycle Cl_x, NO_x, and O_x without loss of O_x. For example,

$$Cl + O_3 \rightarrow ClO + O_2$$
$$ClO + NO \rightarrow Cl + NO_2$$
$$NO_2 + h\nu \rightarrow NO + O$$

$$\boxed{\text{Net: } O_3 + h\nu \rightarrow O_2 + O} \qquad (5.65)$$

Such a sequence of reactions effectively transforms one form of O_x to another with no net loss, while preventing NO_x and Cl_x molecules from participating in normal catalytic cycles. Such interference cycles effectively slow destruction of O_x.

5.11.9 Sources of Bromine

Bromine is another molecule effective for ozone destruction. Bromine accounts for significant ozone loss (about 20–40%) inside the Antarctic ozone hole. The sources of bromine are both anthropogenic and natural. Methyl bromide (CH_3Br) is produced in the troposphere, but it is also the predominant source of bromine in the stratosphere. Methyl bromide is produced by biological processes on both land and in the ocean. Methyl bromide is also used as a fumigant for agricultural purposes, and is released by biomass burning and from cars using leaded fuel. Two other major sources of bromine are Halon-1211 ($CBrClF_2$) and Halon-1301 ($CBrF_3$), used as fire suppressants.

The losses of methyl bromide occur through reactions with water, the hydroxyl radical, chlorine ions, and photolysis by ultraviolet radiation. Methyl bromide has a long lifetime, which allows some of it to be lifted out of the troposphere and into the stratosphere.

Halons 1211 and 1301 are only destroyed by UV photolysis at wavelengths shorter than 280 nm. Hence, the halons can only be photolyzed in both the upper and lower stratosphere. They have very long lifetimes, since it takes quite a while for a molecule to reach these altitudes. The lifting action is again provided by the Brewer-Dobson circulation.

5.11.10 Br_x Catalytic Reactions

Reactive bromine exists in the form of bromine (Br) and bromine monoxide (BrO). Nonreactive bromine species include hypobromous acid (HOBr) and bromine nitrate ($BrONO_2$). These are typically not referred to as "reservoir species" because they are very easily photolyzed, even by visible light, and hence have very short lifetimes. This means that they do not lock up reactive bromine in the same way that $ClONO_2$ locks up reactive chlorine, and so bromine species in the stratosphere tend to exist in reactive forms of Br and BrO.

There are four distinct catalytic cycles for ozone loss. These reaction cycles are illustrated as follows.

5.11.10.1 Br_x–O_x Reaction Cycle

In this cycle, a two-step reaction occurs between reactive bromine and odd oxygen. First, a BrO molecule reacts with a free O atom to form a free Br atom and a diatomic oxygen molecule. Then, the Br atom reacts with an O_3 molecule to reform the BrO molecule and a molecule of diatomic oxygen. The net reaction is to convert two odd oxygen species, an O atom and an O_3 molecule, into two diatomic oxygen molecules. The reactions are

$$BrO + O \rightarrow Br + O_2$$
$$Br + O_3 \rightarrow BrO + O_2$$

$$\boxed{Net: O + O_3 \rightarrow 2\,O_2} \tag{5.66}$$

5.11.10.2 Br_x–Cl_x–O_x Reaction Cycle

In this cycle, four reactions occur involving reactive Br_x and Cl_x species and odd oxygen, in the form of ozone. The net result is to convert two O_3 molecules into three O_2 molecules. That is, two odd oxygen species are lost. The reactions are

$$BrO + ClO \rightarrow Br + ClOO$$
$$ClOO + M \rightarrow Cl + O_2 + M$$
$$Cl + O_3 \rightarrow ClO + O_2$$
$$Br + O_3 \rightarrow BrO + O_2$$

$$\boxed{Net: O + O_3 \rightarrow 2\,O_2} \tag{5.67}$$

5.11.10.3 Br_x–NO_x–O_x Reaction Cycle

There are five reactions among reactive Br_x and NO_x species and O_x (odd oxygen) in the form of O_3 (ozone) in this cycle. Nonreactive bromine nitrate ($BrONO_2$)

is also involved, which is photolyzed by less energetic near-UV and visible light. These photons are less energetic than UV photons, and hence they are able to penetrate into the lower stratosphere, since ozone higher up screens out only the more energetic UV photons. Bromine thus exists mostly in its reactive forms in the lower stratosphere. The net change is as in Eq. (5.67) above, namely, the conversion of two molecules of an odd oxygen species (ozone) into three molecules of O_2. These reactions are

$$BrO + NO_2 + M \rightarrow BrONO_2 + M$$
$$BrONO_2 + h\nu \rightarrow Br + NO_3$$
$$NO_3 + h \rightarrow NO + O_2$$
$$NO + O_3 \rightarrow NO_2 + O_2$$
$$Br + O_3 \rightarrow BrO + O_2$$

$$\boxed{\text{Net: } O + O_3 \rightarrow 2\,O_2} \tag{5.68}$$

5.11.10.4 Br_x–HO_x–O_x Reaction Cycle

There are four reactions among reactive Br_x and HO_x species and O_x, again in the form of O_3. Another nonreactive form of bromine (HOBr) is also involved, which is quickly photolyzed by near-UV and visible light. The net change is as in Eqs. (5.67) and (5.68) above, the conversion of two O_3 molecules into three O_2 molecules, representing the loss of O_x. These reactions are

$$BrO + HO_2 \rightarrow HOBr + O_2$$
$$HOBr + h\nu \rightarrow OH + Br$$
$$OH + O_3 \rightarrow HO_2 + O_2$$
$$Br + O_3 \rightarrow BrO + O_2$$

$$\boxed{\text{Net: } O + O_3 \rightarrow 2\,O_2} \tag{5.69}$$

These chains of reactions make bromine one of the most efficient destroyers of ozone for two reasons. First, catalytic cycles Eqs. (5.67), (5.68), and (5.69), do not require free oxygen atoms to destroy ozone, meaning that these reactions can occur in the lower stratosphere where there are few O atoms available. Second, HOBr and $BrONO_2$ are very easily photolyzed, so that bromine typically exists as reactive species (Br or BrO) rather than as nonreactive species ($BrONO_2$ and HOBr).

5.12 Stratospheric Particles

Stratosphere contains various particles, which exist throughout the region. Only few types of particles in the stratosphere exist on which gases can either be adsorbed (adhesion to the surface of a particle) or absorbed. The most common particles in the stratosphere are sulfate aerosols.

5.12.1 Sulfate Aerosols

Stratospheric sulfate aerosols are submicron-sized particles and are composed of a solution of sulfuric acid and water. In middle latitudes, the temperatures are warm enough to maintain the sulfate aerosol particles in a liquid state. The sulfuric acid comes from carbonyl sulfide (COS) and sulfur dioxide (SO_2) carried into the stratosphere via tropical lifting by the Brewer-Dobson circulation, or by direct injection of SO_2 into the stratosphere from very explosive volcanic eruptions. Because sulfate aerosols are small, they settle out of the stratosphere at a very slow rate of about 100 m year^{-1} for a spherical aerosol with a radius of 0.1 microns. Because of this small settling speed, most sulfate aerosols are carried out of the stratosphere by Brewer-Dobson circulation descent in the higher latitudes.

5.12.2 Chemical Content of PSCs

Polar stratospheric clouds (PSCs) are responsible for the Antarctic ozone hole. They provide the surfaces on which the ozone-destroying reactions occur, as mentioned in Chapter 1, section 1.18.1. PSCs are divided into two classes, Type I and Type II. Both form only under extremely cold conditions that are found only occasionally in the winter polar vortex region of the Arctic and Antarctic stratosphere. They tend to form much more frequently in the more isolated, colder Antarctic vortex, hence the existence of the ozone hole over there and not over the Arctic. The temperature at which they form is referred to as the frostpoint. Type II PSCs are better understood than Type I PSCs.

Type II PSCs are water ice particles that form when the temperature falls below 188 K. They are relatively large, with a diameter of at least 10 microns. Type II PSCs fall rather rapidly at about 1.5 km day^{-1}. This is known as sedimentation, the settling out of small particles from the atmosphere.

The composition of Type I PSCs are still not well understood. It is believed that Type I PSCs are composed of a super-cooled liquid ternary solution of nitric acid, sulfuric acid, and water ice, as well as frozen nitric acid trihydrates (NAT). The frost point for Type I PSCs is 195 K. Type I PSCs are much smaller than Type II PSCs, with particle diameters on the order of 1 micron. Their sedimentation rate is consequently much slower, on the order of 10 m (0.01 km) day^{-1}.

5.13 Tropospheric Chemistry

Tropospheric chemistry is quite complex and plays a crucial role in the radiative balance and thermal structure of the lower atmosphere. The various gases in the troposphere are constantly mixing and reacting with each other. Gases released by the oceans, emitted by living animals, and produced by human activities alter the balance of chemical nature of the atmosphere.

Anthropogenic activities play an increasingly important role in tropospheric chemistry. Fossil fuel burning generates sulfur oxides, which creates sulfuric acid, and exhaust gases from automobiles produce oxides of nitrogen, which contribute to the formation of smog and nitric acid in the troposphere. Volcanic eruptions, forest fires, lightning, and UV radiation from the Sun are the natural contributions to the change in the chemistry of the troposphere. The oceans and the biosphere exchange vast quantities of gases with the atmosphere's lowest layer. The carbon and nitrogen cycles play key roles in these processes.

5.13.1 Sources of Chemical Components in the Troposphere

Carbon dioxide in the troposphere is produced by photosynthesis by plants and microbes, by decomposition of organic matter, and by fossil fuel combustion. CO_2 is the most important greenhouse gas present in the troposphere. Carbon monoxide comes from forestfires, volcanoes, and incomplete combustion in the exhaust gases from automobiles. It is poisonous and can help raise the levels of greenhouse gases via certain chemical reactions. It reacts with oxygen to transform into carbon dioxide.

Hydrocarbons are combinations of hydrogen and carbon atoms and are released by the combustion of fossil fuels. They are one of the constituents of smog. Methane is a powerful greenhouse gas. Although its concentration in the atmosphere is small, it has a substantial impact on Earth's energy balance. Hydrogen peroxide, often present in small quantities in water droplets in the atmosphere, helps create the sulfuric acid in acid rain via reactions with sulfur dioxide.

In the troposphere about 78% of the gas molecules are nitrogen. Nitrogen atoms move through the environment via the nitrogen cycle. Hot combustion, such as in auto exhausts, incorporates nitrogen into nitrogen oxides. Nitrogen oxides form during high-temperature combustion in air, such as in automobile exhausts. They help to form smog and mix with water to create nitric acid, a component of acid rain.

Peroxyacytyl nitrate (PAN) is a noxious and irritating component of smog. It forms in a reaction involving nitrogen dioxide, oxygen, and substances derived from volatile organic compounds (VOCs). Smog was originally a term for a mixture of smoke and fog. Photochemical smog is a toxic atmospheric pollutant often found in urban areas. It consists of nitrogen oxides, ozone, VOCs, and PAN.

Sulfur dioxide and sulfur trioxide are produced by coal and oil burning, volcanoes, and other human and natural sources. Sulfur dioxide combines with water droplets in the air to form sulfuric acid, a component of acid rain.

5.13.2 Tropospheric Ozone

Tropospheric O_3 is both generated and destroyed by photochemistry within the atmosphere. Ozone is a critical compound for oxidation processes in the troposphere,

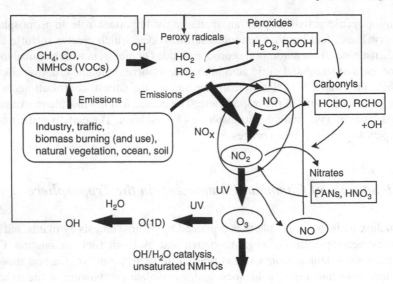

Fig. 5.14 Schematic presentation of the basic cycles of tropospheric ozone chemistry (Courtesy: Sudo Kengo)

and it interacts with source gases, like NO_x, CO, CH_4, H_2O, etc., through its effect on OH, which is the main oxidation agent. The result of the chemical coupling is a pronounced nonlinearity in ozone changes and methane from emission changes. The presence of NO_x ($NO+NO_2$) is crucial for ozone formation in the troposphere. In rural and other unpolluted remote areas with extremely low NO_x levels, precursor emissions lead to ozone loss. In urban areas where the atmosphere is highly polluted and with higher NO_x levels, tropospheric ozone production dominates. Basic concepts of the tropospheric ozone chemistry are illustrated in Fig. 5.14.

The large inhomogeneities in the ozone production and loss processes, combined with the short chemical lifetime of ozone, particularly in the atmospheric boundary layer, leads to large spatial and temporal variations in ozone. In particular, ozone levels often exceed air quality threshold levels in local and regional polluted areas. The presence of halogens in the surface level over oceans in the polar regions leads to efficient depletion of ozone in the boundary layer during spring months.

Tropospheric O_3 is a direct greenhouse gas. The past increase in tropospheric O_3 is estimated to provide the third largest increase in direct radiative forcing since the preindustrial era. In addition, through its chemical impact on OH, it modifies the lifetimes of other greenhouse gases, such as CH_4. Several chemical reactions are involved in the removal of ozone in the troposphere. In addition to chemical reactions, clouds and precipitation through removal of chemical species, and direct surface deposition of ozone affect ozone levels significantly in the troposphere.

Ozone abundances in the lower troposphere vary from less than 10 ppbv over remote oceanic regions where the influence of anthropogenic emissions is negligible. The loss at the surface is significant to more than 100 ppbv over large areas where production is strongly controlled by anthropogenic emission of pollutants.

In the upper troposphere, where ozone transport from the stratosphere is a dominant source, typical mixing ratios are on the order on 100 ppbv. The strong variability in ozone in space and time makes it difficult to determine the tropospheric encumbrance from the exciting surface observations (WMO 2007).

5.13.3 Tropospheric Methane

Methane is well mixed over the entire globe, largely because its lifetime is about a year. Although the global average concentration of methane is slowly increasing, its emission rate is, to a first approximation, equal to the destruction rate at a level between 450 and 550 Tg yr^{-1}. Sources of tropospheric methane fall into two categories (Fig. 5.15). Anthropogenic sources constitute over 80% of the total methane production and include rice paddies, cattle enteric fermentation, gas drilling, coal mining, landfills, and biomass burning. The natural sources of methane are wetlands, swamps, termites, lakes, and oceans. The strength of these sources have been the subject of debate over the last 30 years for the reason that it is extremely difficult to make reliable estimations. In fact, the knowledge of their emission rates is not relevant because major sources already have huge uncertainties associated with them.

Methane escapes from the surface of oceans because it is slightly supersaturated with respect to the boundary layer, due to the presence of anaerobic bacteria. Models for the diffusion process have been proposed but there is some degree of uncertainty

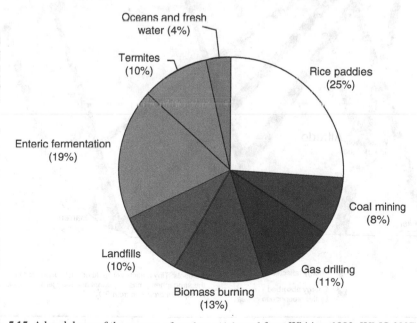

Fig. 5.15 A breakdown of the sources of methane (Adapted from Whiticar 1993; WMO 2007)

associated with extracting an absolute value for the methane flux, because of variations in the film thickness and temperature and the agitation of the sea by the wind (Enhalt 1999).

5.14 Atmospheric Chemistry and Climate

Chemistry and climate strongly influence each other and are mutually related. Atmospheric chemistry is basically the ozone chemistry. Atmospheric parameters, such as temperature, humidity, winds, and the presence of other chemicals in the atmosphere, influence ozone formation, and the presence of ozone, in turn, affects those atmospheric constituents. Figure 5.16 represents a schematic diagram of the interaction of solar UV radiation on atmospheric chemical reactions.

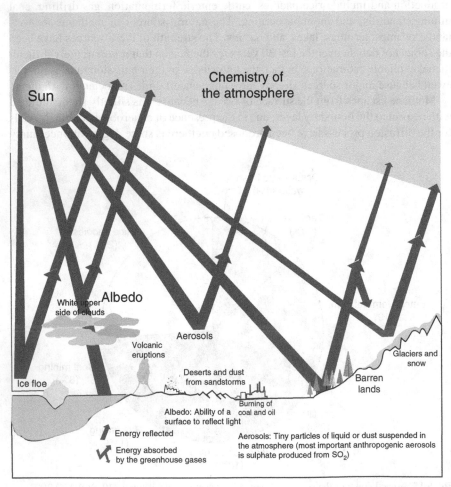

Fig. 5.16 Interaction of solar UV flux on atmospheric chemical actions (Courtesy: UNEP/IPCC 1996)

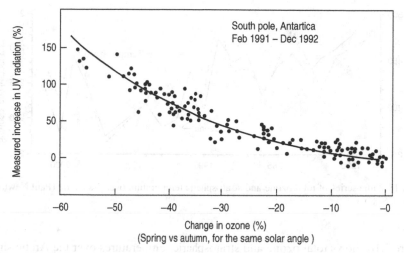

Fig. 5.17 Observational evidence indicating that decrease in stratospheric ozone causes an increase in UV radiation over the Antarctic region (Courtesy: WMO/UNEP 1995)

Ozone changes have an ensuing effect on UV and climate. Observational evidences indicate that surface UV radiation levels have generally increased with a similar geographic pattern to the observed reductions in stratospheric ozone. Surface UV levels are also strongly influenced by cloud cover, surface albedo, and the atmospheric aerosol content. Observation in the Antarctic region during a 2-year period shows that a decrease of 60% ozone causes an increase of 150% of solar UV radiation reaching the Earth's surface (Fig. 5.17). Since ozone is so closely coupled with other atmospheric processes, the complete impact of ozone changes is more complicated. There are many feedback processes, operating in either direction and involving ozone, climate, and UV (Fig. 5.14).

Ozone's impact on climate consists primarily of changes in temperature. The more ozone there is in a given parcel of air, the more heat it retains. Ozone generates heat in the stratosphere, both by absorbing the Sun's ultraviolet radiation and by absorbing upwelling infrared radiation from the troposphere. Consequently, decreased ozone in the stratosphere results in lower temperatures. Observations show that over recent decades, the mid- to upper stratosphere has cooled by 1–6°C. This stratospheric cooling has taken place at the same time that greenhouse gas amounts in the troposphere have risen. The stratospheric cooling and tropospheric warming is seen to be a coupled process.

Ozone loss and low temperature in the stratosphere create a possible feedback loop. The more ozone destruction in the stratosphere, the colder it would get just because there is less ozone. And the colder it would get, the more ozone depletion would occur.

Changes in ozone amounts are closely linked to temperature, with colder temperatures resulting in more polar stratospheric clouds and lower ozone levels (Zeng and Pylee 2003). Atmospheric motions drive the year-to-year temperature changes.

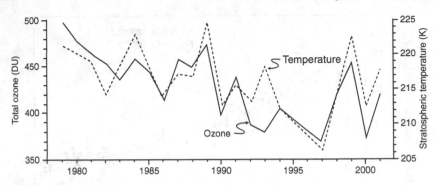

Fig. 5.18 Time series of total ozone and stratospheric temperature over the Arctic (Paul Newman, NASA GSFC)

Figure 5.18 shows total ozone and stratospheric temperatures over the Arctic since 1979. The Arctic stratosphere has cooled slightly since 1979, but the real cause for such a cooling is not yet clear.

In the last three decades, a significant cooling has occurred in the stratosphere. The lower stratospheric cooling is largely due to the observed ozone depletion in the stratosphere. The question whether the cooling will continue or not is uncertain because the expected increase in ozone resulting from decreasing halogens may be offset by dynamical effects. The upper stratospheric cooling will slow down the natural ozone destruction cycles, first compensating any chlorine-catalyzed ozone destruction there and eventually leading to ozone increases compared to pre-ozone hole conditions (Thompson and Solomon 2005; Randel and Wu 2006).

Observational evidences indicate that stratospheric water vapor has increased by about 1% year^{-1} since 1981. This trend is substantially larger than can be attributed to the observed changes in stratospheric methane. Water vapor enters the stratosphere through the tropical tropopause. The trend in water vapor is likely to be caused by changes in the tropical upper troposphere and lower stratosphere (Stenke and Grewe 2005). Water is extremely sensitive to the entry mechanism of air into the stratosphere. A better understanding of how water enters the stratosphere will significantly improve our understanding of the ways in which the tropospheric constituents enter the stratosphere.

The expected decline in halogen loading will result in reduced ozone losses in polar regions. However, climate change will modify polar stratospheric dynamics. The major dynamic factor depends on the stability and characteristics of the polar vortices. Chlorine activation and denitrification are highly temperature-sensitive critical processes.

It is interesting to note that there is no significant trend in total ozone that has been observed in the tropics, so far. However, the processes occurring in the tropics are important in determining stratospheric structure and composition. The tropical tropopause layer largely determines the concentrations of many short-lived species entering the stratosphere as well as the concentration of water vapor. In addition, the tropical troposphere is the critical region for chemical purification of the troposphere, and its structure is sensitive to climate change.

Theoretical and modeling studies indicate that fluctuations in the strength of the stratospheric polar vortices can influence tropospheric weather patterns and sea surface temperatures (Taguchi and Hartmann 2006; Butchart et al. 2006). Future changes in stratospheric composition and circulation could thus have an important influence on the distribution and trends of major patterns of climate variability such as the Arctic oscillation or North Atlantic Oscillation.

Interactions between ozone and climate naturally occur not only in the stratosphere, but also in the troposphere. Tropospheric ozone controls the radiative balance of the earth–atmosphere system by absorbing solar radiation and acts as a greenhouse gas through the absorption and reemission of terrestrial radiation Since ozone is a very good absorber of terrestrial radiation in the 9,600 nm band, an increase in tropospheric ozone has led to trapping of radiation in the lower atmosphere, providing a positive radiative forcing. It has shown that the increase in tropospheric ozone represents the third largest radiative forcing since preindustrial time; therefore it has a significant impact on global warming.

Ozone forms in the troposphere by the action of sunlight on certain chemicals. Chemicals participating in ozone formation include two groups of compounds: nitrogen oxides (NO_x) and volatile organic compounds (VOCs). An increase in temperature accelerates photochemical reaction rates. There is a strong correlation between higher ozone levels and warmer days. With higher temperatures, we can expect a larger number of bad ozone days, when exercising regularly might harm the lungs. However, ozone levels do not always increase with increases in temperature, such as when the ratio of VOCs to NO_x is low.

As the troposphere warms on a global scale, changes in ozone air quality can be expected. Generally speaking, warming temperatures will modify some, but not all, of the complex chemical reactions involved in ozone production in the troposphere, such as those involving methane. Because of the short-lived nature of these chemical constituents and variations across space and time, the uncertainty is too large to make predictions. Scientists can only speculate about specific kinds of change, about the direction of change in a particular location, or about the magnitude of change in ozone amounts that they can attribute to climate.

A warming climate can lead to more water vapor in the lower atmosphere, which would tend to produce more ozone. But cloud cover can also diminish chemical reaction rates because of reduced sunlight and therefore lower rates of ozone formation. Another impact of climate on ozone pollution in the troposphere arises from the probability that higher temperatures will lead to greater demand for air-conditioning and greater demand for electricity in summer. Most of our electric power plants emit NO_x. As energy demand and production rise, we can expect amounts of NO_x emissions to increase, and consequently levels of ozone pollution to rise as well. On the other hand, water vapor is also involved in climate change. A warmer atmosphere holds more water vapor, and more water vapor increases the potential for greater ozone formation. But more cloud cover, especially in the morning hours, could diminish reaction rates and thus lower rates of ozone formation.

It is anticipated that the emission of ozone precursors will increase significantly in the future due to fast industrial development. Therefore, the tropospheric ozone caused due to anthropogenic emissions will continue to contribute to global warming in the 21st century. According to IPCC (2001) the global radiative forcing due to ozone will range from 0.4 to 1.0 Wm^{-2} in 2050 and from 0.2 to 1.3 Wm^{-2} in 2100. The accumulated forcing due to major greenhouse gases based on radiative forcing is given in Fig. 5.19. The maximum and minimum shown in the Fig. 5.19 represent the uncertainity stage. It can be seen that the tropospheric ozone will continue to be the significant greenhouse gas, and even overtake methane and become the second major contributor to global warming after CO_2, by the end of this century.

Modeling studies (Sausen et al. 2002; Joshi et al. 2003) show that as a result of ozone changes, largest radiative forcings are produced for changes at tropopause levels, because of the low temperature in this region. The climate response to a given radiative forcing depends on the altitude at which ozone changes. The GCM study suggested that for the same magnitude of radiative forcing, tropospheric ozone is about 25% less effective in generating a climate response, compared to CO_2.

Future changes in tropospheric chemistry will be sensitive to climate-induced changes in humidity, temperature, and convection. The climate-change-induced increase in humidity is estimated to dampen the increase in tropospheric ozone. The increased humidity will enhance the production and abundance of OH, although the increase in OH is somewhat dampened due to the reduction in ozone. The increased OH concentration will reduce the lifetime of methane and its growth rate.

5.15 Nobel Prize in Atmospheric Chemistry

The Nobel prize in chemistry in 1995 was shared by three scientists: Dr. Mario J. Molina, Professor of Chemistry and Earth, Atmospheric and Planetary Sciences at the Massachusetts Institute of Technology; Dr. F. Sherwood Rowland of the University of California at Irvine; and Dr. Paul Crutzen of the Max Planck Institute for Chemistry in Mainz, Germany, for predicting the man-made chemicals that would destroy the Earth's protective ozone layer. Their research led to an international ban on ozone-depleting chemicals.

Molina joined as a postdoctoral fellow in 1974 and worked in Rowland's lab at the University of California at Irvine. In that year, they published a paper on "Stratospheric sink for chlorofluoromethanes: chlorine atom catalyses destruction of ozone" in *Nature*, outlining the threat posed to the ozone layer by chlorofluorocarbon (CFC) gases in aerosol cans, refrigerators, air conditioners, and styrofoam.

Molina and Rowland (1974) reported that the chemically inert CFCs would be transported to the ozone layer, where intense ultraviolet light would break them down into reactive constituents, notably chlorine atoms. Chlorine atoms already were known to decompose ozone. If mankind continued to use CFCs, they warned, it would dramatically deplete the ozone layer and create a hole in the atmosphere.

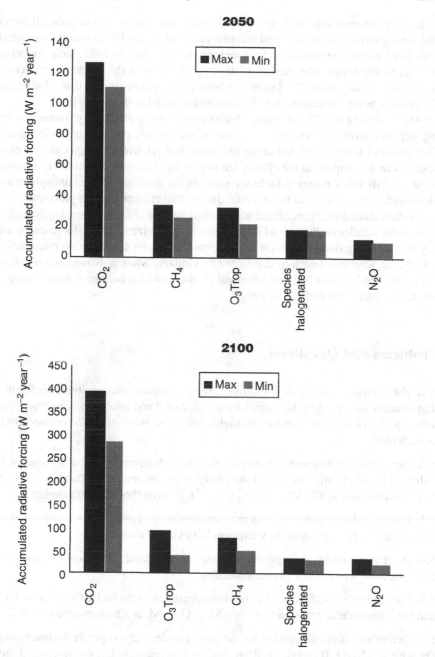

Fig. 5.19 CH4, N2O, and a range of halogenated greenhouse gases (Courtesy: EUR 2003)

In fact, this research work stirred enormous controversy within political arenas and among environmentalists and manufacturers who used CFCs in consumer products. Even many scientists criticized their predictions. But in 1985, their prediction became reality when other scientists discovered a hole in the ozone layer over the south pole. Today, with the benefit of perception, scientists say that Molina and Rowland actually underestimated the real threat posed to the ozone layer.

After winning the Nobel Prize, Molina continued to study why ozone was being depleted over the Antarctic in some of the coldest parts of the stratosphere. That phase of his research led to the discovery that relatively benign chlorine compounds can decompose in ice clouds, leading to the release of chemicals destructive to ozone. His latest research includes work on the interface of the atmosphere and biosphere, which is critical to understanding global climate-change processes.

Crutzen shared the Nobel Prize with Molina and Rowland for showing that nitrogen oxides accelerate the rate of ozone reduction (Crutzen 1979). His research was key to explaining the mechanism behind observed ozone depletion over Antarctica. Jointly, the scientific research conducted by Crutzen, Molina, Rowland, and others led to the 1987 United Nations Montreal Protocol, which banned the most dangerous CFCs, which took effect in 1996.

Problems and Questions

5.1. The Chapman life cycle for ozone results in ozone concentrations higher than that actually observed in the tropics by a factor of 2 and concentrations lower than actually observed in the middle and higher latitudes. What other factors need to be considered?

5.2. According to Chapman's theory, explain what happens to O and O_3 when solar radiation is cut off after sunset? If the catalysts are present, what happens to X and XO? Assume that $k_2[O](M) + k_3[O_3] \geq 10^{-1} s^{-1}$ throughout the stratosphere.

5.3. What is odd oxygen? What is the source of odd oxygen? How is it lost? What is the role of odd oxygen in the Chapman life cycle of ozone?

5.4. What is meant by absorption cross-section of a molecule? Explain absorption cross-section of oxygen or ozone molecules.

5.5. Find the rate coefficients for the bimolecular reaction $NO + O_3 \rightarrow NO_2 + O_2$, and the termolecular reaction $O + O_2 + M \rightarrow O_3 + M$, at a temperature of 198 K.

5.6. Derive an alalytical solution for the loss of molecular oxygen in the reactions of $O_2 + h\nu \rightarrow O + O$. If oxygen is destroyed by this reaction, but not re-created, how long will it take for its concentration to decrease to 10% of its initial value?

5.7. What is the average lifetime for a free oxygen atom and for an ozone molecule in the daytime, middle latitude, lower stratosphere? Given the lifetime of free oxygen and ozone, why isn't the atmosphere completely depleted of ozone?

5.8. Upper stratosphere is the region where all of the major O_x loss pathways involve O atoms. Now let us assume that the Lifetime of O_x and O_3 in the upper stratosphere (say around 40 km) is of the order of a day and a minute, respectively. For what fraction of O atoms are produced from O_3 photolysis reaction with O_2 to re-form ozone molecules.

5.9. Find the lifetime of $ClONO_2$ for background and volcanically perturbed conditions with respect to the reaction on aerosol in the lower stratosphere with surface area density 0.14 mm. Assume that the reactive uptake coefficient for loss of $ClONO_2$ on aerosol is 0.05, and the thermal velocity is 250 ms^{-1}.

5.10. Estimate the e-folding time of methane against loss by OH, if [OH] is 5.0×10^5 molecules cm^{-3} in the chemical reaction $CH_4 + OH \rightarrow CH_3 + H_2O$, where the rate coefficient is 6.2×10^{-15} cm^3 molecule^{-1} at 298 K.

5.11. What are the observed distributions of water vapor and methane with altitude in the stratosphere? Explain these observed distributions of water vapor and methane with altitude in the stratosphere given that both gases have source regions in the troposphere. Why are water vapor and methane important in ozone chemistry?

5.12. What are the reactions and the net effect of the $Br_x - NO_x - O_x$ catalytic reaction cycle? Discuss what happens to the reservoir species $BrONO_2$ in this set of reactions and what does this imply about bromine in the stratosphere?

References

Brasseur GP, Solomon S (2005) Aeronomy of the Middle Atmosphere, Springer, The Netherlands

Butchart H, Scaife AA, Bourqui M, de Grandpre J, Hare SHE, Kettleborough J, Langematz U, Manzini F, Sassi F, Shibata K, Shindell D, Sigmond M (2006) Simulations of anthropogenic change in the strength of the Brewer Dobson Circulation, Clim Dyn, 27, 727–741

Chapman S (1930) A theory of upper atmospheric ozone, Mem Royal Met Soc, 3:103–125

Crutzen PJ (1979) The role of NO and NO_2 in the chemistry of the troposphere and stratosphere, Ann Rev Earth Planet Sci, 7: 443–472

Dessler A (2000) The Chemistry and Physics of Stratospheric Ozone, Academic, New York

Enhalt DH (1999) Gas-phase chemistry in the in the troposphere, in Global Aspects of Atmospheric Chemistry, R. Zellner (ed.), Springer, The Netherlands

European Community Report (2003) Ozone-climate interactions, Scientific Assessment Report No. EUR 20623, Brussels, Belgium

IPCC (Intergovernmental Panel on Climate Change) (2001), Climate Change 2001

Joshi M, Shine KP, Ponater M, Stuber N, Sausen R, Li L (2003) DLR-Institut fur Physik der Atmosphere, Report No. 173, Oberpfaffenhofen, ISSN 0943–4771

Kengo S (2002) Warming and atmospheric composition change interaction, Fourth Assessment Report, IPCC Working Group 1 (http://www.kyousei.aesto.or.jp/~k021open/results/FY2004/figure2004.31.gif)

McElroy MB, Salawitch RJ, Wofsy SC, Logan JA (1986) Reductions of Antarctic ozone due to synergistic interactions of chlorine and bromine, Nature 321: 759–762

Molina MJ, Molina LT (1987) Production of chlorine oxide (Cl_2O_2) from the self reaction of the chlorine oxide (ClO) radical, J Phys Chem, 91: 433–436

Molina MJ, Rowland SF (1974) Stratospheric sink for chlorfluoromethanes: chlorine atom cataly-
    ses destruction of ozone, Nature 249: 810–812
NASA (2007) Studying Earth's Environment from Space (http://www.ccpo.odu.edu/SEES/
    index.html)
Newman PA, Morris G (2003) Stratospheric Ozone; An Electronic Textbook, Studying Earths
    Environment From Space, NASA
Pitts FBJ, Pitts JN Jr (2000) Chemistry of the Upper and Lower Atmosphere, Academic, San Diego,
    CA
Randel WJ, Wu F (2006) Biases in stratospheric and tropospheric temperature trends derived from
    historical radiosonde data, J Clim, 19: 2094–2104
Rasmussen RA, Khalil MAK (1981) Atmospheric methane (CH4) : trends and seasonal cycles,
    J Geophys Res, 86: 9826–9832
Salby ML (1996) Fundamentals of atmospheric chemistry, Academic Press, San Diego, CA
Saltzman ES, Aydin M, De Bruyn WJ, King DB, Yvon-Lewis SA (2004) Methyl bromide
    in pre-industrial air: Measurements from Antarctic ice core, J Geophys Res, 109, doi
    10.1029/2003JD004157
Sausen R, Ponater M, Stuber N, Joshi M, Shine K, Li L (2002) Climate response to inhomo-
    geneously distributed forcing agents, in Non-CO2 Greenhouse Gases Millpress, Rotterdam,
    The Netherlands, ISBN 90-77017-70-4: 377–381
Science and Society, Stratospheric Ozone: Production, Destruction and Trends Lecture
    Notes, Columbia University (http://www.ideo.columbia.edu/edu/dees/V1001/images/chapman.
    profile.gif)
Solomon S (1997a) Chemistry of the Atmosphere: NATO ASI Series, The Stratosphere and its
    Role in the Climate System, GP Brasseur (ed.), Springer, Heidelberg, 54: 219–226
Solomon S (1997b) Chemical Families: NATO ASI Series, The Stratosphere and its Role in the
    Climate System, GP Brasseur (ed.), Springer, Heidelberg, 54: 227–241
SPARC Data Center, Institute for Terrestrial and Planetary Atmospheres State University of
    New York, Stony Brook (http://www.sparc.sunysb.edu)
Stenke A, Grewe V (2005) Simulations of stratospheric water vapor trends: impact of stratospheric
    ozone chemistry, Atmos Chem Phys, 5: 1257–1272
Taguchi M, Hartmann DL (2006) Increased occurrence of stratospheric sudden warmings during
    El Nino as simulated by WACCM, J Clim, 19, 324–332
Thompson DWJ, Solomon S (2005) Interpretation of recent southern hemispheric climate change,
    Science, 296 (5569): 895–899
UNEP/IPCC (1996) The science of climate change, Contribution of Working Group I to
    the Second Assessment Report (http://www.grida.no/climate/vitalafrica/English/graphics/10-
    threefactors.jpg)
United National Environmental Programme (UNEP) (2002) Production and consumptions of
    ozone depleting substances under the Montreal Protocol 1986–2000, Ozone Secretariat,
    Nairobi, Kenya
Wayne RP (1991) Chemistry of Atmosphere, Oxford University Press, New York
Whiticar MJ (1993) Stable Isotopes and global budgets, in Atmospheric methane: sources, sinks,
    and role in global change, MAK Khalil (ed.), NATO ASI Series I, Global Environmental
    Change, 13: 138–167
WMO/UNEP (1995) Scientific Assessment of Ozone Depletion: 1994, Global Ozone Research
    and Monitoring Project – Report No. 37 Executive Summary (http://www.esrl.noaa.gov/
    csd/assessments/1994/common_questions_q_1.gif) (http://www.esrl.noaa.gov/csd/assessments/
    1994/common_questions_q_7.gif)
World Meteorological Organisation (WMO) (2007) Scientific Assessment of ozone Depletion:
    2006, Global ozone Research and Monitoring Project- Report No. 50, Geneva, Switzerland
Zeng G, Pylee JA (2003) Changes in tropospheric ozone between 2000 and 2001 modelled in a
    chemistry-climate model, Geophys Res Lett, 30: 1392, doi: 10.1029/2002GL016708

Chapter 6
Stratospheric Ozone Depletion and Antarctic Ozone Hole

6.1 Introduction

Stratospheric ozone plays a very significant role in the radiation balance of the Earth–atmosphere system and also protects life on the Earth's surface from harmful UV radiation. Changes in stratospheric ozone levels can affect human health and ecosystem as well as the chemistry of the troposphere.

In Chapter 5, we have seen that the atmospheric ozone can be destroyed by a number of free radical catalysts, the most important of which are the hydroxyl radical (OH), the nitric oxide radical (NO), and atomic chlorine (Cl) and bromine (Br). All these have both natural and anthropogenic (man-made) sources. At the present time, most of the OH and NO in the stratosphere is of natural origin, but human activity has dramatically increased the high concentration of carbon dioxide, chlorine, and bromine. These elements are found in certain stable organic compounds, especially chlorofluorocarbons (CFCs), which may find their low reactivity. Once the Cl and Br atoms are liberated from the parent compounds by the action of UV light, it remains in the stratosphere for a longer period and goes on destroying ozone in this region.

A single chlorine atom would keep on destroying ozone for up to 2 years, the timescale required to transport back down to the troposphere, were it not for reactions that remove them from this cycle by forming reservoir species such as hydrogen chloride (HCL) and chlorine nitrate (ClONO$_2$). On a per atom basis, bromine is even more efficient than chlorine at destroying ozone, but there is much less bromine present in the atmosphere. As a result, both chlorine and bromine contribute significantly to the overall ozone depletion. Laboratory studies have shown that fluorine and iodine atoms participate in analogous catalytic cycles. However, in the Earth's stratosphere, fluorine atoms react rapidly with water and methane to form strongly bound HF, while organic molecules which contain iodine react so rapidly in the lower atmosphere that they do not reach the stratosphere in significant quantities.

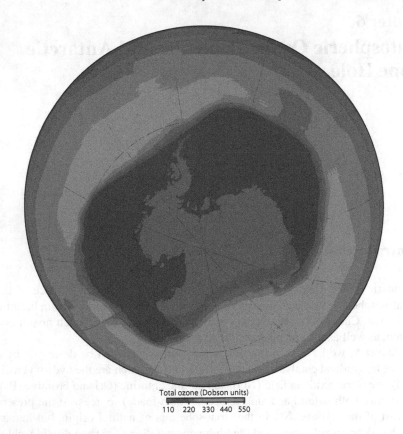

Total ozone (Dobson units)

110 220 330 440 550

Fig. 6.1 Satellite observations of Antarctic ozone hole on 24 September 2006 (Courtesy: NASA)

Figure 6.1 depicts the status of the ozone layer over the south pole during October 2006. Ozone depletion and climate change have usually been thought of as environmental issues with little in common other than their global scope and the major role played in each by CFCs and other halocarbons. With increased understanding of these issues, however, has come a growing recognition that a number of very important linkages exist between them. These linkages will have some bearing on how each of these problems and the atmosphere as a whole will evolve in the future.

6.2 Factors Affecting Stratospheric Ozone Variability

Various natural and anthropogenic factors contribute to the variability in stratospheric ozone depletion. A major issue is the variability in ozone induced by these factors that masks or resembles the expected ozone change due to halogen loading.

6.2.1 Chemical Aspects

Stratospheric ozone is depleted by reactions involving reactive halogen gases, which are produced mostly by photodissociation of halogen source gases. The evolution of stratospheric halogen loading is an obvious factor impacting ozone. Stratospheric halogen loading is important for attributing observed decrease in ozone concentration.

Apart from increasing abundances of ozone-depleting substances (ODS), changes in other gases could affect the evolution of ozone and the timing of ozone recovery by changing the background chemical composition of the atmosphere. In particular, increases in gases producing radicals, such as N_2O, CH_4, molecular hydrogen (H_2), and water (H_2O), catalytically destroy ozone. Catalytic ozone loss in the stratosphere occurs from the reactive nitrogen (NO_x), hydrogen (HO_x), oxygen (O_x), chlorine (ClO_x), and bromine (BrO_x) families. Ozone loss through these families is strongly altitude- and latitude-dependent, with NO_x dominating in the middle stratosphere (25–40 km), and HO_x dominating in the lower and upper stratosphere. Under conditions of high chlorine loading, ClO_x is important in the upper stratosphere (peak near 40 km) and in regions where heterogeneous reaction rates are large, such as in the polar regions during spring.

Figure 6.2 shows the association between chlorine monoxide and ozone between latitudes 63°S and 72°S in the Antarctic region. The Cl atom plays the role of a catalyst in the reaction mechanism scheme. One Cl atom can destroy up to 100,000 O_3 molecules before it is removed by some other reaction. The ClO species is an intermediate because it is produced in the first elementary step and consumed in the second step. The above mechanism for the destruction of ozone has been supported by the detection of ClO in the stratosphere in recent years.

Increase of oxides of nitrogen (NO_x) in the lower stratosphere causes a decrease in HO_x and ClO_x catalyzed losses, along with increases in tropospheric ozone

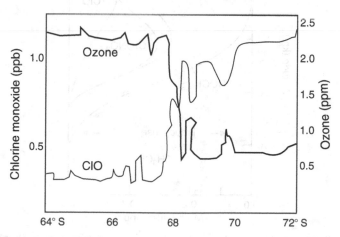

Fig. 6.2 Association between ClO and ozone in the Antarctic region during spring (F. S. Rowland, Courtesy: Encyclopedia of Earth)

production mechanisms. In the middle stratosphere, NO_x-induced changes are reduced by interactions with chlorine species. On the other hand, in some cases coupling between different chemical processes can amplify the effects of source gas emissions, e.g., nitrogen dioxide (NO_2) concentrations over southern middle latitudes have gone up at approximately twice the rate of their source gas N_2O as a result of changes in ozone.

Change in the stratospheric water vapor could affect the ozone concentration in the stratosphere quite significantly. An increase in water vapor would increase HO_x and thus cause ozone decreases in the upper and lower stratosphere. In the polar regions, increases in water vapor would cause an increase in heterogeneous reaction rates and an increase in the surface areas of polar stratospheric cloud (PSC) particles. Both effects are likely to lead to an increase in chlorine activation and ozone loss.

The effects of water vapor increases on ozone (via HO_x), induced by increases in methane, are partially offset by the reaction of methane with atomic chlorine, which deactivates ClO_x and reduces ClO_x-driven ozone loss. This could be important throughout the stratosphere. The coupling of water vapor and methane with ClO_x-induced ozone loss will be eliminated by decreasing ODS levels during the 21st century. An illustration of the distribution of chlorine with various other forms of chemicals released into the atmosphere in the form methyl chloride is depicted in Fig. 6.3. At steady state, the sum of the mixing ratios at all altitudes is constant, which is equal to the mixing ratios of CH_3Cl alone.

Future stratospheric halogen concentrations will depend on imminent emissions of ODSs and on transport into and through the stratosphere. Model simulations

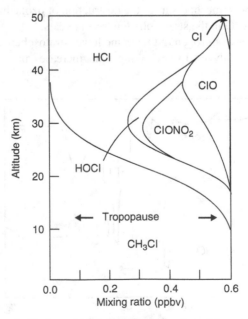

Fig. 6.3 Distribution of chlorine released into the atmosphere from various chemical forms (F. S. Rowland, Courtesy: Encyclopedia of Earth)

suggest that increases in greenhouse gases, such as carbon dioxide (CO_2), nitrous oxide (N_2O), and methane (CH_4), may lead to an increased stratospheric circulation and to reduced transport timescales (Stolarski et al. 2006a; b).

6.2.2 Dynamical Processes

Other than chemical aspects, atmospheric ozone variations can considerably be influenced by two important dynamical transport processes. They are (i) the interannual and long-term changes in the strength of the stratospheric mean meridional (Brewer-Dobson) circulation, which is responsible for the winter–spring buildup of extratropical ozone; and (ii) the changes in tropospheric circulation, particularly changes in the frequency of local nonlinear synoptic wave forcing events, which lead to the formation of ozone mini-holes and associated large increases in tropopause height. It is therefore important to consider interannual changes in both the Brewer-Dobson circulation and the nonlinear synoptic wave forcing when estimating the component of interannual ozone variability and trends that can be attributed to the dynamical transport processes.

When the stratospheric polar vortex is strong, tropospheric wave forcing is weaker, thereby the Brewer-Dobson circulation becomes feeble. As a result, less ozone is transported to the extratropics during winter and spring. In this period, the zonal wind field in the midlatitude lower stratosphere is less cyclonic, implying a greater frequency of anticyclonic, poleward wave-breaking events that lead to ozone mini-holes and localized tropopause height increases. At northern midlatitudes in winter and spring, these two dynamical transport mechanisms tend to reinforce one another.

The QBO also causes ozone values at a particular latitude to expand and contract roughly 3%. Since stratospheric winds move ozone, and do not destroy it, the loss of one latitude is the gain of another and globally the effects cancel out.

Changes in dynamical processes also affect the polar vortex conditions and, as a result, polar ozone loss. The midlatitude ozone is influenced by polar loss via airmass mixing after the polar vortex breakup in early spring. The tropospheric planetary scale waves that are dominantly responsible for driving the Brewer-Dobson circulation are also associated with synoptic wave events and local tropopause height changes.

6.2.3 Stratospheric Temperatures

Stratospheric temperatures depend on stratospheric dynamics, radiation, and composition. At the same time, the rates of chemical reactions, and the formation of PSCs, rely on temperature. Thus, changes in temperature can have a large influence on ozone. Temperature changes need to be accounted for when attributing observed

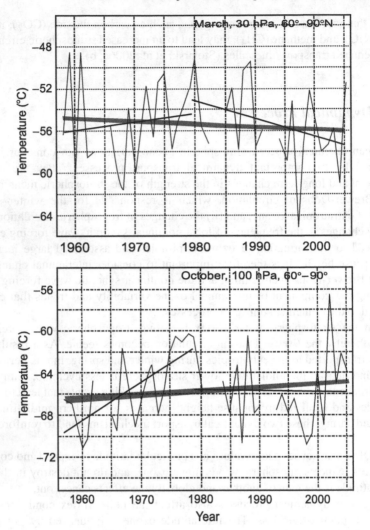

Fig. 6.4 Springtime temperatures at 30 hPa level in the Arctic (60–90°N) during March and Antarctic (60–90°S) during October (Labitzke and Kunze 2005)

ozone variations to changes in halogen loading and when predicting future ozone levels (Pawson et al. 1998). This is especially important for attribution in polar regions, where interannual variations in ozone are closely coupled to variations in polar temperatures. The size of the ozone hole depends on the temperature in the vortex area (60–70°S) over the Antarctic region.

Figure 6.4 illustrates the springtime temperature at 30 hPa level over the Arctic in March and the Antarctic in October for the period 1958–2005, derived from NCEP–NCAR reanalysis data (Labitzke and Kunze 2005). It can be seen that during the last five decades, the Arctic temperature in the lower stratosphere is decreasing, whereas an increase in temperature trend is noted in the Antarctic. But it is interesting to note that the temperatures in both Arctic and Antarctic lower stratosphere reversed their trends during the late 1970s. The ozone hole was detected in the same period. Similar reversal in trend was also observed in several tropospheric climatic factors, such as ENSO, TBO, etc.

Future changes in temperature, and hence ozone, are not likely to be uniform throughout the stratosphere. Cooling due to increased CO_2 and other greenhouse gases is expected to slow down gas-phase ozone loss reactions. When stratospheric chlorine contents decrease to pre-1980 levels, and if there are no other changes, the cooling will lead to an increase in ozone to values higher than in 1980. However, increases in greenhouse gases will also alter the chemical composition of the stratosphere and possibly the Brewer-Dobson circulation. These effects are likely to affect ozone. The impact of stratospheric cooling on ozone might be the opposite in polar regions.

6.2.4 Atmospheric Transport

Atmospheric transport is a major factor contributing to stratospheric ozone variability. The changes in the stratospheric meridional circulation in the stratospheric polar vortices and in tropospheric weather systems have a strong influence on stratospheric ozone and can produce variability on a wide range of timescales. Some of these changes can be linked to waves propagating from the troposphere, but internal stratospheric dynamics also play a role. Transport processes in the stratosphere and troposphere are dealt with in detail in Chapter 7. Changes in temperatures and transport not only complicate the detection and attribution of recovery milestones, but also affect ozone projections over the rest of this century.

6.2.5 The Solar Cycle

Solar activity in the 11-year cycle has a direct impact on the radiation and ozone budget of the middle atmosphere. During years with maximum solar activity, the solar UV irradiance is enhanced, which leads to additional ozone production and heating in the stratosphere and above. By modifying the meridional temperature gradient, the heating can alter the propagation of planetary and smaller-scale waves that drive the global circulation. Although the direct radiative forcing of the solar cycle in the upper stratosphere is relatively weak, it could lead to a large indirect dynamical response in the lower atmosphere through a modulation of the polar night jet and the Brewer-Dobson circulation (Kodera and Kuroda 2002). Such dynamical changes can feedback on the chemical budget of the atmosphere because of the temperature dependence of both the chemical reaction rates and the transport of chemical species.

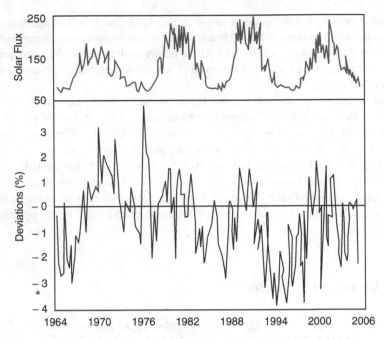

Fig. 6.5 Solar variability and total ozone in the tropics (Adapted from WMO 2007)

When attributing recent changes in stratospheric ozone to changes in ODSs, it is important to consider ozone variations related to the 11-year solar cycle because the timing of the recent maximum in solar activity, between 1999 and 2003, was around the time when *equivalent effective stratospheric chlorine* (EESC) peaked in the stratosphere. Observations continue to indicate a statistically significant solar variation of ozone, with ozone in phase with solar activity. This suggests that an increase in solar activity during the 1999–2003 solar maximum will have contributed to the slowing of the decline and increase of ozone (Dameris et al. 2005). Proper attribution of the cause of the ozone changes in recent years requires the separation of ozone increases due to changes in solar irradiance from those due to changes in halogen levels.

Figure 6.5 shows the deseasonalized, area-weighted total ozone deviations from five datasets for the latitude bands 25°S–25°N. The amplitude of ozone changes due to solar activity varies with altitude and latitude. In the upper stratosphere, ozone during solar maximum is 2–5% higher than in solar minimum, with an uncertainty of around 2%. Sensitivity studies indicate that current estimates of the solar cycle effect on ozone are probably sufficiently accurate to allow the separation of halogen decrease-related ozone increases from solar cycle effects in the upper stratosphere.

Depending on latitude and location, total column ozone is between 2 and 10 Dobson units (DU) higher during solar maximum, both in observations and model simulations, with uncertainty ranging from 2 to over 5 DU (Reinsel et al. 2005,

Steinbrecht et al. 2006). One reason for this large uncertainty in the magnitude of the solar cycle variation in total ozone is the fact that the two solar maxima before 1999–2003 coincided with large volcanic eruptions. It is difficult to separate the impacts of eruptions and solar cycle on observed ozone. There were no major volcanic eruptions during the 1999–2003 solar maximum.

6.2.6 Volcanic Eruptions

Volcanic eruptions can have a large impact on stratospheric ozone by changing heterogeneous chemistry, thermal structure, and circulation in the stratosphere. Because of this, it is necessary to consider volcanic eruptions both when interpreting observed changes and when making projections of future changes of ozone. There have been no large volcanic eruptions since the 1991 Mt. Pinatubo eruption, and the stratospheric aerosol loading in recent years has remained at low, nonvolcanic levels.

However, the impact of the Mt. Pinatubo eruption still needs to be considered when attributing changes in ozone in the last decade of the 20th century to changes in ODSs. The Mt. Pinatubo eruption contributed to a large decline in northern hemisphere ozone, which was followed by an increase in ozone as stratospheric aerosols decayed back to low, nonvolcanic levels (see Fig. 6.6). This decrease in aerosol levels occurred at around the same time that the growth in EESC slowed and reached its peak value.

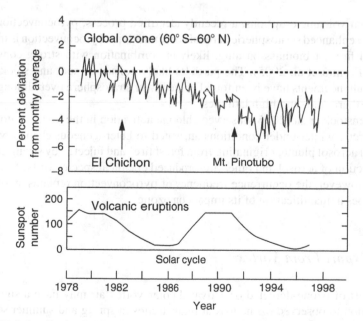

Fig. 6.6 Global ozone trend and major volcanic eruptions (Adapted from WMO 2007)

Outside the polar regions, the primary effect of an increased rate of heterogeneous reactions is to cause a reduction of nitrogen oxides. In the current high-chlorine conditions, this causes an increase in reactive chlorine and increased ozone depletion, as observed following the Mt. Pinatubo eruption. However, in low-chlorine conditions, a large volcanic eruption could cause a small ozone increase due to the suppression of nitrogen oxides. Hence, a large, Pinatubo-like eruption within the next 20 years, when there will still be significant amounts of halogens in the stratosphere, may lead to an increase in ozone destruction by ODSs and a temporary delay in ozone recovery.

6.2.7 Aerosol Effects

It is well known that the presence of enhanced aerosols in the stratosphere can cause significant chemical ozone loss through heterogeneous chemical reactions. Sulfate aerosols in the lower stratosphere provide surfaces for the activation of chlorine. The distribution of sulfate surface area depends on background sulfur emissions and volcanic eruptions. Thus, the Mt. Pinatubo period to the present gives a good span of potential heterogeneous effects, barring a huge future volcanic eruption. The reactions do not suggest significant missing processes or erroneous rates.

6.2.8 Pyroconvection

Apart from volcanic eruptions, a recently identified process, pyroconvection, may also cause enhanced stratospheric aerosol concentrations. Pyroconvection is induced by boreal fires or biomass burning, likely in combination with strong convective activity (Fromm et al. 2005). The particles likely consist of soot and smoke. The aerosol enhancements have been found in the lower stratosphere over all longitudes in the northern mid- and high latitudes.

Blumenstock et al. (2006) observed chlorine activation in the lower stratosphere in late Arctic winter under conditions attributed to heterogeneous chemistry on an enhanced aerosol plume, originating from forest fires and injected by strong convection. Injection of aerosol and other fire-produced chemical species will affect ozone locally. However, the occurrence frequency of pyroconvection remains an open issue, as does a quantification of its impact on ozone.

6.2.9 Export From Vortex

The export of ozone-depleted or activated polar vortex air may have a significant contribution to observed ozone loss at midlatitudes in spring and summer seasons.

This effect exists in both hemispheres but is expected to be larger in the southern hemisphere due to the larger and more regular ozone depletion in the Antarctic vortex. Although transport is involved in this process, the ultimate cause is chemical O_3 loss in the polar region by Cl and Br species. The mass of missing ozone in the ozone hole has the same order of magnitude as the mass deficit over southern mid- and high latitudes in summer, and illustrates the strong correlation between loss in the ozone hole and summer.

6.3 Basis of Ozone Depletion

The amount of ozone in the stratosphere is determined by a balance between photochemical production and recombination. The overall effect is to increase the rate of recombination, leading to an overall decrease in the amount of ozone. For this particular mechanism to operate, there must be a source of O atoms, which is primarily the photodissociation of ozone. This mechanism is only important in the upper stratosphere where such atoms are abundant. More complicated mechanisms have been discovered that lead to ozone destruction in the lower stratosphere as well. Concentrations of ozone in the stratosphere fluctuate naturally in response to variations in weather conditions and amounts of energy being released from the Sun, and to major volcanic eruptions (Solomon et al. 2005a, b).

6.3.1 Special Features of Polar Meteorology

In order to understand how so much destruction of ozone occurs during spring over the Antarctic region, we need to go through the possibilities of the occurrence of ozone loss. First, we will look at the way in which the polar atmosphere behaves and the peculiar features of the meteorology in the stratosphere.

The role of sunlight in ozone depletion is the reason why the Antarctic ozone depletion is greatest during spring. During winter, there is no light over the pole to drive the chemical reactions. As a result, a strong circumpolar wind, known as polar vortex, forms in the middle to lower stratosphere. The polar vortex isolates the air over the polar region. During the spring the Sun appears after the winter, providing energy to drive photochemical reactions and releasing the trapped compounds.

Since there is no sunlight during polar night, the air within the polar vortex become very cold. Once the air temperature goes to below about 195 K, polar stratospheric clouds (PSC) can form. PSCs initially form as nitric acid trihydrate. As the temperature becomes colder, larger droplets of water-ice with nitric acid dissolved in them can form. However, their exact composition is still the subject of intense scientific investigation. These PSCs are crucial for ozone loss to occur.

Changes in stratospheric meteorology cannot explain the ozone hole (Brasseur and Solomon 2005). Measurements show that wintertime Antarctic stratospheric temperatures of past decades have not changed significantly. Ground, aircraft, and satellite measurements have provided clear evidence of the importance of the chemistry of chlorine and bromine originating from man-made compounds in depleting Antarctic ozone in recent years.

6.3.2 Chemical Processes Leading to Polar Ozone Depletion

It is now accepted that chlorine and bromine compounds in the atmosphere cause the ozone depletion observed in the "ozone hole" over Antarctica and over the north pole. However, the relative importance of chlorine and bromine for ozone destruction in different regions of the atmosphere has not yet been clearly explained. Nearly all of the chlorine, and half of the bromine in the stratosphere, where most of the depletion has been observed, comes from human activities (Schoeberl et al. 2006).

Figure 6.7 shows a schematic illustration of the life cycle of the CFCs: how they are transported into the upper stratosphere/lower mesosphere; sunlight breaks down the compounds and then their breakdown products descend into the polar vortex. The main long-lived inorganic carriers of chlorine are hydrochloric acid (HCl) and

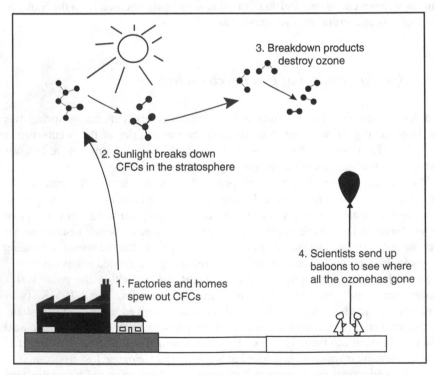

Fig. 6.7 Schematic presentation of life cycle and transportation of CFC in the atmosphere (Adapted from Eduspace, European Space Agency)

chlorine nitrate ($ClONO_2$). These form from the breakdown products of the CFCs. Dinitrogen pentoxide (N_2O_5) is a reservoir of oxides of nitrogen and also plays an important role in the chemistry. Nitric acid (HNO_3) is significant in that it sustains high levels of active chlorine.

6.3.3 Production of Chlorine Radicals

One of the most important points to realize about the chemistry of the ozone hole is that the key chemical reactions are unusual. They cannot take place in the atmosphere unless certain conditions are present. The main feature of this unusual chemistry is that the chlorine reservoir species HCl and $ClONO_2$ are converted into more active forms of chlorine on the surface of the polar stratospheric clouds.

6.4 Anthropogenic Contribution in Ozone Depletion

Human activities produce the emission of ozone-depleting gases, containing chlorine and bromine atoms, into the atmosphere and ultimately lead to stratospheric ozone depletion. The source gases that contain only carbon, chlorine, and fluorine are called chlorofluorocarbons (CFCs). Along with CFCs, carbon tetrachloride (CCl_4) and methyl chloroform (CH_3CCl_3) have been the most important chlorine-containing gases that are emitted by human activities which destroy stratospheric ozone (see Fig. 6.8).

The chlorine-containing gases have been used in many applications, including refrigeration, air-conditioning, foam blowing, aerosol propellants, and cleaning of metals and electronic components. These activities have typically caused the emission of halogen-containing gases to the atmosphere. Halogenated hydrocarbon gases (halons) and methyl bromide (CH_3Br) also deplete ozone significantly. Halons are widely used to protect large computers, military hardware, and commercial aircraft engines (Ramaswamy et al. 2006). Because of these uses, halons are often directly released into the atmosphere. Methyl bromide, used primarily as an agricultural fumigant, is also a significant source of bromine to the atmosphere.

After emission, halogen source gases are either naturally removed from the atmosphere or undergo chemical conversion. The time to remove or convert about 60% of a gas is often called its atmospheric lifetime. Lifetimes vary from less than 1 year to 100 years for the principal chlorine- and bromine-containing gases. Gases with the shortest lifetimes (e.g., the hydrochlorofluorocarbons (HCFCs), methyl bromide, methyl chloride, and the very short-lived gases) are substantially destroyed in the troposphere, and therefore only a fraction of such emitted gases contribute to ozone depletion in the stratosphere. The long-lived species, CFC11 and CFC12, reach stratospheric altitudes and contribute substantially to the ozone depletion .

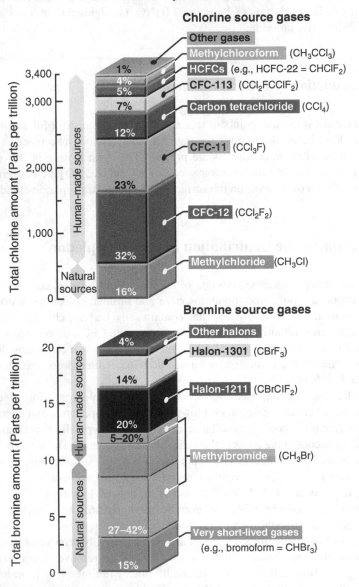

Fig. 6.8 Percentage contribution of primary source gases of chlorine and bromine in the stratosphere (Adapted from WMO 2007)

Human emissions of the principal chlorine- and bromine-containing gases have increased substantially since the middle of the 20th century. The result has been global ozone depletion, with the greatest losses occurring in polar regions.

6.4.1 Chlorine Compounds

Most of the chlorine in the stratosphere as a result of human activities is released at the ground level. Figure 6.9 describes the various forms of chlorine from anthropogenic and natural sources entering into the stratosphere.

The compounds containing chlorine that dissolve in water cannot reach stratospheric altitudes in significant amounts because they are "washed out" of the atmosphere in rain or snow. For example, large quantities of chlorine are released from evaporated ocean spray as sea salt particles. However, because sea salt dissolves in water, this chlorine is taken up quickly in clouds or in ice, snow, or rain droplets and does not reach the stratosphere. Another ground-level source of chlorine is from its use in swimming pools and as household bleach. When released, this chlorine is rapidly converted to forms that dissolve in water and therefore are removed from the lower atmosphere. Such chlorine never reaches the stratosphere in significant amounts.

Volcanoes can emit large quantities of hydrogen chloride, but this gas is rapidly converted to hydrochloric acid, which dissolves in rain water, ice, and snow, and does not reach the stratosphere. Even in explosive volcanic plumes that rise high in the atmosphere, nearly all of the hydrogen chloride is removed by precipitation before reaching stratospheric altitudes. The exhaust from the space shuttle and from some rockets does inject some chlorine directly into the stratosphere, but the quantities are very small.

The major ozone-depleting human-produced halocarbons, such as chlorofluorocarbons (CFCs) and carbon tetrachloride (CCl_4), do not dissolve in water, or

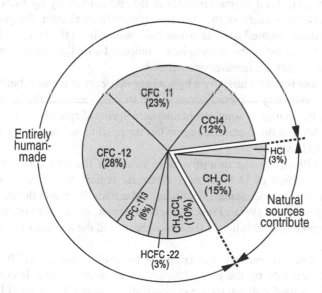

Fig. 6.9 Primary sources of chlorine entering the stratosphere in the early 1990s (Adapted from WMO 2007)

react with other natural surfaces, and are not broken down chemically in the lower atmosphere. These halocarbons and other human-produced substances containing chlorine do reach the stratosphere.

6.4.2 Chlorofluorocarbons in the Stratosphere

Chlorofluorocarbon (CFC) molecules are indeed several times heavier than air. CFCs reach the stratosphere because the Earth's atmosphere is always in motion and mixes the chemicals added into it. Measurements from balloons, aircraft, and satellites confirm that the CFCs are actually present in the stratosphere. Air motions mix the atmosphere to altitudes far above the top of the troposphere much faster than molecules can settle according to their weight. Gases such as CFCs that do not dissolve in water are relatively nonreactive in the troposphere, are mixed relatively quickly, and reach the stratosphere regardless of their weight.

6.5 Antarctic Ozone Hole

The Antarctic ozone hole is a region of extreme ozone loss that has been appearing annually since the 1970s. Ozone amounts over Antarctica drop dramatically in the course of a few weeks. The hole begins to develop each September and ceases by early October, and it starts recovering in November and subsequently disappears by early December. Total ozone amounts in this period fall by up to 50% inside the hole. Observations taken in the Antarctic region from aircraft, the ground, and satellites have demonstrated that the ozone hole results from the increased amounts of chlorine and bromine in the stratosphere, combined with the unique atmospheric conditions of the southern hemisphere winter.

The ozone hole is not technically a hole where no ozone is present, but is actually a region of exceptionally depleted ozone in the stratosphere over the Antarctic that happens at the beginning of southern hemisphere spring (September–October). The present ozone level in the Antarctic region has dropped to as low as 33% compared to that of pre-1975 values.

Figure 6.10 illustrates the monthly mean values of the total amount of ozone recorded in the month of October over Antarctic region. It can be seen that the total ozone during October started declining in the mid-1970s and the decrease became alarming after the 1980s. Present levels of depletion have served to highlight a surprising degree of instability of the atmosphere, and the amount of ozone loss is still increasing.

The ozone hole was weakly observable as far back as the mid-1970s, and became easily observable by the early 1980s as it grew in severity. It constitutes a 60% reduction in total column ozone concentrations, and a 100% local loss in the 12–20 km layer. The hole is associated with the Antarctic polar vortex and extremely cold winter temperatures that occur inside the vortex.

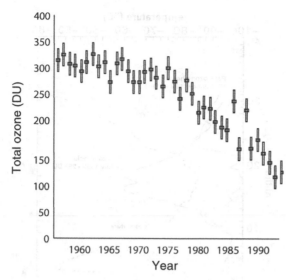

Fig. 6.10 Monthly mean total ozone over Antarctica during October (F. S. Rowland, Courtesy: Encyclopedia of Earth)

Figure 6.11 shows the vertical profiles of ozone over the south pole on 28 July 1999, representing normal profile and on 13 October 1999, when the ozone hole became well established. On a normal day, the ozone concentration is higher between 12 and 24 km in the polar lower stratosphere. But during the time of ozone hole formation, nearly complete ozone depletion occurs between 13 and 23 km, where extremely low temperatures support the heterogeneous photochemical destruction of ozone. But, above and below these heights the air temperature is not low enough for this type of ozone destruction, and ozone amounts remain virtually unchanged.

Measurements by the TOMS instrument aboard the Nimbus-7 satellite showed that the depletion of ozone during the southern hemisphere spring occurred over the entire Antarctic continent, centered on the south pole. Because of the visual appearance of this Antarctic low ozone region, the phenomenon was quickly dubbed the "Antarctic ozone hole."

The depth and area of the ozone hole are primarily governed by amounts of chlorine and bromine in the Antarctic stratosphere. Very low temperatures are needed to form polar stratospheric clouds (PSCs). Chlorine gases react on these PSCs to release chlorine into a form that can easily destroy ozone. The chlorine and bromine chemical catalytic reaction to destroy ozone needs sunlight (see Chapter 5 for details). Hence, the ozone hole begins to grow as the Sun is rising over Antarctica at the end of winter.

The ozone hole begins to develop in late August and reaches its largest area in depth in the middle of September to early October. In the 1980s, the hole was small because chlorine and bromine levels over Antarctica were low. Year-to-year variations in area and depth are caused by the interannual variations in temperature. Colder conditions result in a larger area and lower ozone values in the center of the hole.

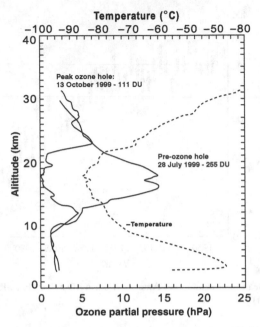

Fig. 6.11 Vertical profiles of ozone over the south pole (Adapted from NASA 2003)

Any place where the concentration drops below 220 DU is considered part of the ozone hole. Average ozone concentrations in the ozone hole are around 100 DU. Reduction of stratospheric ozone is harmful to the biosphere.

6.5.1 Discovery of Antarctic Ozone Hole

During the 1970s, it was recognized that the ozone is naturally low over Antarctica as a result of the weaker poleward and downward circulation in the southern hemisphere. In 1985, Joesph Farman, Brian Gardiner, and Jonathan Shanklin of the British Antarctic Survey published a paper (Farman et al. 1985) showing that the total ozone content of the atmosphere during Antarctic spring had decreased considerably after 1979. The amounts measured were much less than even the naturally occurring low amounts over Antarctica in the southern spring. This observational evidence on the spectacular seasonal drop in ozone over Antarctica during the spring was a surprise to the atmospheric scientists.

In the 1980s the first measurements of this loss were actually documented. In 1984, when the British first reported their findings, October ozone levels were about 35% lower than the average for the 1960s. When the first measurements were taken, the drop in ozone levels in the stratosphere was so dramatic that at first the scientists thought their instruments were faulty. The US satellite Nimbus-7 quickly confirmed the results, and the term Antarctic ozone hole entered popular language.

Earlier it was believed that the physical and photochemical processes controlling ozone production and loss are well understood. Computer model simulations of ozone production, transport, and loss processes agreed well with the observed data. Based on the measurements from ground-based Dobson spectrophotometer at Halley Bay (76°S, 27°W), Farman et al. (1985) reported that the Antarctic total ozone is depleted by about 50% in early spring between 1975 and 1984, and the large losses are primarily confined to the spring season (September–October).

The observed average amount of ozone during September, October, and November over the British Antarctic Survey station at Halley, Antarctica, first revealed notable decreases in the early 1980s, compared with the preceding data obtained starting in 1957. The ozone hole is formed each year when there is a sharp decline up to 60% in the total ozone over most of Antarctica for a period of about 3 weeks to months during spring in the southern hemisphere. Late summer (January–March) ozone amounts did not show similar sharp decline in the 1980s and 1990s. Total ozone observations taken from three other stations in Antarctica, including the Indian station, Maitri, and from satellite-based instruments, revealed similar decrease in springtime amounts of total ozone. Balloon-borne ozone instruments show dramatic changes in the way ozone is distributed with altitude.

Figure 6.12 shows the vertical profiles of ozone over the Antarctic station Syowa during the month of October in the pre-ozone hole period (1968–1980) and the severe ozone hole period (1991–1997). In the spring season, the vertical profile in the pre-ozone hole era resembles very much that of any other season, having maximum ozone concentration between 13 and 23 km, with peak values of the order of 15 milli pascals in the 14–16 km altitude, whereas, the ozone concentration decreased to lowest minimum value of less than 2 milli pascals in the 14–16 km region during the severe ozone hole period. Nearly 80% decrease in ozone concentration is noted at 14–16 km region between the pre-ozone hole period and the severe ozone hole time during spring season over Antarctica.

Before the stratosphere was affected by human-produced chlorine and bromine, the naturally occurring springtime ozone levels over Antarctica were about 30–40% lower than springtime ozone levels over the Arctic. This natural difference between Antarctic and Arctic conditions was first observed in the late 1950s by Dobson. It stems from the exceptionally cold temperatures and different winter wind patterns within the Antarctic stratosphere as compared with the Arctic. This is not at all the same phenomenon as the marked downward trend in ozone over Antarctica in recent years. The ozone hole appeared first over the colder Antarctic because the ozone-destroying chemical process works best in cold conditions. The Antarctic continent has colder conditions than the Arctic, which has no landmass. As the years have gone by the ozone hole has increased rapidly and is as large as the Antarctic continent.

Average area of the Antarctic ozone hole, where the total ozone is less than 220 DU, detected by the SBUV on NIMBUS-7 and SBUV-7 instruments on NOAA polar orbiting satellites between 1 October and 30 November from 1979 to 2006, is depicted in Fig. 6.13. The ozone hole area has increased more than 16 million square kilometers within a short span of 20 years from 1979 to 1998. The hole lasts for only

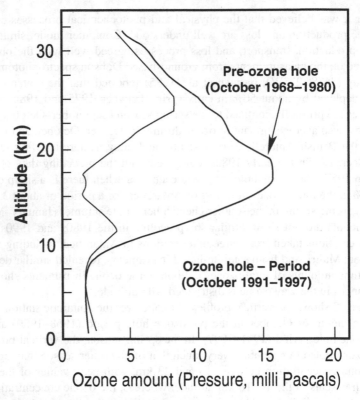

Fig. 6.12 Springtime depletion of the ozone layer over Syowa, Antarctica (Adapted from WMO 2007)

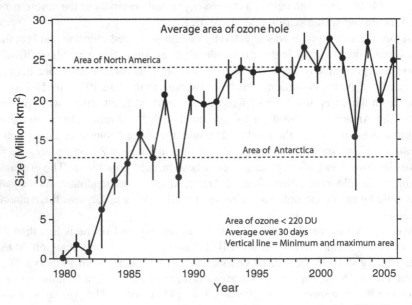

Fig. 6.13 Monthly variations of zonal wind in the troposphere and stratosphere at 60° S (Courtesy: NASA)

2 months, but its timing could not be worse. Just as sunlight awakens activity in dormant plants and animals, it also delivers a dose of harmful ultraviolet radiation. After 8 weeks, the hole leaves Antarctica, only to pass over more populated areas, including the Falkland Islands, South Georgia, and the tip of South America. This biologically damaging, high-energy radiation can cause skin cancer, injure eyes, harm the immune system, and upset the fragile balance of an entire ecosystem. The ozone hole can be as big as 1.5 times the United States.

However, less well-known is that ozone depletion has been measured everywhere outside the tropics, and that it is, in fact, getting worse. In the middle latitudes, ozone levels have fallen about 10% during the winter and 5% in the summer. Since 1979, they have fallen about 5% per decade when averaged over the entire year (WMO 2007). Depletion is generally worse at higher latitudes, i.e., farther from the equator.

6.5.2 Theories of Ozone Hole

Earlier Farman et al. (1985) have attempted to explain the reasons for the sudden formation of ozone hole, and the causes that changed the atmosphere over Antarctica to produce such a large, sudden loss of ozone on a seasonal basis. Since there was no apparent change in the meteorological parameters over the Antarctic region and only weak transport effects at the altitude of the hole, it is postulated that the sudden loss in ozone each October may possibly be due to the apparent increase in halocarbon amounts since the 1960s and the extremely cold temperature in the lower stratosphere above Antarctica. The increasing chlorine concentrations with the very cold temperatures over Antarctica were enhancing ozone loss.

The ozone loss rates computed by using Farman's approach were much too small to explain the large ozone losses seen during September. The computer models showed that amounts of free oxygen atoms, necessary for the catalytic destruction of ozone, were too low in the 15–24 km altitudes range where most of the ozone loss was taking place. Hence, Farman's theory was found to be incorrect.

Thereafter, three theories emerged to explain the Antarctic ozone hole. They are the dynamical theory, the nitrogen oxide theory, and the heterogeneous chemistry theory.

6.5.3 Dynamical Theory

The dynamical theory proposed that the atmospheric circulation over Antarctica had changed in such a way that air from the troposphere, where there is little ozone, was being carried into the polar lower stratosphere, and hence the observed reductions. If ozone-poor air from the troposphere was indeed being transported into the lower stratosphere, then other long-lived trace gases should also be measurably increasing in the lower stratosphere.

Nitrous oxide (N_2O) is emitted into the troposphere by biological processes, and is destroyed in the stratosphere by either UV radiation photolysis or by a reaction with excited O atoms (WMO 1995). The loss of N_2O takes place in the upper stratosphere, since O atoms are generally produced by the photolysis of O_2, which requires UV wavelengths under 240 nm. Such energetic UV radiation cannot penetrate into the troposphere because of the screening by ozone molecules. Thus, N_2O has fairly high amounts in the troposphere (between 300 and 310 ppbv) and low amounts in the upper stratosphere. This general profile of N_2O has been confirmed by satellite, balloon, and aircraft observations.

The dynamical theory predicts that Antarctic N_2O amounts should be high if the air was transported upward from the troposphere into the lower stratosphere where ozone was low.

The dynamical theory of the ozone hole proposed that the Antarctic circulation associated with the Brewer-Dobson circulation was changing, and that ozone-poor air from the troposphere was being transported upward into the lower stratosphere. Evidence that disproved the dynamical theory came from N_2O (and other long-lived trace gas) observations. Later it was found that air inside the lower stratosphere of the Antarctic polar vortex had indeed descended from the middle and upper stratosphere, in line with the Brewer-Dobson circulation. This meant that ozone amounts should have been higher.

The observations show that N_2O is substantially lower than amounts characteristic of the troposphere (300–310 ppbv) in the region inside the polar vortex where the Antarctic ozone hole appears in the spring. These N_2O observations, as well as observations of other long-lived trace gases, demonstrate that air inside the lower stratospheric Antarctic polar vortex had indeed descended from the middle and upper stratosphere, otherwise N_2O amounts would have been much higher. Furthermore, the air ought to have contained higher ozone amounts, since the air was brought down by the Brewer-Dobson circulation from higher altitudes and lower latitudes. The dynamical theory for ozone loss was thus proved to be incorrect.

6.5.4 Nitrogen Oxide Theory

The nitrogen oxide theory of the ozone hole was proposed by Callis and Natarajan (1986), and it projected that large amounts of NO_x compounds were being produced as a result of the sunspot maximum in 1979. This NO_x would be photochemically produced as a result of increased energetic UV light in the middle to upper stratosphere of the tropics and transported into the polar lower stratosphere by the Brewer-Dobson circulation. Measurements of NO_2 (nitrogen dioxide) at high latitudes indicate very low concentrations during the spring period of the ozone hole.

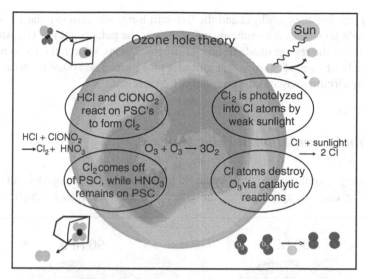

Fig. 6.14 Schematic diagram of ozone hole theory (Adapted from NASA 2003)

6.5.5 Heterogeneous Chemistry Theory

The third theory of the ozone hole involves heterogeneous chemical reactions on the surfaces of solid (frozen) particles that formed in the cold lower stratosphere of the Antarctic vortex. These reactions would free chlorine from "inactive" forms into "reactive" forms, where the chlorine could destroy ozone in the catalytic cycles, as discussed in Chapter 5, section 5.11. The theory proposed that reactions which normally do not occur in gas phase might be greatly enhanced if chlorine-containing compounds such as $ClONO_2$ (chlorine nitrate) and HCl (hydrochloric acid) could collect on the surfaces of these particles and then react to release the chlorine into a reactive form that could cause large ozone losses. This is called the heterogenous chemistry theory, and it turned out to be convincing one for explaining Antarctic ozone losses. The ozone hole theory is schematically represented in Fig. 6.14.

Further study in this area has shown that the heterogeneous chemistry theory is consistent with observations. The dynamics and nitrogen oxide theories are found to be deceptive.

6.6 The Antarctic Polar Vortex

The polar stratospheric regions of both hemispheres are surrounded by a narrow band or stream of fast-moving winds very high up blowing from west to east. Similar to the upper tropospheric jet stream, this jet stream develops along a zone of a tight temperature gradient. In this case, it is the temperature gradient that develops along the line between sunlight and the 6-month long, wintertime polar night. This high-altitude jet stream is commonly referred to as the polar night jet. The Antarctic polar vortex is the region inside of this jet stream. During winter, the jet stream can reach speeds of over of 50 m s^{-1} at altitudes of 21 km. The Antarctic polar vortex completely circumnavigates the continent of Antarctica.

6.6.1 Wind Circulation in Polar Vortex

The westerly circulation of the Antarctic polar vortex is strongest in the upper stratosphere and strengthens over the course of the winter (see Fig. 6.15). The

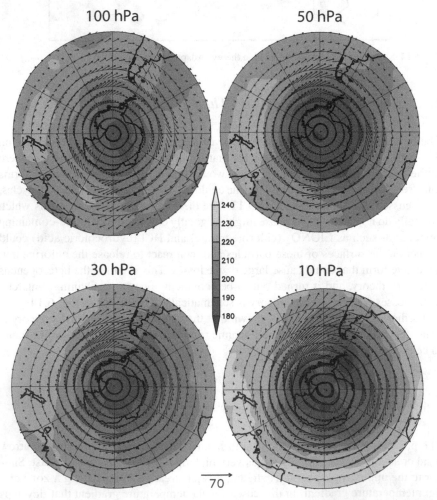

Fig. 6.15 Wind circulation pattern around the Antarctic polar vortex at 100, 50, 30 and 10 hPa levels on 14 August 2001(NCEP/NCAR Reanalysis data)

polar night jet is important because it acts as a barrier to transport between the southern polar region and the southern midlatitudes. It is a barrier because it effectively blocks any mixing between air inside and outside the vortex during the winter. Thus, ozone-rich air in the midlatitudes cannot be transported into the polar region.

The isolation of polar air allows the ozone loss processes to proceed unimpeded with no replenishment by intrusions of ozone-rich air from midlatitudes. This isolation of the polar vortex is a key ingredient to polar ozone loss, since the vortex region can evolve without being disturbed by the more conventional chemistry of the midlatitudes. The polar night jet over the Arctic is not as effective in keeping out intrusions of warmer, ozone-rich midlatitude air. This is because there is more wave activity and hence more north–south mixing of air in the northern hemisphere than in the southern hemisphere.

6.6.2 Polar Night Jet and Polar Vortex

The development of the Antarctic polar vortex and the southern polar night jet is illustrated in Fig. 6.16 through the altitude-versus-time contour plot of mean zonal wind at 50 hPa level obtained from ERA40 Reanalysis data at 60°S latitude region. As the amount of sunlight decreases and temperatures drop in the southern polar region, the night jet winds increase. At higher altitudes, the polar vortex begins to develop in the March–April (early autumn) period and is fully developed by May,

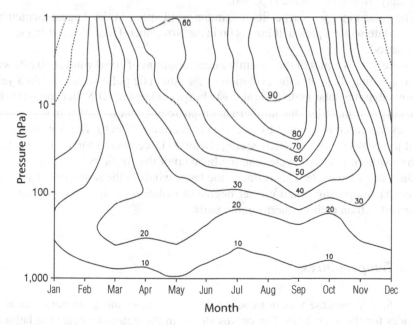

Fig. 6.16 Tropospheric and stratospheric mean zonal wind distribution in an year at 60°S latitude based on climatological wind data obtained from ERA40 Reanalysis

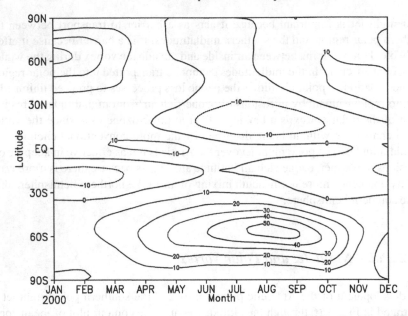

Fig. 6.17 Latitudinal variation of mean zonal wind in an year at 50 hPa level from ERA40 Reanalysis data

corresponding to the onset of the period of complete polar night darkness. At lower altitudes, the vortex develops more slowly, not becoming fully developed until the June–July (first half of winter) period.

The polar night jet reaches its maximum wind speed in the August–September (mid- to late winter) period. It breaks up in the November–December (spring second half) period.

Figure 6.17 is a plot of the monthly mean variations of zonal wind at 50 hPa with latitude, which illustrates the evolution of the Antarctic polar vortex within a year. It shows that the polar night jet is almost always centered at 60°S. In contrast to this southern polar night jet , the northern hemisphere polar jet is weaker in mid-winter, and has decreased in strength by late winter (February–March). The Antarctic polar night jet breaks up in mid to late spring (October–December), nearly 2 months later in the southern seasonal cycle than the breakup of the Arctic polar night jet in the northern seasonal cycle. This is due to the faster winds of the southern polar jet and the colder temperatures and greater degree of isolation of air inside the Antarctic polar vortex than their northern counterparts.

6.6.3 Temperatures

Figure 6.18 shows the variations in the 15–27 km layer mean temperatures at all latitudes for the year 2006. The curves shown in the figure delineate the latitudes

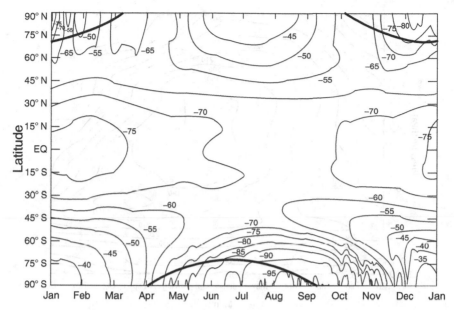

Fig. 6.18 Latitudinal variation of temperature at 50 hPa level during the year 2006 (Source: NCEP/NCAR Reanalysis data)

that are not sunlit during winter. Latitudes poleward of this line are in complete darkness. Only when and where the cold polar air is sunlit, photochemical reactions depleting ozone can take place in the presence of chlorine and other halons.

Temperatures are below 192 K over a deep layer (15–27 km), and extending from the south pole to 70°S. These cold temperatures develop during the polar night because of the lack of sunlight, which causes the air to cool radiatively toward its equilibrium temperature, which without sunlight is quite cold. The lack of north–south mixing due to the isolation of the polar vortex region allows this cooling to occur throughout winter long without any intrusions of warmer air.

6.6.4 South Polar Cold Temperatures During the Winter

The development of the cold temperatures over the south polar region is illustrated in Fig. 6.19 using a zonal mean plot of the temperatures at 80°S as a function of altitude. The polar region cools over the course of the fall period at all altitudes. This cooling is exceptionally strong at the higher altitudes in early fall (40–48 km), with warming beginning in the June–July period. The coldest temperatures (i.e., temperatures less than 192 K) first appear in July at approximately 24 km (30 hPa). These cold temperatures begin to appear at lower altitudes later in the season. Hence, at higher stratospheric altitudes, the coldest period is in early winter, while at lower stratospheric altitudes, the coldest temperatures occur in late winter. The temperatures rapidly warm during the breakup of the polar vortex. This breakup occurs

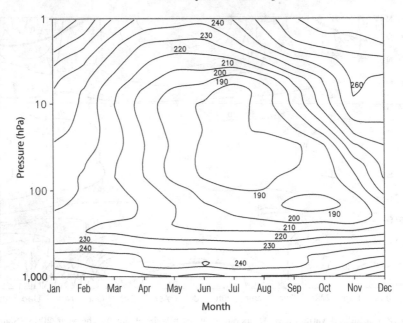

Fig. 6.19 Mean temperature distribution in the troposphere and stratosphere at 80°S (Data Source: ERA40 Reanalysis)

earlier at the higher altitudes, and takes place the latest at the lower altitudes. Because the ozone hole is observed at the lower altitudes below 30 km, the temperature region of greatest concern is between 10 and 30 km.

Temperatures inside the southern polar vortex are colder than inside the northern polar vortex (Ramaswamy et al. 2001). The cold temperatures inside of the polar vortex are crucial to the large polar ozone losses. This is because the formation of polar stratospheric clouds (PSCs) requires such cold temperatures, and the PSCs are the key to the loss process. PSCs have been observed in both the Arctic and Antarctic stratosphere for quite some time, though the colder Antarctic leads to more frequent PSC formation.

6.6.5 Potential Vorticity

In order to identify stratospheric air that is inside or outside of the polar vortex, the potential vorticity, from weeks to days, is used. The temperature and wind speeds of a particular air parcel might vary in time, whereas the potential vorticity remains almost the same from day to day. Hence, potential vorticity is a key tool for following the motion of air in the stratosphere. Motions of air in the stratosphere are mostly in the horizontal, and the vertical motions are weak. Horizontal motions often occur on the isentropic surfaces.

A set of images of potential vorticity at 10 day intervals between August 1 and October 10, 2001 at 70 hPa level is illustrated in Fig. 6.20. This period encompasses

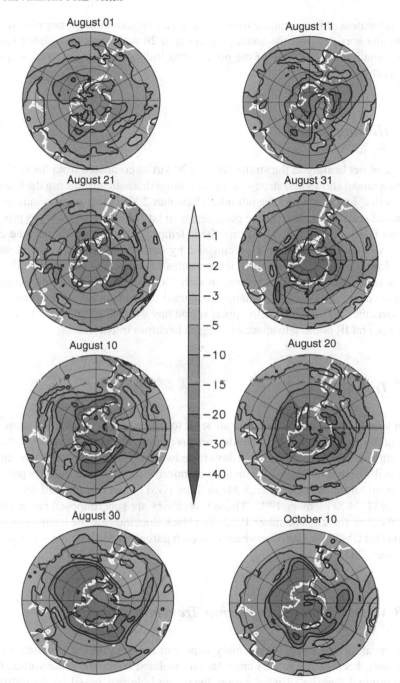

Fig. 6.20 Potential vorticity at 70 hPa level in the southern hemisphere at 10-day intervals between August 1–October 10, 2001. (Data source: NCEP/NCAR Reanalysis)

the development of the Antarctic ozone hole. It can be seen in large, negative values of potential vorticity over the south pole in Fig. 6.20 with somewhat higher values in the midlatitudes. The edge of the polar vortex indicates the location of the polar night jet stream.

6.6.6 Heating

Average of net heating in the stratosphere at 20 km as computed from the Goddard radiation model shows that stratosphere cools quite dramatically, during the March–April period in the south polar latitudes (Newman 2003). This corresponds to the absence of solar radiation over the polar region in late March. However, this process does not continue indefinitely. Stratospheric temperatures eventually become cold enough so that longwave radiation emission by carbon dioxide, ozone, and water vapor slows down. By August, the temperatures have become so cold that the net cooling is near zero. Sunlight returns in early October as spring arrives, warming the stratosphere. Net cooling remains small until after the polar vortex breakup in late November, by which time the polar region has warmed to relatively high temperatures, and IR cooling to space once again becomes important.

6.6.7 Transport

When viewed from space, Antarctic air tends to move in a clockwise direction, because of the prevailing westerly winds over this region. The air at 20 km level circles the south pole about once every 4–6 days in midwinter. This basic background circulation is illustrated in Fig. 6.21 with a set of trajectories initialized on 20 September 1992 at midnight (00) Greenwich Mean Time (GMT) and run forward for 3 days to 00 GMT 24 September 1992. These trajectories are superimposed on an image of total ozone for 21 September 1992. The black dots indicate the location of each parcel at 00 GMT. We can observe how far each parcel moved over the 4-day period (Newman 2003).

6.6.8 Vertical Motions and Ozone Transport

The vertical motion of air is extremely important for studying the evolution of the ozone hole, because air parcels are relatively isolated inside the polar vortex. One of the original theories of what causes the ozone hole was based on the transport of low ozone air from the troposphere into the lower stratosphere. Rosenfield et al. (2002) have shown that the air below 20 km inside the ozone hole during September has descended from altitudes near 25 km over the course of the southern hemisphere

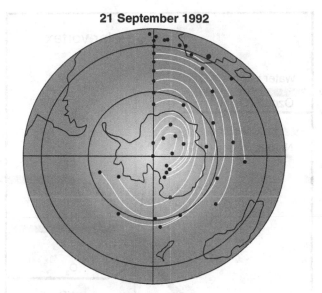

Fig. 6.21 Air parcel trajectories inside the polar vortex, 20 August to 14 September 1992, with total ozone (Adapted from NASA 2003)

winter. Thus, the theory of transport of ozone-poor tropospheric air upward into the stratosphere is ruled out. This descent has also been determined via observations of the descent rates of long-lived tracers in the UARS satellite data (Schoeberl et al. 2006).

6.6.9 Chemistry of the Polar Vortex

Figure 6.22 shows that some of the most complex atmospheric chemical reactions occur during the 4–6-week duration of the ozone hole. The polar vortex seals the Antarctic atmosphere during this period, creating what is essentially an extraordinary chemical reaction bowl. It can be seen that very dramatic changes occur in the chemical composition of the stratosphere as one races from outside the vortex to the inside. The concentrations of many chemicals drop dramatically, including water vapor, nitrogen oxides, and ozone. At the same time, the concentrations of other chemicals, like chlorine monoxide, increase dramatically (Maduro and Schauerhammer 1992).

The walls of the polar vortex act as the boundaries for these extraordinary changes in chemical concentrations. Now the polar vortex can be considered as a sealed chemical reactor bowl, containing a water vapor hole, a nitrogen oxide hole, and an ozone hole, all occurring simultaneously (Labitzke and Kunze 2005). This chemical condition exists only in the short-lived Arctic polar vortex.

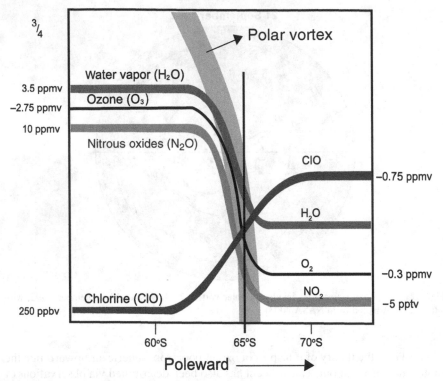

Fig. 6.22 Concentration of chemicals inside and outside of the polar vortex (Maduro and Schauerhammer 1992, Courtesy: John Wiley & Sons)

6.7 Structure and Dynamics of the Antarctic Ozone Hole

The ozone hole's horizontal and vertical structure includes (i) the day-to-day variability of the ozone hole; (ii) the growth of the ozone hole during the southern hemispheric mid- to late winter (August–September) time period; and (iii) the decay of the ozone hole during the mid- to late spring (November–December) time period (Newman 2003).

6.7.1 Horizontal Structure

The horizotonal structure of the ozone hole during the month of October exhibits several features. During this time the ozone hole is almost symmetric about the south pole, which is usually located near its midpoint. There is an underlying *wave one* structure in the ozone field that becomes evident when the zonal mean field is removed. The ozone hole is quite mobile, sometimes rotating along an elongated axis toward the east. Each of these aspects and their implications are discussed below.

6.7.1.1 Symmetric

The ozone hole is almost symmetric about the south pole. Total ozone inside the collar region during October over Antarctica is slightly offset towards the south Atlantic Ocean. The difference between the former and later periods is that ozone amounts are much lower over Antarctica in the later years, corresponding to the annual appearance of the ozone hole. The ozone hole has deepened more dramatically over the years, but the approximate geographical configuration has not really changed.

6.7.1.2 Wave One Structure

There appears a north–south wave structure in the total ozone field. The ozone is always higher at 60°S than it is over Antarctica (80°S to the pole). October fields are dominated by a single low in the south Atlantic region, and a single high near 150°W. This high/low structure is known as a wave one pattern. The wave one pattern has its maximum amplitude near 60°S and it falls off to a near-zero amplitude near 40°S.

The global total ozone amounts are low over Antarctica, show a midlatitude maximum, and are low in the tropics. Prior to 1980, the October average amounts in the polar region were greater than 280 DU. These amounts have decreased in the late 1980s and 1990s to about 120 DU (see Fig. 6.23).

6.7.1.3 Mobility and Eastward Rotation

The ozone hole tends to be highly dynamic and develops elongation of the hole that slowly rotates eastward. For example, an ozone hole that is generally centered

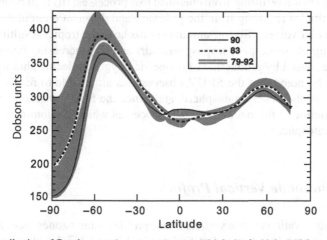

Fig. 6.23 Distribution of October zonal mean total ozone with latitude (Adapted from WMO 2007)

over the pole on a day in October is elongated toward the tip of South America. By the following day, this elongation axis is oriented toward the South Atlantic. The hole continues to rotate in a clockwise sense over a period of about 8 days. Such elongations result in quasi-periodic passages of extremely low ozone amounts over sites on the edge of the hole, such as the Antarctic Peninsula.

6.7.2 Vertical Ozone Structure

The vertical structure of the ozone hole can be studied by examining how the ozone varies with height by the different ways it is measured, such as ozone mixing ratio, ozone density, and partial pressure (the fraction of air pressure due to the presence of ozone molecules alone). The results from these different ways of measuring ozone allow us to see important features in the vertical distribution of stratospheric ozone.

6.7.2.1 Mixing Ratio and Density

Ozone is produced in the stratosphere via the photolysis of oxygen molecules by energetic UV light. As a result, ozone mixing ratios are largest in the tropical middle stratosphere. The Brewer-Dobson circulation transports these high ozone concentrations through the winter hemisphere, toward the pole, descending from the upper to the middle and lower stratosphere.

Zonal mean graph of ozone observations from the Nimbus-7 SBUV instrument in October 1987 is shown in Fig. 6.24a for the ozone mixing ratio and Fig. 6.24b for the ozone density. Figure 6.24a shows that most of the ozone is contained in the lower stratosphere between 18 and 28 km. Over Antarctica, both the ozone density and mixing ratios were extremely low during 1987. The important features of these Nimbus-7 SBUV images are (i) low ozone amounts in the lower stratosphere over the Antarctic region resulting from chemical loss processes; (ii) high ozone amounts in the midlatitudes resulting from the poleward and downward circulation near the edge of the polar vortex; (iii) low amounts of ozone in the tropics resulting from the upward lifting of ozone-poor tropospheric air; and (iv) increasing ozone amounts with altitude caused by production of ozone via oxygen molecule photolysis.

It should be noted that the SBUV observations are unable to resolve the vertical structure of the lower stratosphere, and hence the SBUV instrument, is not an adequate monitor of the ozone hole loss processes which predominantly occur in the lower stratosphere.

6.7.3 Ozonesonde Vertical Profiles

Long-term observations of the vertical ozone profiles using ozonesondes at the south pole (90°S) during October and at Sodankyla, Finland (67°N), during March are

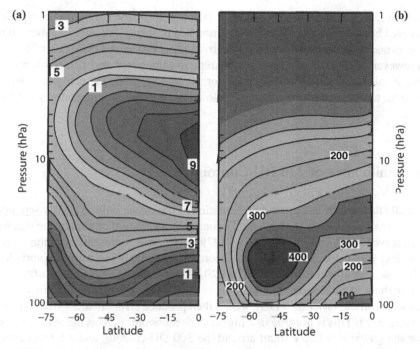

Fig. 6.24 Zonal mean image of ozone observations from the Nimbus-7 SBUV instrument in October 1987 (Adapted from NASA 2003)

shown in Fig. 6.25. It is evident in Fig. 6.25 that the ozone hole is largely confined to the 14–22 km region, and virtually more than 95% of the ozone was destroyed in this region during the 1992–2001 decade. Before the decade of the appearance of the ozone hole (1962–1971), maximum concentration of ozone over Antarctica

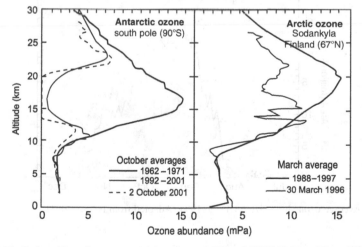

Fig. 6.25 Vertical profiles of ozone partial pressure at 90°S and 67°N (Adapted from WMO 2007)

was between 14 and 18 km region. in 2 October 2001, the ozone was completely destroyed between 14 and 20 km over Antarctica in spring. Average October values in the ozone layer were reduced by 90% from pre-1980 values (WMO 2007).

However, March ozone values in the Arctic in some years are often below normal average values as shown in Fig. 6.25 for 30 March 1996. In such years, winter minimum temperatures are generally below PSC formation temperatures for long periods.

6.8 Significance of 220 DU Contour

The 220 DU contour is a good representation of the ozone hole. It clearly separates the low total ozone from the high total ozone. It is an amount of total ozone that was not observed over Antarctica prior to 1979, and hence represents a region of real ozone loss with respect to the historic record. It is relatively insensitive to variations in absolute instrument calibration. The 220 DU typically exists within a sufficiently tight gradient of total ozone that the effects of calibration errors in instrument measurements (on the order of 5 DU), which then produce errors in areal size estimate, are minimized. This is not true of a higher ozone amount, such as the 300 DU level, since the gradient is fairly small around the 300 DU contour, and a 5 DU calibration error can easily produce large errors in the estimate of area inside the 300 DU contour (Newman 2003).

Figure 6.26 illustrates the average Antarctic ozone hole for the period 1979–1994 and its evolution from July to December. The black line represent the area inside the 220 DU ozone contour for 1992 and the black dots represents the area inside the 220

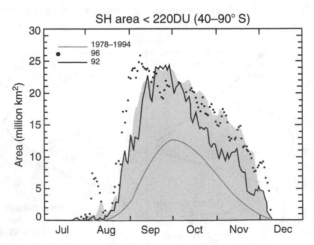

Fig. 6.26 Average ozone hole size for 1979–1994 (Adapted from NASA 2003)

DU ozone contour for 1996. The average ozone hole size for 1979–1994 is plotted as thin line, while the gray shading displays the range of area values observed on each day over the same 16-year period.

6.9 Severity of Ozone Depletion

The severity of the ozone depletion can be described from various metrics that capture different aspects of the ozone hole, such as ozone hole area, ozone minimum, ozone mass deficit, and date of ozone hole appearance and disappearance. Figure 6.27a, b, and c displays ozone hole area, ozone hole minimum, and ozone mass deficit, respectively.

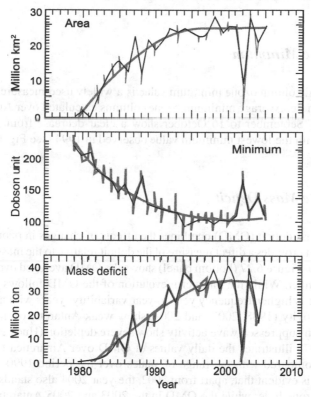

Fig. 6.27 Area of the Antarctic ozone hole for 1979–2005 (top); an average of daily minimum ozone values over Antarctica during the period from 21 September to 16 October (middle); ozone mass deficit (OMD) average over the 21–30 September period (Adapted from WMO 2007)

6.9.1 Ozone Hole Area

Ozone hole area is a primary estimate of the severity of the Antarctic ozone hole based on its geographic area (WMO 2007). The ozone hole area calculated from the area contained by total column ozone values less than 220 DU during 21–30 September. The value 220 DU was chosen to define the ozone hole because it is almost always a middle value in a strong gradient of total ozone, and because it is lower than pre-1980 observed ozone values.

The average ozone hole area currently reaches approximately 25 million square kilometers each spring, while the single largest daily value reached nearly 30 million square kilometers in September 2000 (Schoeberl et al. 1996; Newman et al. 2004). Ozone depletion can be enhanced by volcanic perturbations as could be seen in the very strong ozone depletion (deep ozone holes) observed in the 1990s after the Mt. Pinatubo eruption in 1994. The area growth of the ozone hole slowed during the mid-1990s, with a large dip in 2002 (see Fig. 6.27a).

6.9.2 Ozone Minimum

The daily total column ozone minimum value is a widely used measure of the state of the ozone hole. Average minimum ozone columns calculated over Antarctica for the period 21 September to 16 October show a clear decrease from 1979 to the mid-1990s, with the lowest minimum value observed in 1994 (see Fig. 6.27b).

6.9.3 Ozone Mass Deficit

The ozone mass deficit (OMD) combines the effects of changes in ozone hole area and depth, and provides a direct measure of the deficit relative to the mass present for a value of 220. Figure 6.27 (bottom panel) shows the OMD averaged over the period 21–30 September. While the long-term evolution of the OMD follows the halogen loading, there is higher-frequency year-to-year variability; years with anomalously high wave activity (1988, 2002, and 2004) show weak Antarctic ozone depletion, and years with suppressed wave activity show severe depletion (Huck et al. 2005).

Figure 6.27c illustrates the daily values of OMD over Antarctica for the years 2002–2005 compared with the range of values over the period 1990–2001. From Fig. 6.27c, it is evident that, apart from 2002, the year 2004 also stands out as having a weak ozone hole, while the OMD in the 2003 and 2005 Antarctic winters is more comparable to those observed during the 1990s. Although lower stratospheric (50 hPa) minimum temperatures were below average over much of the 2004 Antarctic winter, they increased and remained near-average after mid-August (Santee et al. 2005). The lower stratosphere warmed rapidly in September, stopping further heterogeneous processing of vortex air by the end of the month. This resulted in a slow increase in OMD and a leveling off in late September. At the end of September, a large increase in mixing accompanied by a weakening vortex transport barrier signaled vortex erosion, leading to the breakup.

6.10 Annual Cycle of Antarctic Ozone

Global distribution of annual cycle of total ozone is displayed in Fig. 6.28. The figure shows Nimbus-7 TOMS total ozone as a function of time and latitude. The data are averaged over both time (1979–1992) and longitude. The gaps seen in the polar regions during the winter months result from the inability of TOMS to make measurements during polar night, since the TOMS observations require solar UV light for its ozone measurement technique (Newman 2003).

Figure 6.28 shows ozone amounts over the course of an entire year for all latitudes outside of the polar night. As it is based on 14 years of data, one can see whether the average amount of ozone changes over the course of the year at different latitudes. It is evident from Fig. 6.28 that the ozone amounts vary little over the course of the year in the tropics, while they vary considerably over both polar regions.

Ozone amounts are extremely low over Antarctica during October, with a collar of high ozone just north of there in the 40°–70°S region, and relatively low amounts throughout the tropics. The southern hemisphere ozone high collar region is almost always present, though the amounts decrease in the southern summer. It reaches its highest amounts in late October as a result of the continual accumulation of ozone in the lower stratosphere that is driven by the poleward and downward transport of the Brewer-Dobson cell.

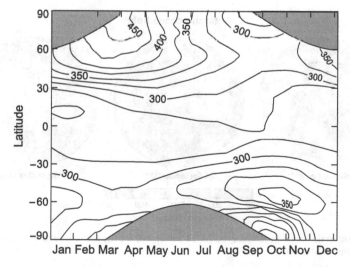

Fig. 6.28 Latitudinal distribution of annual total ozone (Adapted from NASA 2003)

6.11 Anomalous 2002 Antarctic Ozone Hole

The anomalous Antarctic ozone hole which occurred in 2002 showed features that astonished the scientific community. In this year, the hole had much less area and much less ozone depletion when compared with values in several preceding years. Figure 6.29 displays the ozone hole area and minimum ozone values of the 2001, 2002, and 2003 of the Antarctic ozone hole. The smaller area noted in 2002 is unexpected because the conditions required to deplete ozone, namely low temperatures and available reactive halogen gases, were not expected to have large year-to-year variations.

In 2002, the ozone was depleted in August and early September, but the hole broke apart into two separate depleted regions during the last week of September (see Fig. 6.30). The depletion in these two regions is significantly less than was observed inside either the 2001 or 2003 ozone holes, but still substantially greater than that observed in the early 1980s. The anomalous behavior in 2002 occurred because of specific atmospheric air motions that sometimes occur in polar regions, not large decreases in reactive chlorine and bromine amounts in the Antarctic stratosphere.

The Antarctic stratosphere was warmed by very strong, large-scale weather systems in 2002 that originated in the lower atmosphere (troposphere) at midlatitudes in late September. During this period, Antarctic temperatures are generally very low and ozone destruction rates are near their peak values. These tropospheric systems traveled poleward and upward into the stratosphere, upsetting the circumpolar wind flow and warming the lower stratosphere where ozone depletion was in progress. The higher than normal impact of these weather disturbances during the critical time period for ozone loss reduced the total loss of ozone in 2002.

Large Antarctic ozone depletion returned in 2003 through 2005, in a manner similar to that observed from the mid-1990s to 2001. The high ozone depletion

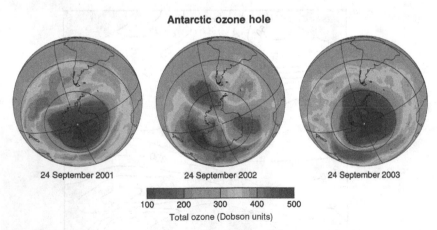

Fig. 6.29 Antarctic ozone hole observed on 24 September for the previous and next years of 2002 (Adapted from WMO 2007)

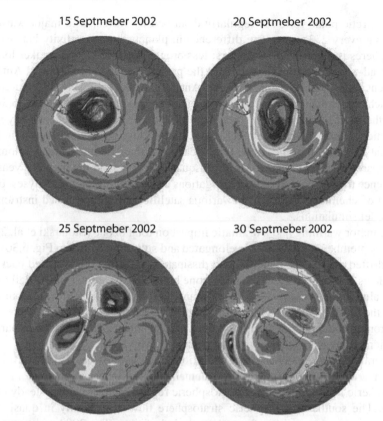

Fig. 6.30 TOMS total ozone maps for 4 days during September and October 2002. The white space around the south pole is polar night, where no measurements are made (Adapted from Stolarski et al. 2005, Courtesy: American Meteorological Society)

found since the mid-1990s, with the exception of 2002, is expected to be typical of coming years. A significant, sustained reduction of Antarctic ozone depletion, defined as ozone recovery, requires the removal of halogen from the stratosphere.

6.11.1 Major Stratospheric Warming in Southern Hemisphere

For the first time in the history of observational records, a major stratospheric warming occurred in the southern hemisphere (SH) in September 2002, which broke the polar vortex and affected the ozone hole (WMO 2007). This major warming caused dramatic stratospheric circulation changes. Warmings normally occur in the northern hemisphere due to the effect of planetary waves propagating up from the troposphere.

The Arctic polar vortex is regularly disturbed by waves, with major warmings occurring every 2–3 years. The difference in planetary wave activity between the hemispheres is due to various factors: less orographic forcing and weaker longitudinal land–sea contrast in the SH, and the presence of the cold elevated Antarctic continent at the pole. As a consequence, Antarctic winter stratospheric temperatures are much colder than the Arctic and exhibit much less variability (see Fig. 6.30).

Furthermore, temperature records from 1940 do not show evidence of any major Antarctic warmings (Naujokat and Roscoe 2005). This unprecedented event induced a dramatic reduction of the ozone hole area to less than 5 million square kilometers as compared with more than 20 million square kilometers in the previous years. Its occurrence triggered numerous investigations using meteorological analyses, observations of chemical species from various satellite and ground-based instruments, and model simulations.

The major warming had a dramatic impact on total ozone (Stolarski et al. 2005). On 23 September, the ozone hole elongated and split in two pieces (Fig. 6.30). One piece drifted over South America and dissipated, while the other drifted back over the pole as a significantly weakened ozone hole. The 2002 total ozone daily minimum value did not reach values lower than 150 DU, as compared with around 100 DU in the preceding years. Higher total ozone values were observed in the polar region from mid-September to mid-October. Ozone hole metrics all show remarkable deviations from averages over the last decade (WMO 2007).

Several studies examined the conditions that led to the major warming in 2002 in order to explain this unprecedented event. Numerical simulations of the stratospheric flow show distinct stratospheric regimes that are either steady or wavering. The southern hemispheric stratosphere flow is generally in quasi-steady state or in wavering regimes for a short period of time. The 2002 winter was in a wavering regime beginning in June. This vacillation induced a systematic weakening of the polar night jet that ultimately allowed a strong pulse of planetary wave to propagate into the stratosphere.

The warming in 2002 was unprecedented in Antarctic meteorological observations. Warming events are difficult to predict because of their complex formation conditions.

6.12 Arctic Ozone Hole

Arctic stratosphere also shows significant ozone depletion in recent years in the late winter or in early spring (February–March). But the intensity of ozone depletion is not as severe as that observed in Antarctic. Arctic ozone depletion is more variable from year to year. Even though some reduction in Arctic ozone is observed in every winter/late spring, a complete depletion each year over a broad vertical layer is not found in the Arctic.

The ozone content in the Arctic stratosphere is dependent on chemical and dynamical conditions and shows great interannual variability. The variability in

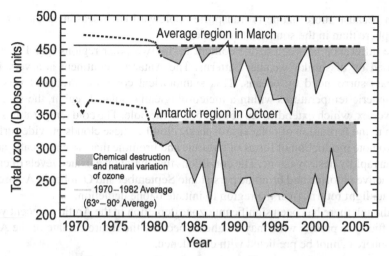

Fig. 6.31 Springtime mean total ozone over Arctic and Antarctic (Adapted from WMO 2007)

amplitude of PSC formation potential, derived from stratospheric temperature analyses, in the Arctic has increased over the last 40 years, with 1995/1996, 1999/2000, and 2004/2005 standing out as cold winters with high PSC formation potential, whereas 1998/1999, 2001/2002, and 2003/2004 were warm winters with very low PSC formation potential. This variability in dynamical conditions has a correlated impact on ozone transport and chemistry and is therefore reflected in the total ozone column over the polar regions.

Figure 6.31 shows a comparison of average springtime total ozone values found in Arctic and Antarctic regions for the last three decades based on satellite and ground-based observations. Decreases from the pre-ozone hole average values (1970–1982) were observed in the Arctic beginning in the 1980s, even though the decreases were not large, when similar changes were occurring in the Antarctic. Each point represents the monthly mean average in March in the Arctic and October in the Antarctic. The decrease in ozone in the Arctic region, which was 5–6% in the 1980s has reached a maximum of about 30% in 1995 but remained smaller than those found in the Antarctic since the mid-1980s. It can be seen that significant ozone depletion as occurred in most of the years in the Arctic and all years in the Antarctic after 1982. The largest average depletion of ozone has occurred in the Antarctic since 1990. Changes in ozone are due to the combination of both chemical rearrangements and natural variations.

The interannual changes in the Arctic and Antarctic average ozone values reflect annual variations in meteorological conditions that affect the extent of low polar temperatures and the transport of air into and out of the polar stratosphere. Essentially all of the decrease in the Antarctic and usually most of the decrease in the Arctic each year are attributable to chemical destruction by reactive halogen gases. Average total ozone values over the Arctic are larger at the beginning of each winter season because more ozone is transported poleward each season in the northern hemisphere than in the southern hemisphere.

The difference between ozone content in the two polar regions (see Fig. 6.31) is caused by dissimilar weather patterns. The Antarctic continent is a very large landmass surrounded by oceans. This symmetrical condition produces very low stratospheric temperatures within a meteorologically isolated region, the so-called polar vortex, which extends from about 65°S to the pole. The cold temperatures lead in turn to the formation of polar stratospheric clouds. These clouds provide surfaces that promote production of forms of chlorine and bromine that are chemically active and can rapidly destroy ozone. The conditions that maintain elevated levels of chemically active chlorine and bromine persist into September and October in Antarctica, when sunlight returns over the region to initiate ozone depletion.

Although there has been significant ozone depletion in the Arctic in recent years, it is difficult to predict what may lie ahead, because the future climate of the Arctic stratosphere cannot be predicted with confidence.

6.13 Montreal Protocol

In 1985, a treaty called the Vienna Convention for the Protection of the Ozone Layer was signed by 20 nations in Vienna. The signing nations agreed to take appropriate measures to protect the ozone layer from human activities. The Vienna Convention supported research, exchange of information, and future protocols. In response to growing concern, the Montreal Protocol on Substances that Deplete the Ozone Layer was signed in 1987 and, following country ratification, entered into force in 1989. The Protocol established legally binding controls for developed and developing nations on the production and consumption of halogen source gases known to cause ozone depletion. National consumption of a halogen gas is defined as the amount that production and imports of a gas exceed its export to other nations.

6.13.1 Amendments and Adjustments

As the scientific basis of ozone depletion became more certain after 1987 and substitutes and alternatives became available to replace the principal halogen source gases, the Montreal Protocol was strengthened with Amendments and adjustments. These revisions put additional substances under regulation, accelerated existing control measures, and prescribed phaseout dates for the production and consumption of certain gases. The initial Protocol called for only a slowing of chlorofluorocarbon (CFC) and halon production. The 1990 London Amendments to the Protocol called for a phaseout of the production and consumption of the most damaging ozone-depleting substances in developed nations by 2000 and in developing nations by 2010. The 1992 Copenhagen Amendments accelerated the date of the phaseout to 1996 in developed nations. Further controls on ozone-depleting substances were agreed upon in later meetings in Vienna (1995), Montreal (1997), and Beijing (1999).

6.13.2 Montreal Protocol Projections

Future stratospheric abundances of effective stratospheric can be calculated based on the provisions of the Montreal Protocol. The concept of effective stratospheric chlorine accounts for the combined effect on ozone of chlorine- and bromine-containing gases. The results are shown in Fig. 6.32 for the cases: (i) no Protocol and continued production increases of 3% per year; (ii) continued production and consumption as allowed by the Protocols original provisions agreed upon in Montreal in 1987; (iii) restricted production and consumption as outlined in the subsequent Amendments and adjustments as decided in London in 1990, Copenhagen in 1992, and Beijing in 1999; and (iv) zero emissions of ozone-depleting gases starting in 2007.

Figure 6.32 shows past and projected stratospheric abundances of chlorine and bromine without the Protocol, under the Protocol's original provisions, and under its

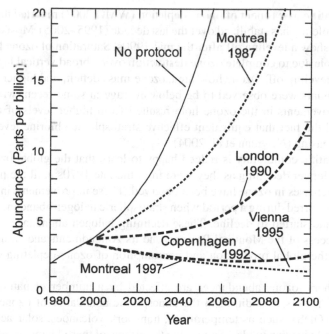

Fig. 6.32 Effect of the International Agreements on ozone-depleting stratospheric chlorine/bromine (Adapted from WMO 2007)

subsequent agreements. Without the Montreal Protocol and its Amendments, continuing use of chlorofluorocarbons (CFCs) and other ozone-depleting substances would have increased the stratospheric abundances of chlorine and bromine tenfold by the mid-2050s compared with the 1980 amounts. Such high chlorine and bromine abundances would have caused very large ozone losses, which would have been far larger than the depletion observed at present.

In 1987, the recognition of the potential for chlorine and bromine to destroy stratospheric ozone led to the Montreal Protocol on substances that deplete the ozone layer, as part of the 1985 Vienna Convention for the Protection of the Ozone Layer, to reduce the global production of ozone-depleting substances. Subsequently, global observations of significant ozone depletion have prompted amendments to strengthen the treaty. The 1990 London Amendment calls for a ban on the production of the most damaging ozone-depleting substances by 2000 in developed countries and by 2010 in developing countries. The 1992 Copenhagen Amendment changed the date of the ban to 1996 in developed countries. Further restrictions on ozone-depleting substances have been agreed upon in Vienna (1995) and Montreal (1997).

6.14 Present Status of Ozone Depletion

Recent scientific assessment of ozone depletion (WMO 2007) reported that Antarctic ozone depletion has stabilized over the last decade (1995–2005). Most ozone hole diagnostics show a leveling off after the mid-1990s. Saturation of ozone loss inside the ozone hole due to complete ozone destruction over a broad vertical layer plays a role in this leveling off. Ozone hole area, ozone mass deficit, and higher minimum column amounts were observed to be below average in some recent winter years. These improvements in the ozone hole resulted from higher levels of dynamical forcing, and the fact that equivalent effective stratospheric chlorine levels are not further decreasing (Newman et al. 2004).

The scientific community is indeed happy to learn that the global ozone levels are now no longer declining as they were from the late 1970s until the mid-1990s, and some increases in ozone have been observed. These improvements in the ozone layer have occurred during a period when stratospheric halogen abundances reached their peak and started to decline. These declining halogen abundances clearly reflect the success of the Montreal Protocol and its Amendments and adjustments in controlling the global production and consumption of ozone-depleting substances (ODSs).

Stratospheric ozone abundances are affected by a number of natural and anthropogenic factors in addition to the atmospheric abundance of ozone-depleting substances (ODS), such as temperatures, transport, volcanoes, solar activity, and hydrogen and nitrogen oxides. Separating the effects of these factors is complex because of nonlinearities and feedbacks in the atmospheric processes affecting ozone.

Figure 6.33 describes the time series of various factors and the total ozone for the last four decades.

Global mean total ozone column values for the period (2002–2005) were approximately 3.5% below compared to the 1964–1980 average values. However, 2002–2005 mean values are similar to that of the 1998–2001, indicating that ozone is no longer decreasing. These changes are seen uniformly in all available global datasets.

Total amount of ozone over the tropics (25°–25°S) for the past 25 years remains unchanged. Averaged total ozone for the period 2002–2005 for the northern hemisphere and southern hemisphere midlatitudes (35°–60°) are comparable to the 1998–2001 mean values, but about 3% and 5.5% respectively lower than the 1964–1980 average values. Changes in the seasonal ozone since pre-1980 period over northern midlatitudes are larger in spring, while those southern mid-latitudes are nearly the same throughout the year (WMO 2007).

In the vertical, upper and lower stratospheric ozone declined during 1979–1995, but has been relatively constant during the last decade. The net ozone decrease was ∼10–15% over midlatitudes; smaller but significant changes occurred over the tropics. Available observational evidences, such as Umkehr, lidar, and microwave ozone measurements along with stratospheric aerosol and gas experiment (SAGE I+II) and solar backscatter ultraviolet (SBUV(/2)) satellite instruments confirm these findings.

In the lowermost stratosphere, between 12 and 15 km in the northern hemisphere, a strong decrease in ozone was observed from ozonesonde data between 1979 and 1995, followed by an overall increase from 1996 to 2004, leading to no net long-term decrease at this level. These changes in the lowermost stratosphere have a substantial influence on the column. The southern hemisphere midlatitude data do not show a similar increase since 1995 at these altitudes.

Significant ozone decreases of ∼3% between the tropopause and 25 km are found in the SAGE satellite measurements between 1979 and 2004 at 25°S–25°N. Since no change is found in total ozone over the tropics, this could be explained by significant increases in tropospheric ozone in the tropics. While regional increases in tropical tropospheric ozone have been seen, not all tropical regions or datasets show an increase.

6.15 Ozone Layer in Future

The process of recovery of the ozone layer will depend not only on the decline of ozone-depleting substances, but also on many other factors. Although some of these factors can be accounted for empirically, projecting future ozone coupling between the different chemical, dynamical, and radiative processes involved requires the use of models that include these interdependencies to make well-founded projections. In particular, models used for prognostic studies should incorporate the effects of changes in temperature and transport that are likely to occur due to the rise in concentrations of greenhouse gases.

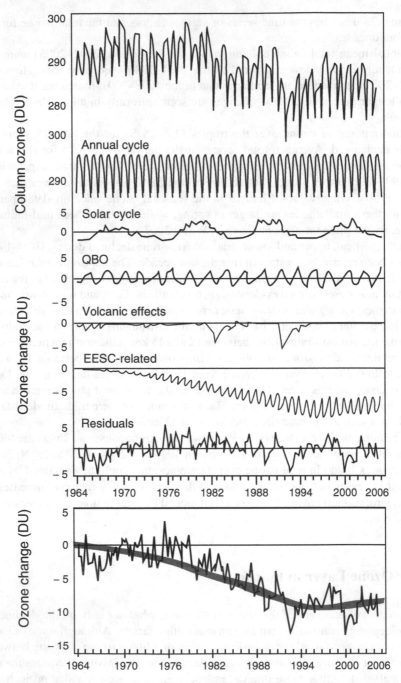

Fig. 6.33 Ozone variations for 60°S–60°N from 1964–2005 and deseasonalized, area-weighted total ozone deviations adjusted for solar, volcanic, and QBO effects (Adapted from WMO 2007)

The projections of total column ozone are examined for three periods: (i) the beginning of the century (2000–2020), when EESC is expected to start to decrease or continue to decrease; (ii) mid-century (2040–2050 in extrapolar regions, 2060–2070 in polar regions), when EESC is expected to fall below 1980 values; (iii) end of the century (2090–2100), when factors other than ODSs are expected to control stratospheric ozone.

Confidence in projections near the beginning of the century is higher than near the middle or end of the century because the former can be supported by observations, empirical studies, and extrapolations, while the latter are more influenced by uncertainties in the emission scenarios and other boundary conditions. In general, a separation of the different factors contributing to the ozone variability in the model output has not been performed. Therefore the modeled ozone time series cannot be used for attribution of ozone changes to variations in ODSs.

The expected slow improvement of Antarctic ozone over the next decade means that variability will continue to complicate detection of the first and second stages of ozone hole recovery, even after accounting for temperature variations. The return of (i) the relationship between temperature and ozone, and (ii) the variance in ozone abundances to historical values may provide early signals of the beginning of recovery inside the ozone hole. However, after a stage of recovery has occurred, it is unclear how long it will take to achieve the detection of the stage using either diagnostic.

6.15.1 Recovery Stages of Global Ozone

The recovery process for global ozone is schematically shown in Fig. 6.34. As the overall decline in ODS and gases continues in response to Montreal Protocol provisions, global ozone is expected to recover, approaching or exceeding pre-1980 values. Ozone-recovery attributable to decreases in ozone-depleting gases can be described as a process involving three stages: (1) the initial slowing of ozone decline, identified as the occurrence of a statistically significant reduction in the rate of decline in ozone; (2) the onset of ozone increases (turn around), identified as the occurrence of statistically significant increases in ozone above previous minimum values; and (3) the full recovery of ozone from ozone-depleting gases, identified as when ozone is no longer significantly affected by ozone-depleting gases from human activities.

6.15.2 Natural Factors

Stratospheric ozone is influenced by two important natural factors, namely, changes in the output of the Sun and volcanic eruptions. The solar effect on ozone is expected

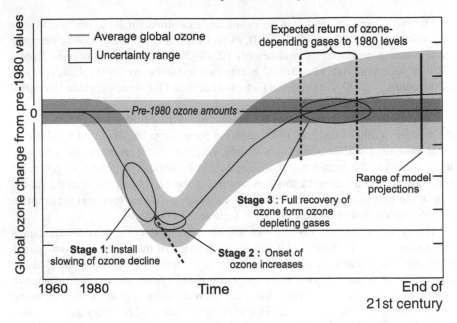

Fig. 6.34 Recovery stages of global ozone in the next 100 years (Adapted from WMO 2007)

to be predictable based on the 11-year cycle of solar output. Volcanic eruptions are particularly important because they enhance ozone depletion caused by reactive halogen gases, but cannot be predicted. The occurrence of a large volcanic eruption in the next decades when effective stratospheric chlorine levels are still high may obscure progress in overall ozone recovery by temporarily increasing ozone depletion.

The ozone depletion caused by human-produced chlorine and bromine compounds is expected to gradually disappear by about the middle of the 21st century as these compounds are slowly removed from the stratosphere by natural processes. This environmental achievement is due to the landmark of international agreement to control the production and use of ozone-depleting substances.

However, the future state of the ozone layer depends on more factors than just the stratospheric concentrations of human-produced chlorine and bromine. It will also be affected to some extent by the changing atmospheric abundances of several other human-influenced constituents, such as methane, nitrous oxide, and sulfate particles, as well as by the changing climate of the Earth. As a result, the ozone layer is unlikely to be identical to the ozone layer that existed prior to the 1980s. Nevertheless, the discovery and characterization of the issue of ozone depletion from chlorine and bromine compounds and a full global compliance with the international regulations on their emissions will have eliminated what would have been a major deterioration of the Earth's protective ultraviolet shield.

Questions and Exercises

6.1. Discuss the principal steps in the depletion of stratospheric ozone by anthropogenic activities.

6.2. Stratospheric ozone is called as 'good ozone' and tropospheric ozone is known as 'bad ozone'. Why? What happens when tropospheric ozone increases?

6.3. What are natural sources of halogens to the atmosphere? Are these more or less important than industrial sources? Explain

6.4. What is the Antarctic ozone hole? How is it defined? How does it compare to the annual spring low total ozone amounts over Antarctica?

6.5. When did the ozone hole first appear? How big is the ozone hole, and is it getting bigger? Will the ozone hole keep growing?

6.6. Describe the characteristics of the Antarctic ozone hole in respect of the longitudinal and latitudinal symmetry, spring-time ozone amounts as a function of time and its mobility with respect to the pole.

6.7. In assessing the severity of the Antarctic ozone loss surface areas with ozone amounts less than 220 DU can be monitored. Why is the 220 DU contour a reasonable delineation of the ozone hole?

6.8. Discuss the effect of solar activity and volcanic eruptions on ozone depletion.

6.9. What emissions from human activities lead to ozone depletion? Discuss the reactive halogen gases that destroy stratospheric ozone.

6.10. Chloroflurocarbons and other halogen source gases are heavier than air and they are produced in the troposphere. How they reach the stratosphere and deplete ozone in this region?

6.11. Why has ozone hole appeared over the Antarctica (southern hemispheric polar region) when ozone depleting gases are mainly produced from countries in the northern hemisphere?

6.12. Ozone in the stratosphere is destroyed due to human activities. Whether human could replace the loss of global stratospheric ozone by producing ozone and transporting into the stratosphere? If so, what are the major difficulties and consequences?

6.13. What are the characteristics of the polar night jet and polar vortex. Explain the impact of the polar night jet on stratospheric polar ozone concentrations?

6.14. Why do extremely cold temperatures develop in the polar vortex during the winter season? Do the coldest temperatures appear at all altitudes at the same time during the winter season? Explain.

6.15. Why would you expect the Antarctic polar vortex to be highly contained? What evidence shows this containment? What impact does the potential vorticity have on the ozone hole?

References

Brasseur G, Solomon S (2005) Aeronomy of the Middle Atmosphere, 2nd edition, Springer, Dordrecht

Blumenstock T, Kopp G, Hase F, Hochschild G, Mikuteit S, Raffalski U, Ruhnke R (2006) Observation of unusual chlorine activation by ground-based infrared and microwave spectroscopy in the late Arctic winter 2000/01, Atmos Chem Phys, 6: 897–905

Callis, LB, Natarajan M (1986) Ozone and nitrogen dioxide changes in the stratosphere during 1979–84, J Geophys Res, 91:10771–10796

Dameris M, Grewe V, Ponater M, Deckert R, Eyring V, Mager F, Matthes S, Schnadt C, Stenke A, Steil B, Bruhl C, Giorgetta MA (2005) Long-term changes and variability in a transient simulation with a chemistry climate model employing realistic forcing, Atmos Chem Phys, 5: 2121–2145

Eduspace, The ozone hole, European Space Agency (http://eduspace.esa/int/subdocument/images/ozone-gen.jpg)

Huck PE, McDonald AJ, Bodeker GE, Struthers H (2005) Interannual variability in Antarctic ozone depletion controlled by planetary waves and polar temperatures, Geophys Res Lett, 32: doi 10.1029/2005GL022943

Farman JC, Gardiner BG, Shanklin JD (1985) Large losses of total ozone in Antarctica reveal seasonal ClOx/NOx interaction, Nature, 315: 207–210

Fromm M, Bevilacqua R, Servranckx R, Rosen J, Thayer JP, Herman J, Larko D (2005) Pyrocumulonimbus injection of smoke to the stratosphere: Observations and impact of a super blowup in northwestern Canada on 3–4 August 1998, J Geophys Res, 110: D08205, doi 10.1029/2004-JD005350

Kodera K, Kuroda Y (2002) Dynamical response to the solar cycle, J Geophys Res, 107: 4749, doi 10.1029/2002JD002224

Labitzke K, Kunze M (2005) Stratospheric temperatures over the Arctic: Comparison of three data sets, Meteorol Z, 14: 65–74

Langematz U, Kunze M (2006) An update on dynamical changes in the Arctic and Antarctic stratospheric polar vortices, Clim Dyn, 27: 647–660, doi 10.1007/s00382-006-0156-2

Maduro RA, Schauerhammer R (1992) The holes in the ozone scare: the scientific evidence that the sky is not falling, In 21st Century Science Associates, John Wiley & Sons, Washington DC

NASA (2003) Studying Earth's Environment from Space (http://www.ccpo.odu.edu/SEES/index.html)

Naujokat B, Roscoe HK (2005) Evidence against an Antarctic stratospheric vortex split during the periods of pre-IGY temperature measurements, J Atmos Sci, 62: 885–889

Newman PA (2003) The Antarctic ozone hole, Chapter 11: Stratospheric Ozone: An Electronic Text, NASA, GSFC

Newman PA, Kawa SR, Nash ER (2004) On the size of the Antarctic ozone hole, Geophys Res Lett, 31: doi 10.1029/2004GL020596

Newman PA, Nash SR, Kawa ER, Montzke SA (2006) When will the Antarctic ozone hole recover? Geophys Res Lett, 33: doi. 10.1029/2005GL025232

Pawson S, Labitzke K, Leder S (1998) Stepwise changes in stratospheric temperature, Geophys Res Lett, 25: 2157–2160

Reinsel GC, Miller AJ, Weatherhead EC, Flynn LE, Nagatani R, Tiao GC, Wuebbles DJ (2005) Trend analysis of total ozone data for turnaround and dynamical contributions, J Geophys Res, 110: D16306, doi 10.1029/2004JD004662

Rosenfield JE, Frith SM, Stolarski RS (2005) Version 8 SBUV ozone profile trends compared with trends from a zonally averaged chemical model, J Geophys Res, 110: D12302, doi 10.1029/-2004JD005466

Ramaswamy V, Chanin ML, Angell JK, Barnett J, Gaffen D, Gelman ME, Keckhut P, Koshelkov Y, Labitzke K, Lin JJR, ONeill A., Nash J, Randel WJ, Rood R, Shine K, Shiotani M, Swinbank R (2001) Stratospheric temperature trends: Observations and model simulations, Rev Geophys, 39: 71–122

Ramaswamy V, Schwarzkopf MD, Randel WJ, Santer BD, Soden BJ, Stenchikov GL (2006) Anthropogenic and natural influences in the evolution of lower stratospheric cooling, Science, 311: 1138–1141

Rosenfield JE, Douglass AR, Considine DB (2002) The impact of increasing carbon dioxide on ozone recovery, J Geophys Res, 107: 4049, doi 10.1029/2001JD000974

Rowland RF (2007) Stratospheric ozone depletion by chlorofluorocarbons (Nobel Lecture), Appeared in Encyclopedia of Earth (ed; C. J. Cleveland)

Santee MI, Manney GL, Livesey NJ, Froidevaux L, MacKenzie IA, Pumphrey HC, Read WG, Schwartz MJ, Waters JW, Harwood RS (2005) Polar processing and development of the 2004 Antarctic ozone hole: First results from MLS on aura, Geophys Res Lett, 32, doi 10.1029/2005GL022582

Schoeberl MR, Douglass AR, Kawa SR, Dessler A, Newman PA, Stolarski RS, Roche AE, Waters JW, Russel III JM (1996) Development of the Antarctic ozone hole, J Geophys Res, 101: 20909

Schoeberl MR, Kawa SR, Douglass AR, McGee TJ, Browell EV, Waters J, Livesey N, Read W, Froidevaux L, Santee ML, Pumphrey HC, Lait LR, Twigg L (2006) Chemical observations of a polar vortex intrusion, J Geophys Res, 111: D20306, doi 10.1029/2006JD007134

Solomon S, Garcia RR, Rowland FS, Wuebbles DJ (2005a) On the depletion of Antarctic ozone, Nature, 321: 755–758

Solomon S, Portmann RW, Sasaki T, Hofmann DJ, Thompson DWJ (2005b) Four decades of ozonesonde measurements over Antarctica, J Geophys Res 110: D21311, doi 10.1029/2005JD005917

Steinbrecht W, Hassler B, Bruhl C, Dameris M, Giorgetta MA, Grewe V, Manzini E, Matthes S, Schnadt C, Steil B, Winkler P (2006) Interannual variation pattern of total ozone and lower stratospheric temperature in observations and model simulations, Atmos Chem Phys, 6: 349–374

Stolarski RS, McPeters RD, Newman PA (2005) The ozone hole of 2002 as measured by TOMS, J Atmos Sci, 62: 716–720 (http://www.eoearth.org/Rowland_nobel_lecture_fig05.gif)

Stolarski RS, Douglass AR, Steenrod S, Pawson S (2006) Trends in stratospheric ozone: Lessons learned from a 3D Chemical Transport Model, J Atmos Sci, 63: 1028–1041

Stolarski RS, Douglass AR, Newman PA, Pawson S, Schoeberl MR (2006) Relative contribution of greenhouse gases and ozone changes to temperature trends in the stratosphere: A chemistry-climate model study NASA Report Document ID: 20070008218

WMO (1995) Scientific Assessment of Ozone Depletion: 1994, Global Ozone Research and Monitoring Project Report No. 37, Geneva, Switzerland

WMO (World Meteorological Organization), Scientific Assessment of Ozone Depletion: 2006 (2007) Global Ozone Research and Monitoring Project Report No. 50, Geneva, Switzerland

Chapter 7
Transport Processes in the Stratosphere and Troposphere

7.1 Introduction

Transport of air across the tropopause plays an important role in determining the chemical composition, and hence radiative properties, of both the troposphere and stratosphere. Quantifying this transport presents a significant challenge on account of the many multiscale processes involved from the global scale mean meridional circulation, through intermediate advective and convective processes, to molecular diffusion. It has long been recognized that tropospheric air enters the stratosphere principally in the tropics, and moves poleward in the stratosphere.

To understand the large-scale circulation in the troposphere and stratosphere, it is useful to look at transport processes averaged around a latitude circle. Ozone production mainly takes place in the tropical stratosphere as the direct solar radiation photodissociates oxygen molecules (O_2) into oxygen atoms (O), which quickly react with other O_2 molecules to form ozone (O_3). But most ozone is found in the higher latitudes rather than in the tropics, i.e., outside of its natural tropical stratospheric source region. This higher-latitude ozone results from the slow atmospheric circulation that moves ozone from the tropics where it is produced into the middle and polar latitudes. This slow circulation is known as the Brewer-Dobson circulation.

7.2 The Brewer-Dobson Circulation

Figure 7.1 shows the zonally averaged circulation in the middle atmosphere superimposed on top of an annual average ozone density. The Brewer-Dobson circulation is represented by the thick arrows. The figure also shows the seasonally averaged ozone density from north pole to south pole.

The Brewer-Dobson circulation is a slow circulation pattern, first proposed by Brewer to explain the lack of water in the stratosphere. He presumed that water vapor is freeze-dried as it moves vertically through the cold equatorial tropopause

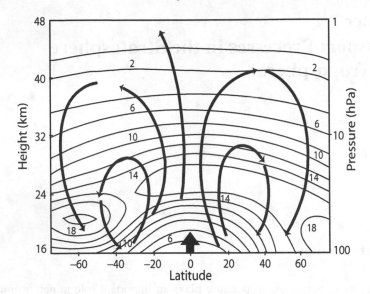

Fig. 7.1 Zonal mean middle atmospheric circulation and annual average ozone density (DU/km), from Nimbus-7 SBUV Observations during the period 1982–89 (Courtesy: NASA)

(see Fig. 7.1). Dehydration can occur in this region by condensation and precipitation as a result of cooling to temperatures below 193 K. The lowest values of water are found just near the tropical tropopause. Later Dobson suggested that this type of circulation could also explain the observed high ozone concentrations in the lower stratosphere polar regions which are situated far from the photochemical source region in the tropical middle stratosphere. The Brewer-Dobson circulation additionally explains the observed latitudinal distributions of long-lived constituents like nitrous oxide and methane.

This conceptual model has since been refined but not drastically altered. That Brewer-Dobson circulation is controlled by stratospheric wave drag, quantified by the Eliassen-Palm flux divergence, sometimes lays claim to the extratropical pump (as shown in Fig. 7.2), with the circulation at any level being controlled by the wave drag above that level. However, the wave drag can be difficult to compute accurately and it is common to diagnose the mean circulation from the diabatic heating. It is possible to estimate the net mass flux across a given isentropic surface from the diabatic heating.

On the other hand, transport of material along isentropic surfaces, such as that between the tropical upper troposphere and the lowermost stratosphere, is more difficult to quantify, especially the net transport of a given species that results from the two-way mixing. Observations show that the composition of the lowermost stratosphere varies with season, and suggest a seasonal dependence in the balance between the downward transport of air of stratospheric character and the horizontal transport of air of upper-troposphere character. For any time period, the integrated mass flux to the troposphere at middle and high latitudes is the sum of the mass flux

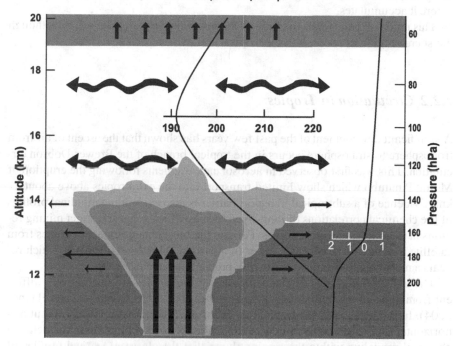

Fig. 7.2 Transport processes between stratosphere and troposphere by means of Brewer-Dobson circulation (Adapted from University of Washington)

across the 380 K potential temperature surface, the net mass transported between the tropical upper troposphere and the lowermost stratosphere, plus (minus) the mass decrease (increase) of the lowermost stratosphere (Appenzeller et al. 1996). The first quantity is easy to compute, but the last two quantities are sensitive to small-scale processes, including synoptic-scale disturbances and convection.

The classical picture of the stratosphere–troposphere coupling has evolved over the last few years. Such developments and tuning are essential for a good description of processes that are important for the stratosphere–troposphere coupling.

7.2.1 Low Ozone in Tropical Stratosphere

The Brewer-Dobson circulation consists of three basic parts. The first part is rising tropical motion from the troposphere into the stratosphere. The second part is pole-ward transport in the stratosphere. And the third part is descending motion in both the stratospheric middle and polar latitudes, though there are important differences. The middle-latitude descending air is transported back into the troposphere, while the polar-latitude descending air is transported into the polar lower stratosphere, where it accumulates.

This model explains why tropical air has lower ozone than polar air, even though the source region of ozone is in the tropics.

7.2.2 Circulation in Tropics

A significant development of the past few years has shown that the ascent of air from troposphere to stratosphere occurs in the tropical branch of the Brewer-Dobson circulation. This was first observed in aerosol measurements following the eruption of Mount Pinatubo which show limited transport into the extratropics above about 22 km. Evidence of a subtropical transport barrier is also evident in latitudinal profiles of the chemical correlations of long-lived species which suggest distinct mixing regions in the tropics and extratropics. Perhaps the most notable evidence comes from satellite measurements of water vapor in the tropical lower stratosphere, which reveals a meridionally confined, vertically propagating annual cycle.

The transport and mixing characteristics of the tropics are qualitatively different from those of the midlatitudes (Appenzeller et al. 1996, Bonazzola and Haynes 2004). In the ideal case, the tropical air is assumed to advect upward without any horizontal exchange to midlatitudes, and is called a tropical pipe. However, detailed chemical modeling within the tropics shows that the observed vertical profiles of chemical species cannot be reconciled with a strict interpretation of the tropical pipe; some dilution from midlatitudes is required, especially below 22 km. Several studies (Rind et al. 2001, Cordero et al. 2002) to quantify the amount of dilution, by balancing the dilution against vertical upwelling and chemical processing show that nearly 50% of the air in the tropical upwelling region around 22 km is of midlatitude origin.

Numerous transport processes involved in the chemistry and dynamics of the troposphere and stratosphere are schematically represented in Fig. 7.3. The air that is slowly lifted out of the tropical troposphere into the stratosphere is very dry, with low ozone, and high CFC levels. This tropical lifting circulation out of the lower stratosphere is quite slow, on the order of 20–30 m day^{-1}. Most of the air rising into the stratosphere at the tropopause never makes it into the upper stratosphere. Between 16 and 32 km, the air density decreases by about 90%. This means that of the mass coming into the stratosphere at 16 km, approximately 90% will move toward the middle latitudes rather than be carried up to 32 km.

7.2.3 Ozone Transport

Air in the troposphere has relatively low ozone concentrations, except in highly polluted urban environments. Even polluted regions have relatively low ozone concentrations when compared to stratospheric levels. As this ozone-clean air moves

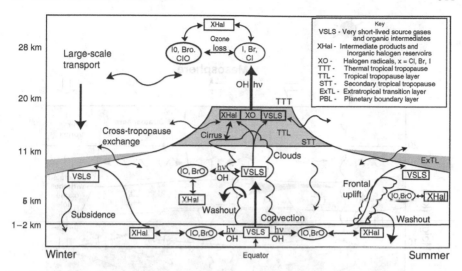

Fig. 7.3 Various transport processes involved in the chemistry and dynamics of the troposphere and stratosphere (Courtesy: Cox and Haynes 2003; WMO 2007)

slowly upward in the tropical stratosphere, ozone is being created by the slow photochemical production caused by the interaction of solar UV radiation and molecular oxygen.

Ozone is created in the tropical region by the overhead Sun during the day all year long. There is sufficiently enough energetic UV light to split the molecular oxygen, O_2, and form ozone. It normally takes more than 6 months for air near the tropical tropopause (16 km) to rise up to about 27 km (Cox and Haynes 2003, WMO 2007).

Even though ozone production is small and slow in the lower tropical stratosphere, the slow lifting circulation allows enough time for ozone to build up. The ozone density is maximum near 27 km, which is generally referred to as the ozone layer.

7.2.4 Circulation in Extratropics

Any global budget of chemical species in the extratropical lower stratosphere must take account of transport into and out of this region. The long radiative timescales in this part of the atmosphere mean that the steady-state, downward controlled limit is not achieved on seasonal timescales, and therefore likely to be a lag between the extratropical wave drag and the downwelling through the extratropical tropopause (Perlwitz and Harnik 2004). In addition, the mass of the stratosphere itself undergoes a seasonal breathing as the tropopause moves up and down.

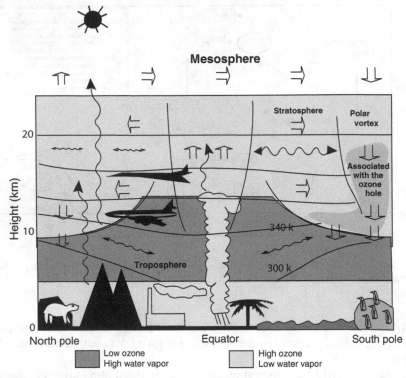

Fig. 7.4 Diagram illustrating the transport processes in the atmosphere from tropics to poles (Courtesy: WMO/UNEP)

In the stratosphere, the Brewer-Dobson circulation carries air from the equator to the poles (see Fig. 7.4). About 30°N and 30°S, the circulation becomes downward as well as poleward. This poleward and downward circulation tends to increase ozone concentrations in the lower stratosphere of the middle and high latitudes. This increase of ozone at lower altitudes observed in the higher latitudes is due to the result of the Brewer-Dobson circulation.

Another cause for the increase in ozone amounts in the high-latitude lower stratosphere is that the ozone molecule gets a longer lifetime in this region. The ozone is produced by molecular oxygen photolysis, and it is destroyed in catalytic reactions. Since there are very few oxygen atoms in the lower stratosphere (because most of the UV necessary to produce them is absorbed at higher altitudes), the lifetime of ozone is very long. Thus, ozone is not easily destroyed in the lower stratosphere. As a result, ozone can accumulate as the Brewer-Dobson circulation moves air poleward from the tropical source region into higher latitudes and downward into lower altitudes.

7.2.5 CFC Transport

Another result of this mass circulation is that most CFCs are carried from the troposphere into the stratosphere in the tropics, and are then recycled back into the troposphere in the middle to high latitudes.

Figure 7.5 gives a schematic representation of the CFC transport from troposphere to stratosphere and between tropics and midlatitudes. Since it is the intense UV in the upper stratosphere that breaks down CFCs, and since very few CFC molecules make it to the upper stratosphere, the lifetime of CFCs is quite long. It is estimated that the timescale needed to reduce CFC-12 by 63% is approximately

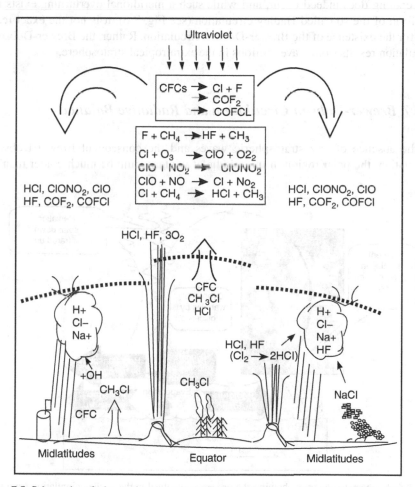

Fig. 7.5 Schematic of the processes that transport chlorine to the stratosphere (Courtesy: WMO/UNEP)

120 years. This lifetime results from the very slow circulation and the decrease of density, which both significantly impact the rate at which CFCs reach the upper stratosphere and are broken down by UV light.

7.2.6 Existence of Brewer-Dobson Circulation

The mechanism behind the Brewer-Dobson circulation is quite intricate and interesting. One expects that the circulation results from solar heating in the tropics, and cooling in the polar region, causing a large meridional overturning of air as warm tropical air rises and cold polar air sinks (Austin and Li 2006). While this heating and cooling does indeed occur, and while such a meridional overturning exists in the form of the so-called Hadley circulation (see Fig. 7.6), it is not the exact reason for the existence of the Brewer-Dobson circulation. Rather, the Brewer-Dobson circulation results from wave motions in the extratropical stratosphere.

7.2.7 Brewer-Dobson Circulation and Radiative Balance

In the absence of any stratospheric waves and the consequent Brewer-Dobson circulation, the polar region in the middle of winter would be much colder than it

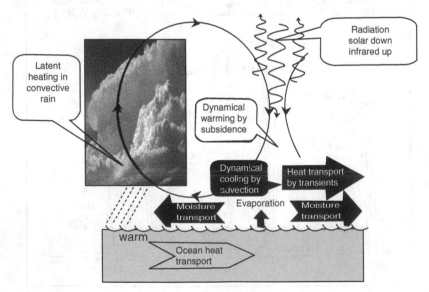

Fig. 7.6 A schematic diagram showing the processes involved in the Hadley circulation from the standpoint of the heat budget, especially in the subtropics (Courtesy: Trenberth and Stepaniak 2003, Courtesy: American Meteorological Society)

actually is. Calculations show that without the waves and resulting Brewer-Dobson circulation, the polar stratosphere would be phenomenally cold. It is estimated that at 30 km polar temperature would be about 160 K, as opposed to the measured 200 K.

The temperature field derived without any waves or circulation is known as the radiative equilibrium. In the winter lower stratosphere, the circulation in the polar region is downward. This adiabatic descent (compression) results in temperatures that are warmer than the corresponding radiative equilibrium value, producing a slightly warmer radiative balance.

7.2.8 Hemispheric Differences in Brewer-Dobson Circulation

It has already been seen that the Brewer-Dobson circulation features rising motions around the equator, poleward transport, and sinking motions in the higher latitudes. This is the general pattern. However, there are significant hemispheric differences in the strength and behavior of the Brewer-Dobson cell. These are evident from the large-scale features of the zonal mean distributions of ozone, trace gases, temperature, and zonal wind in the stratosphere.

The midwinter Brewer-Dobson circulation cells in the southern and northern hemispheres are quite different. The dissimilarity arises from the hemispheric differences in planetary wave forcing coming out of the troposphere. Large scale topographical features like the Rockies and the Himalaya–Tibet complex are mostly in the northern hemisphere. The southern hemisphere has significantly less land than the northern hemisphere and is almost entirely ocean from 55°S south to the Antarctic continent. The weak winter wave activity in the southern hemisphere means that the Antarctic polar vortex is much more isolated than its Arctic counterpart, and as a result, temperatures in the Antarctic polar vortex get extremely cold. These wave forcing differences have a profound influence on transport and result in the observed hemispheric differences in distribution of ozone and other stratospheric trace constituents.

Because of the prominent topography and land–ocean contrasts in the northern hemisphere, the northern stratosphere has more frequent and intense planetary wave activity during the northern winter than the southern hemisphere stratosphere during the southern winter. This stronger wave activity in the northern hemisphere leads to a stronger Brewer-Dobson circulation in the northern midwinter than during the southern midwinter. These horizontal mixing processes in the southern hemisphere are confined to the subtropics and middle latitudes, seldom reaching the Antarctic polar region. By contrast, in the northern hemisphere, mixing processes often extend into the Arctic polar region, owing to the significant planetary wave activity and resultant Brewer-Dobson circulation (Eichelberger and Hartmann, 2005).

7.3 Atmospheric Waves and Tracer Transport

The Brewer-Dobson circulation is induced by the growth and dissipation of large-scale atmospheric waves. These waves can transmit heat and momentum across large distances on a global scale. They can displace trace gas constituents, either temporarily or permanently. Therefore, the transport of ozone and other trace elements are mainly controlled by the atmospheric waves and the processes involved developing and dissipating them.

The waves in the stratosphere are largely composed of those generated in the troposphere which propagates upward. There are two types of waves that are critical for the mixing processes and the Brewer-Dobson circulation: the Rossby waves and gravity waves. Gravity waves are important for understanding the entire circulation of the stratosphere and mesosphere, whereas Rossby waves are principally responsible for the existence of Brewer-Dobson circulation (Scott and Polvani 2004). Rossby waves induced by topography, known as the stationary planetary waves propagate vertically and break in the stratosphere, causing the wintertime stratospheric sudden warmings in the polar regions that lead to the Brewer-Dobson circulation.

Because of the hemispheric asymmetries of the wave-forcing mechanisms, the wave energy is significantly larger in the northern stratosphere than in the southern stratosphere. In the southern hemisphere, there are only small planetary wave features. The polar vortex appears to be centered right on the south pole. This inter-hemispheric asymmetry has a profound effect on the distribution of ozone and other trace gases in the two hemispheres. It is interesting to note that in the summer hemisphere, there are virtually no stratospheric waves in the geopotential field (Gettleman and Foster 2002).

Outside the equatorial region, large-scale waves have a major influence in the winter hemisphere stratosphere. These intense northern hemisphere waves produce a stronger Brewer-Dobson circulation, a weaker polar vortex, warmer polar temperatures in response to the stronger Brewer-Dobson circulation, and an earlier breakup of the polar vortex. In addition, because of the stronger Brewer-Dobson circulation, more ozone is transported to the northern polar lower stratosphere, leading to higher ozone levels in the northern hemisphere. This hemispheric asymmetry in wave activity thus has a profound effect on the distribution and mixing processes of ozone and other trace gases between the two hemispheres.

Figure 7.7 shows that the years when planetary waves in the northern hemisphere are unusually weak, an ozone hole can form over the Arctic. During the period of strong planetary wave activity, high-latitude stratosphere becomes warmer than normal and ozone loss becomes minimum. In the year of weak planetary wave activity, the polar stratosphere is colder than the normal conditions and causes high depletion of ozone.

High mountains and land–sea boundaries combine to generate vast undulations in the atmosphere due to planetary scale waves, which act to heat polar air. Planetary scale waves are so large that some of them wrap around the whole Earth displacing air north and south as they travel around the globe.

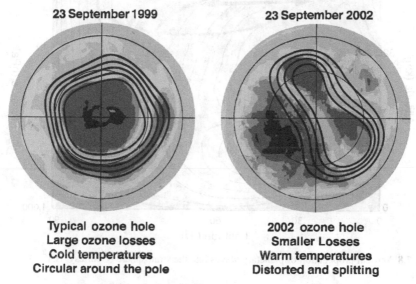

Fig. 7.7 Diagram illustrating planetary longwave activity and ozone depletion (Courtesy: NASA)

Stronger planetary waves in the northern hemisphere warm the Arctic strato-sphere and suppress ozone destruction. Land forms in the southern hemisphere also produce planetary waves, but they tend to be weaker because there are fewer tall mountain ranges and more open oceans around Antarctica.

7.3.1 Movement

After being generated in the troposphere, the planetary waves propagate vertically into the stratosphere along the axis of the jet core, eventually bending toward the tropics (see Fig. 7.8). As these waves move upward into the stratospheric layers, they grow in amplitude because of decreasing air density in the stratosphere. The wave decelerates the polar night jet by depositing easterly momentum into the fast-moving westerly jet. The Brewer-Dobson circulation is induced by this wave deceleration and the accompanying stratospheric sudden warming.

7.3.2 Wave Growth and Dissipation

The growth and dissipation of atmospheric waves result in meridional exchange or transport of air in the stratosphere. Periods of rapid growth of extratropical planetary

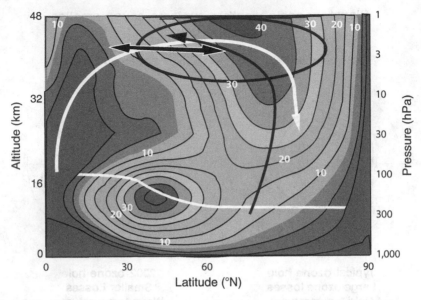

Fig. 7.8 Vertical propagation of planetary waves into the stratosphere (Courtesy: NASA)

waves, known as wave transience, are most common in northern high latitudes during the northern winter. This rapid wave growth can lead to sudden and dramatic changes in the temperature and circulation structure in the stratosphere. This is the stratospheric sudden warming phenomenon, explained earlier in Chapters 1 and 4 of this book.

Wave growth also occurs in the southern high latitudes during the southern late winter and spring, although it is less dramatic than its northern hemisphere counterpart. It is this wave transience in the southern hemisphere that is responsible for the breakup of the polar vortex and the Antarctic ozone hole during early spring.

Wave dissipation has two main causes: thermal dissipation and wave-breaking. Thermal dissipation occurs through radiative processes, while wave-breaking refers to a rapid mixing of air parcels from different regions (Sherwood and Dessler 2001).

7.3.2.1 Thermal Dissipation

Thermal dissipation refers to the process of wave dissipation in which radiative heating and cooling lessen the temperature differences that are associated with Rossby wave formation. These Rossby waves have associated large-scale areas of warm and cold temperature perturbations. The warm regions will radiatively cool to space at greater rates than the colder regions and restore the atmosphere to a more uniform temperature field. In general, this thermal damping process becomes more significant with increasing altitude in the stratosphere.

7.3.2.2 Wave-Breaking

Wave dissipation or damping also occurs by a process referred to as wave-breaking. Atmospheric waves grow to large amplitudes and break, thereby causing rapid meridional mixing. This process is particularly evident in the winter middle latitudes. Waves propagate vertically from the troposphere into the stratosphere, and then equatorward into the subtropics. As the wave moves upward, the density of the atmosphere decreases, and the amplitude of the wave consequently grows. This eventually leads to wave-breaking, in which air parcels undergo large and rapid latitudinal excursions causing them to undergo strong, irreversible, meridional mixing. As a result, the long-lived tracers become thoroughly mixed throughout the subtropics and lower middle latitudes (Trenberth and Stepaniak 2003).

The strength of the mixing is directly related to the strength of the waves (see Fig. 7.8). This means that greater mixing occurs during the northern midwinter than during the southern midwinter. The well-mixed region occurs on the equatorward side of the polar night jet. This is known as the *surf zone*. The wave dissipation occurs as energy is transferred from the larger wave scales to smaller wave scales, which are thermally dissipated. The planetary wave-breaking generally occurs when a wave propagates into a region where the wave speed matches the mean flow.

Wave-breaking processes occur not only for stratospheric planetary waves, but also for very small-scale gravity waves. Breaking gravity waves are important in the mesosphere where gravity wave amplitudes become large enough to generate convective instability, overturning, and rapid vertical mixing of air parcels. These gravity waves are also thought to decelerate the mean flow in the upper stratosphere and mesosphere, and hence also affect the mean circulation.

7.3.3 Wave Transport

Wave growth and dissipation generates meridional mass transport by two processes. First, the Brewer-Dobson circulation results from these waves because of the transfer of easterly momentum and energy via waves from the troposphere to the stratosphere, which act as a break on the westerly polar night jet, creating first a radiative and then a mass imbalance. Second, meridional exchange of mass occurs as waves dissipate in the atmosphere, producing a meridional stirring of air. This stirring or mixing tends to occur approximately on isentropic surfaces.

7.3.4 Wave-Mixing

Planetary- or synoptic-scale waves can cause irreversible changes in ozone concentration. The high ozone levels over the polar region are within the polar vortex which persists throughout the winter. These high concentrations are maintained by the combination of the downward component of the Brewer-Dobson circulation, which brings ozone-rich air downward and poleward, and wind patterns which isolate the vortex from middle latitude air. The wave-mixing thus causes stirring up of air parcels and redistributing trace gas constituents (Shepherd 1997).

7.3.5 Wave Influence on Mean Circulation

These planetary wave-induced ozone changes that are illustrated in Figs. 7.7 and 7.8 create a high degree of ozone variability across the globe, but this variability does not contribute to the global average of ozone on timescales of months or longer. Rather, the global average of ozone is primarily driven by the Brewer-Dobson circulation, which produces the observed ozone gradient of elevated extratropical latitude stratospheric ozone and low tropical stratospheric ozone (Stohl et al. 2003). The wave-mixing causes the gradient to vanish. These wave events continually pull low ozone air from the tropics into the extratropical region and high ozone extratropical air into the tropical region.

7.4 Other Factors

While most interest has focused on lateral inmixing into the tropics from midlatitudes, one expects this process also to be associated with lateral outmixing. There is clear evidence of outmixing events in the middle stratosphere, corroborated by tracer advection calculations, and aerosol measurements also suggest significant outmixing in the lower stratosphere. Measurements of water vapor, carbon dioxide, and ozone in the midlatitude lower stratosphere also point to a direct transport pathway from the tropical upper troposphere/lower stratosphere into the midlatitude lower stratosphere, which amounts to a *short-circuiting* of the Brewer-Dobson circulation. In fact, the lowest part of the tropical stratosphere is revealing itself as a particularly complex part of the atmosphere, which plays a critical role in determining the global structure of stratospheric transport and mixing.

7.5 The Quasi-Biennial Oscillation and the Brewer-Dobson Circulation

Atmospheric weather and wave dynamics vary from year to year. These differences produce an interannual variability in the wave activity that affects the Brewer-Dobson circulation. One of the principal sources of year-to-year variability in the total ozone distribution is the quasi-biennial oscillation (QBO).

Total ozone distribution is affected by the QBO for two reasons: (i) the QBO affects the stratospheric temperature structure, which in turn affects the photochemical balance of the upper stratosphere (see Chapter 5); and (ii) the QBO directly modifies the Brewer-Dobson circulation.

Since the QBO is due to the internal dynamics of tropical waves rather than the annual change of seasonal cycle, the period of this wind oscillation is highly variable with periods ranging from 22 to 34 months. The zonal winds blowing in one direction descend in altitude in time and are replaced by winds blowing in the opposite direction. There are westerly and easterly wind regimes associated with the QBO.

The QBO phenomenon changes the overall equatorial zonal wind field. These changes in wind direction produce temperature anomalies, which in turn modify the Brewer-Dobson circulation.

7.5.1 QBO Circulation

The temperature anomalies associated with the QBO winds induce a modification to the normal Brewer-Dobson circulation. This QBO circulation is superimposed on the normal Brewer-Dobson circulation. Depending on which phase, this circulation will either be speeded up or weakened.

The QBO in its descending easterly phase maintains colder temperatures between the overlying easterlies and underlying westerlies (see Fig. 7.9). The result is that the infrared cooling to space will be lesser than normal in the QBO cold region. Because the heating from solar UV is approximately constant, the weakened cooling to space means that the total heating in the tropics is somewhat larger. This greater heating in the tropics results in a speeding up of the normal Brewer-Dobson circulation lifting in the tropics.

Conversely, the QBO in its descending westerly phase maintains warmer temperatures between the overlying westerlies and underlying easterlies (see Fig. 7.9). The result is that infrared cooling to space will be greater than normal in the QBO warm region. Because the heating from solar UV is approximately constant, the greater cooling to space means that the total heating in the tropics is somewhat smaller (Shepherd 2002). This lesser heating in the tropics results in a slowing up of the normal Brewer-Dobson circulation lifting in the tropics.

These downward and upward motions associated with the QBO at the equator are balanced by upward and downward motion in the subtropics, respectively. This circulation cell, is connected by poleward or equatorward motions, and is called the QBO-induced meridional circulation.

The subtropical branch of the QBO-induced circulation cell is located approximately between 15°N and 15°S. The QBO-induced circulation has an influence on trace gas constituents in the tropical stratosphere. QBO signals in ozone, methane, hydrogen, fluorine, and nitrous oxide have been reported from long-term satellite observations.

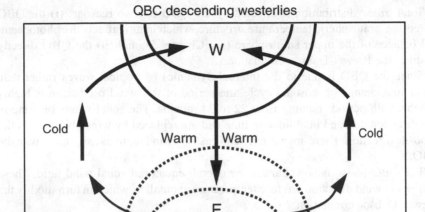

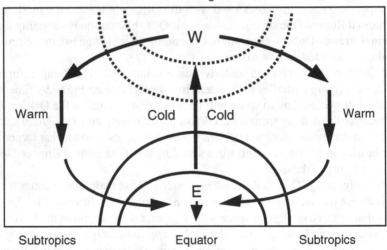

Fig. 7.9 Schematic diagram of the QBO descending westerlies (top panel) and descending easterlies (bottom panel) phases (Courtesy: NASA)

7.5.2 Ozone Transport: Influence of QBO

The QBO represents an important source of ozone variability throughout the lower stratosphere. The equatorial zonal wind QBO is confined to the equatorial stratosphere, a region where ozone is controlled by both transport and photochemistry. Below 30 km in the tropics, ozone is primarily under dynamical control, and thus is affected by the QBO-induced circulation that exists atop the Brewer-Dobson circulation. Above 30 km, ozone increasingly comes under photochemical control, and thus responds to the QBO-induced temperature anomalies rather than transport effects.

7.5.2.1 Tropical Ozone and QBO

The descending westerlies of the QBO are associated with a vertical circulation pattern that produces downward motion in the tropics and upward motion in the subtropics, weakening the normal Brewer-Dobson circulation in the tropics. Because the upward motion of air is slowed down and as the vertical gradient of ozone mixing ratio is positive in the lower stratosphere (i.e., increasing ozone with altitude), ozone production can proceed for longer periods. The result is a positive column ozone anomaly in the tropics and a negative anomaly in the subtropics. In the descending easterly phase of the QBO when the Brewer-Dobson circulation in the tropics is enhanced, ozone production has less time to occur, and the column ozone anomalies are reversed, resulting in a negative ozone anomaly in the tropics and a positive ozone anomaly in the subtropics.

It can be seen in Fig. 7.10 that there exists alternating high and low ozone values in the 20–30 km altitude range, corresponding to the westerly and easterly shear zones of the QBO, respectively. Above 30 km, the ozone variability is controlled

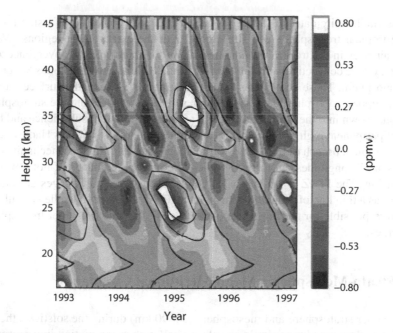

Fig. 7.10 Vertical variation of stratospheric ozone anomalies for the 4°S to 4°N region. The data is based on the UARS HALOE instrument; the contours superimposed on the ozone anomalies are the HRDI zonal winds

more by temperature-dependent photochemical processes than transport. Thus, the ozone variation from 35 to 45 km is a response to the temperature variations due to both the QBO and the semiannual oscillation.

7.5.2.2 Extratropical Ozone QBO

Observations of QBO signals in dynamical variables and constituent fields, such as column ozone and water vapor, exist in the extratropical stratosphere. However, unlike the situation in the tropical stratosphere where the mechanism for the observed ozone anomalies is fairly well understood, there is no well-accepted explanation for how the equatorial QBO anomaly is transmitted to extratropical latitudes. In fact, there is even debate as to how large an extratropical QBO signal exists. Estimates of the magnitude of the midlatitude total ozone QBO range from 5 to 20 DU. Although this is still an area of active research, it is clear that the QBO has an important influence on the extratropical circulation and various constituent fields.

7.6 Tropospheric Meridional Circulation

While the Brewer-Dobson circulation is the most important circulation for understanding stratospheric ozone, other circulations impact the stratosphere in minor ways.

The Hadley circulation is a tropospheric circulation consisting of rising motion in the tropical troposphere and subsidence over the extratropical regions. Warm, moist air rises in the tropical troposphere along the intertropical convergence zone (ITCZ), where convectively driven cumulonimbus (thunderstorm) clouds tower into the atmosphere. These convective towers pump materials from the surface into the tropical upper troposphere, where they can slowly be carried into the stratosphere. Air sinks down into the cooler subtropical regions, resulting in subsidence and belts of semipermanent high pressure. This circulation pattern is known as Hadley cell.

The tropical branch of the Hadley circulation is not a continuous mechanism that occurs at all longitudes but rather a series of concentrated bands of hot towers of convection. The ITCZ is observed in the global visible cloud images around the equator as a thin line of thunderstorms within 10° of the equator. Hadley circulation is also responsible for the lower tropospheric easterlies and the upper tropospheric westerlies.

7.7 Strato-Mesospheric Mean Meridional Circulation

In the upper stratosphere and mesosphere (30–90 km) during the solstices, the circulation is dominated by a single circulation cell with rising motion in the summer hemisphere and sinking motion in the winter hemisphere with a corresponding summer-to-winter hemisphere meridional drift required by mass continuity near

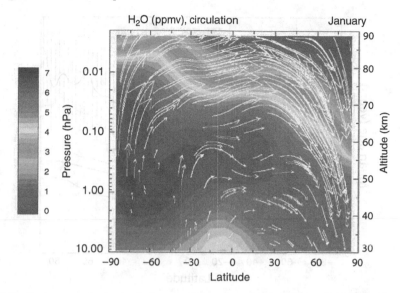

Fig. 7.11 Strato-mesospheric circulation superimposed on the water vapor distribution (Courtesy: NASA)

the stratopause (see Fig. 7.11). As a result, the summer polar mesosphere is much colder than its radiative equilibrium, while the winter polar mesosphere is much warmer. This circulation pattern is generally driven by small-scale gravity waves. These waves occur during all seasons and at all latitudes in the mesosphere.

Water vapor has a photochemical source in the upper stratosphere (approximately 50 km), and a sink in the upper mesosphere above about 80 km. Figure 7.11 shows the mean mesospheric circulation as it transports high water vapor values from the upper stratosphere to the upper mesosphere near 80 km in the middle to high latitudes of the summer hemisphere. Low water vapor values are subsequently transported from the lower thermosphere above 90 km down to the lower mesosphere/upper stratosphere at middle to high latitudes of the winter hemisphere.

7.8 Age of Air in the Stratosphere

Stratospheric ozone is destroyed by catalytic reactions with trace gases such as chlorine. The chlorine in the stratosphere is primarily released from human-produced CFCs via UV photolysis. However, CFC-12 is photolyzed by wavelengths less than 240 nm. Since ozone and oxygen molecules absorb the overwhelming majority of this radiation, CFCs must reach very high altitudes before they can be UV photolyzed by radiation of this wavelength and release their chlorine into radical forms that are then free to attack ozone. The slow Brewer-Dobson circulation cell takes at least 1–2 years to move this air from the troposphere to the upper stratosphere,

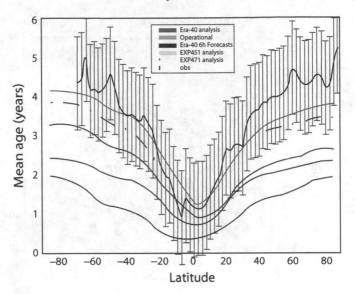

Fig. 7.12 Mean age of air at 20 km altitude from model simulations (Figure courtesy of Beatriz Monge Sanz, University of Leed)

where this material remains for at least 1–2 years. The longer the residence time of CFCs in the stratosphere, the more chlorine can be liberated from the CFC into the radical species, and hence the more ozone that a radical chlorine atom can destroy.

Transport models can be used to estimate the average age of air for the stratosphere. This age is the average amount of time it takes for air parcels to be transported from the ground to a specific latitude and altitude region in the stratosphere. The modeled age is shown in Fig. 7.12.

The air parcels in the troposphere are very quickly mixed due to convection and weather systems, so that the age of air distribution is relatively uniform in this region. This age diagram is based on an idealized tracer with a tropospheric mixing ratio that increases linearly in time. The age distribution reflects the influence of the Brewer-Dobson circulation. Air enters the stratosphere through the tropical tropopause, and ascends through the tropical stratosphere. Nearly 90% of the air entering the stratosphere at 16 km never makes it to the top of the stratosphere around 32 km. Most of the air from the tropics finds its way into the extratropical lower and middle stratosphere, where it enters the descending branch of the Brewer-Dobson circulation. Air that does make it to the top of the stratosphere (Holton et al. 1995) undergoes vigorous overturning in the mesosphere by the circulation that exists there. This circulation is induced by gravity wave-breaking.

It takes air parcels 4.5–5 years to be transported from the ground to the lower mesosphere. Air parcels return to the stratosphere via wintertime descend at middle and high latitudes. The tropical lower stratospheric air is relatively juvenile, since it has been directly lifted out of the troposphere by the Brewer-Dobson circulation. The vertical profiles of the mean air derived from carbon dioxide measurements and

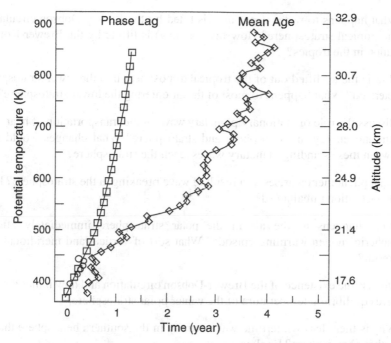

Fig. 7.13 Vertical profiles of "mean age" in the tropical stratosphere, derived from CO_2 measurements, and "phase lag" time, derived from H_2O measurements (Waugh and Hall 2002, Courtesy: American Geophysical Union)

phase lag derived from water vapor measurements in the tropical stratospheric region are illustrated in Fig. 7.13. The young tropical stratospheric air contrasts with lower stratospheric air at middle and high latitudes, where ages can exceed 5 years. Older air has a large abundance of chlorine radicals and small abundance of CFCs because it has had a longer period for the CFCs to be destroyed by UV photolysis (Finlayson-Pitts and Pitts 2000). In comparison, tropical lower stratospheric air has low concentrations of chlorine radicals and higher CFC amounts. While it takes about 4–5 years for materials to reach the upper stratosphere from the troposphere, only a small fraction of all tropospheric air ever makes it that high. The time needed to cycle most of the tropospheric air through the upper stratosphere is in fact many decades.

Questions and Exercises

7.1. What is the Brewer-Dobson Circulation? Discuss the significance of this circulation pattern in the transportation processes from tropics to higher latitudes.

7.2. How is the Brewer Dobson circulation maintained? Discuss the difference in the characteristics of Dobson circulation in northern and southern hemispheres.

7.3. What happens to water vapor as it is lifted by the Brewer-Dobson circulation into the tropical stratosphere? How fast (or slow) is lifting by the Brewer-Dobson circulation in the tropics?

7.4. How is the air lifted out of the tropical troposphere into the lower stratosphere characterized? What happens to most of the air entering the lower stratosphere?

7.5. Discuss the role of stationary planetary wave in the transportation of heat, momentum and energy in troposphere and stratosphere? What changes would take place when these standing planetary waves reach the stratosphere?

7.6. What phenomenon arises as a result of wave breaking in the stratosphere? How is such a situation imbalanced?

7.7. What happens to the air in the polar stratosphere immediately after a stratospheric sudden warming episode? What sort of vertical and meridional motions result?

7.8. Discuss the existence of the Brewer-Dobson circulation and how it modifies the radiative equilibrium temperature of the winter polar stratosphere.

7.9. Why is there less wintertime wave activity in the southern hemisphere than in the northern hemisphere? Explain.

7.10. What are the results of stronger wintertime wave activity in the northern hemisphere for ozone distribution in the Arctic lower stratosphere?

7.11. What are the two main causes of wave dissipation in the stratosphere? Give a brief description of these two causes.

7.12. What is the well-mixed region where wave breaking occurs called? Where does it occur?

7.13. How does the QBO modify the Brewer-Dobson Circulation in the tropics?

7.14. What sort of ozone anomaly results in the tropics and in the subtropics during descending westerly and easterly phases of the equatorial QBO?

References

Austin J, Li F (2006) On the relationship between the strength of the Brewer Dobsun circulation and the age of stratospheric air, Geophys Res Lett, 33, doi 10.1029/2006GL026867

Appenzeller C, Holton JR, Rosenlof KH (1996) Seasonal variation of mass transport across the tropopause, J Geophys Res 101: 15071–15078

Brasseur G, Solomon S (2005) Aeronomy of the Middle Atmosphere, 3rd edition Springer, Dordrecht

Bonazzola M, Haynes PH (2004) A trajectory based study of the tropical tropopause region, J Geophys Res, 109 doi 10.1029/2003JD004536

Butchart N, Scaife AA, Bourqui M, de Grandpre J, Hare SHE, Kettleborough J, Langematz U, Manzini E, Sassi F, Shibata K, Shindell D, Sigmond M (2006) Simulations of anthropogenic change in the strength of the Brewer Dobson circulation, Clim Dyn, 27, 727–741, doi 10.1007/s00382-006-0612-4

Cordero E, Newman PA, Weaver C, Fleming E (2002) Stratospheric dynamics and transport of ozone and other tracer gases, Chapter 6: Stratospheric Ozone An Electronic Text, NASA, GSFC

Cox ME, Haynes P (2003) Scientific assessment of ozone depletion: 2002, WMO Report No. 47., Geneva, Switzerland

Eichelberger SJ, Hartmann D (2005) Changes in the strength of the Brewer Dobson circulation in a simple AGCM, Geophys Res Lett, 33, doi 10.1029/2005GL022924

Finlayson-Pitts BJ, Pitts JN Jr (2000) Chemistry of the upper and lower atmosphere, Academic, London, 2000

Gettleman A, Foster PMdeF (2002) A climatology of the tropical tropopause layer, J Meteorol Soc Japan, 80, 911–924

Holton JR, Haynes PH, McIntyre ME, Douglas AR, Rood RB, Pfister L (1995) Stratosphere troposphere exchange, Rev. Geophys 33: 403–439

NASA (2003) Studying Earth's Environment from Space (http://www.ccpo.odu.edu/SEES/index.html)

Perlwitz J, Harnik N (2004) Downward coupling between the stratosphere and troposphere: The relative role of wave and aonal mean processes, J Clim, 17: 4902–4909

Rind D, Lerner J, McLinden C (2001) Changes of tracer distributions in the doubled carbon dioxide climate, J Geophys Res, 106: 28061–28080

Scott RK, Polvani LM (2004) Stratospheric control of upward wave flux near the tropopause, Geophys Res Lett, 31, doi 10.1029/2003GL017965.1

Shepherd TG (1997) Transport and mixing in the lower stratosphere: a review of recent developments, SPARC Newsletter 9, July 1997

Shepherd TG (2002) Issues in stratospheric tropospheric coupling, J Meteorol Soc Japan, 80: 769–792

Sherwood SC, Dessler AE (2001) A model for transport across the tropopause, J Atmos Sci, 58: 765–779

Stohl A, Wernli H, James P, Bourqui M, Forster C, Liniger MA, Seibert P, Sprenger M (2003) A new perceptive of the stratosphere-trposphere exchange, Bull Amer Meterol Soc, 84, 1565–1573

Trenberth KE, Stepaniak DP (2003) Seamless poleward atmospheric energy transports and implications for the Hadley circulation, J Climate 16: 3705–3721

Waugh DW, Hall TM (2002) Age of stratospheric air: Theory, observations, and modeling, Rev Geophys, 40, doi, 10.1029/2000R000101

WMO (2007): Scientific Assessment of Ozone Depletion: 2006, Global ozone research and monitoring project Report No. 50, Geneva, Switzerland

Chapter 8
Stratosphere–Troposphere Exchange

8.1 Introduction

Stratosphere–troposphere exchange (STE) is a part of the general circulation of the atmosphere that transports air and atmospheric constituents across the tropopause. Stratosphere and troposphere are the two adjoining regions of different characteristics which are coupled radiatively, chemically, and dynamically. Radiation from the Sun warms up the land, sea surface, and air. Heating is greatest in the tropics, and less in the mid- and high latitudes. This means that convection is strongest in the tropics and air rises to higher altitudes compared to elsewhere on Earth.

Above the tropopause, absorption of solar radiation by ozone leads to a warming of the stratosphere. This warming is highest over the tropics, lower in the polar regions and goes to zero during the polar winter. In the troposphere the atmospheric stability is much smaller than that in the stratosphere and therefore mixing is much more rapid in the troposphere than in the stratosphere. The tropopause thus acts as a barrier to upward transport of air and pollutants. Figure 8.1 shows that the warm air rises in the tropics which cools and moves toward the poles.

Even though the dynamics of the troposphere and stratosphere are inseparable, these two regions can be considered as two isolated boxes. Vertical air exchange in the troposphere takes hours to days whereas mixing in the stratosphere takes months to years. The vertical transport of air and chemical species through the depth of the troposphere can occur on timescales as short as a few hours via moist convection, and on timescales of days via baroclinic eddy motions in middle latitudes. On the other hand, the vertical transport through a similar altitude range in the stratosphere takes months to a year or even more in the lower stratosphere and this vertical transport must be accompanied by radiative heating or cooling.

The difference between the vertical transport timescales in the stratosphere and troposphere is mainly due to the rapid increase in ozone mixing ratio and the rapid decrease in water vapor mixing ratio with altitude observed just above the tropopause. Based on the differing chemical compositions, it is possible to distinguish the air parcels in the troposphere and stratosphere.

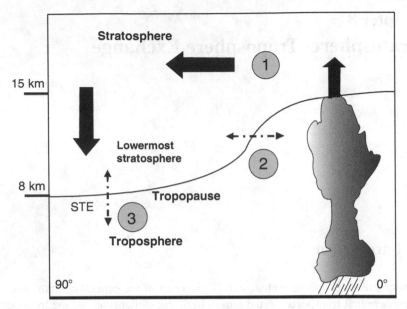

Fig. 8.1 The direction of global circulation and stratosphere troposphere exchange

The mesoscale stratosphere–troposphere exchange (STE) has direct implications on the distribution of atmospheric ozone, in particular, the decrease of lower stratospheric ozone and the increase of tropospheric ozone (Hartmann et al. 2000). STE also impacts the distribution of aircraft emissions and the vertical structure of aerosols and greenhouse gases .

Various observational techniques have shown that anthropogenic and natural trace chemical species are transported are from the troposphere to the stratosphere and initiate the chemical reactions responsible for the ozone depletion. On the other hand, downward transport from the stratosphere not only removes the chemical species including those involved in stratospheric ozone depletion, but also represents a significant input of ozone and other reactive species into the tropospheric chemical system.

The transport of anthropogenic gases, like chlorofluorocarbons, from the troposphere into the stratosphere, affects the chemical balance in both regions and provides the catalysts necessary for stratospheric ozone destruction. Stratospheric–tropospheric exchange also controls the rate of transport between source and sink regions for both tropospheric and stratospheric source gases. A consequence of this is the long lag time between the release of tropospheric trace gases and stratospheric ozone reduction.

Therefore, STE comprises not only the transport across the tropopause, but also the rate at which tropospheric material is supplied to, and removed from, the regions in the stratosphere in which there are chemical sources and sinks, for whichever chemical species are of interest. It is necessary, therefore, to take account of the

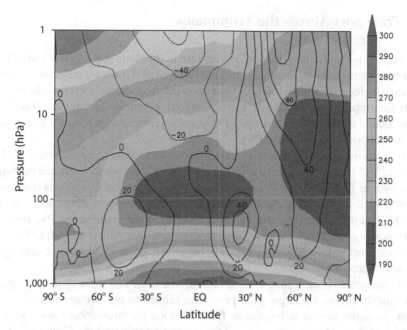

Fig. 8.2 Climatological mean (1957–2006) temperature and wind distribution for January (Data source : NCEP/NCAR Reanalysis data)

species, photochemical sensitivities at different altitudes and latitudes, and of the global-scale circulation, including the spatio-temporal structure of transport within the stratosphere.

There are slight differences between the circulation of air in the northern and southern hemispheres. In the north, the distribution of oceans and landmasses are more inhomogeneous than in the south and the polar vortex is weaker. In addition, seasons have to be taken into account. The variations in temperature and wind patterns between the two hemispheres influence the Brewer-Dobson circulation.

Atmospheric cross section of the temperature and wind distribution from south pole to north pole during the month of January is illustrated in Fig. 8.2. During the northern hemispheric winter (January), the cold tropopause over the tropics and the formation of a polar vortex over the Arctic region can be seen. The polar vortex height extends from the tropopause to 25 km and spreads upto 60 km in the arctic region.

Earlier studies of stratosphere–troposphere exchange of mass and chemical species have mainly emphasized the synoptic- and mesoscale mechanisms of exchange. However, Holton et al. (1995) have presented the detailed global-scale aspects of exchange, such as the transport across an isentropic surface (potential temperature about 380 K) that in the tropics lies just above the tropopause, near the 100 hPa pressure level. Exchange of air between the stratosphere and the troposphere can occur if layers of constant (potential) temperature cross the tropopause or if there are disturbances and convective transport occurs in the midlatitudes.

8.2 Transport Across the Tropopause

The tropopause is characterized by an increase in the static stability in moving from the troposphere to the stratosphere. It often behaves as if it were a material surface to varying degrees of approximation. The tropical tropopause corresponds roughly to an isentropic or stratification surface, whose potential temperature $\theta = 380$ K in the annual mean. The extratropical tropopause corresponds roughly to a surface of constant potential vorticity (PV).

The extratropical tropopause is found from observation to be remarkably close to the 2 PVU potential vorticity surface, where PVU denotes the standard potential-vorticity unit (1 PVU $= 10^{-6}$ m^2 s^{-1} K kg^{-1}). The tropopause slopes downward and poleward, intersecting isentropic surfaces and is usually distinguished on each such intersecting isentropic surface by a band of sharp gradients of PV, with values typically near 2 PVU, whose instantaneous configuration may well be strongly deformed irreversibly by synoptic-scale baroclinic eddies and other extratropical weather-related disturbances.

For timescales greater than several months, the mass flux through the tropopause is ultimately driven by large-scale processes related to the Brewer-Dobson circulation. To show this, it is helpful to further divide the stratosphere into the *overworld*, *middle world*, *underworld*, and the lowermost stratosphere. Figure 8.3 shows schematically these layers of the stratosphere and the dynamical processes that occur in each. This approach places stratosphere–troposphere exchange in the framework of the general circulation, and helps to clarify the roles of the different mechanisms involved, and the interplay between large and small scales.

Tropopause has a quasi-material nature, in which the reversible deformations of its shape, however large and rapid, are of little significance to get transport by

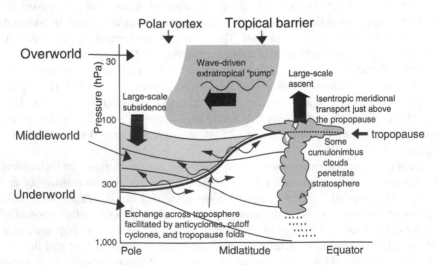

Fig. 8.3 Schematic representation of stratosphere–troposphere exchange through tropopause (Adapted from UGAMP)

themselves. Transport along isentropic surfaces occurs in adiabatic motion, whereas transport across the isentropes requires diabatic processes. Since the tropopause intersects the isentropes, transport can occur in either way, and is likely to occur in both ways.

In the region of upper troposphere and lower stratosphere consisting of isentropic surfaces that intersect the tropopause, air and chemical constituents can be irreversibly transported as adiabatic eddy motions. This leads to large latitudinal displacement of the troposphere, followed by irreversible mixing on small scales. The irreversible transport across the tropopause may be associated with the synoptic-scale baroclinic eddies or with global-scale diabatic ascent or descent.

The region where the isentropic surfaces lie entirely in the stratosphere is referred to as the *overworld* (see Fig. 8.3). Air parcels in this region cannot reach the troposphere directly. However, air parcels from the stratosphere descend first slowly across the isentropic surfaces, a process that must be accompanied by diabatic cooling and reach the troposphere.

Underworld is the region where the isentropic surfaces lie entirely in the troposphere. The tropospheric air parcel reaches the stratosphere by first rising across the isentropic surfaces. The isentropic surface bounding the overworld and the lowermost stratosphere generally has a potential temperature around 380 K, depending on cloud top heights. Holton et al. (1995) used 380 K isentrope to designate the lower boundary of the overworld, but the actual boundary may vary.

The *middle world*, the region that lies between the tropopause and the 380 K θ surface, is a convergence zone between tropospheric and stratospheric air (e.g., Hoskins 1987; Holton et al. 1995). Mixing ratios of ozone and water vapor can span an order of magnitude or more in this region purely due to transport. Therefore, examining transport into the middle world will offer valuable insight into future changes in ozone at mid- and high latitudes.

For chemical purposes, the lowermost stratosphere must be sharply distinguished from the tropospheric part of the middle world, where moist convection strongly transports material across the isentropic surfaces. Similarly, the lowermost stratosphere must be distinguished from the rest of the stratosphere, being the only part of the stratosphere accessible from the troposphere through transport along isentropic surfaces.

The middle world serves as a critically important, seasonally variable conduit that governs coupling of the tropical tropopause level (TTL) and subtropical jet structure to the middle and high latitudes. In the late winter, this segment of the stratosphere links the tropics above 390 K with high latitudes and couples the tropical upper troposphere through the subtropical jet structure (that dictates isentropic exchange between the troposphere and the stratosphere) at surfaces below 350 K. Key reversals in net meridional flow into and through the middle world occur in the spring and fall such that concerted flow from the tropics northward in winter give way to regions of equatorward flow in early summer. The equatorward transport in the midlatitude summer middle world is a result of monsoon circulation.

The current thinking on the general transport through the stratosphere is that it is driven by breaking waves slowing down the mean zonal flow, resulting in a poleward drift of the air parcels, which must descend by mass continuity. The wave-induced forces drive a kind of global-scale extratropical fluid-dynamical suction pump, which withdraws air upward and poleward from the tropical lower stratosphere and pushes it poleward and downward into the extratropical troposphere. The resulting global-scale circulation drives the stratosphere away from radiative equilibrium conditions. Wave-induced forces may be considered to exert a nonlocal control, mainly downward in the extratropics but reaching laterally into the tropics, over the transport of mass across lower-stratospheric isentropic surfaces. This mass transport is for many purposes a useful measure of global-scale stratosphere–troposphere exchange, especially on seasonal or longer timescales. Because the strongest wave-induced forces occur in the northern hemisphere winter season, the exchange rate is also a maximum at that season.

The global exchange rate is not determined by details of near-tropopause phenomena such as penetrative cumulus convection or small-scale mixing associated with upper-level fronts and cyclones. These smaller-scale processes must be considered, however, in order to understand the finer details of exchange. Moist convection appears to play an important role in the tropics in accounting for the extreme dehydration of air entering the stratosphere. Stratospheric air finds its way back into the troposphere through a vast variety of irreversible eddy exchange phenomena, including tropopause folding and the formation of so-called tropical upper-tropospheric troughs and consequent irreversible exchange.

Transport in the overworld is controlled by the Brewer-Dobson circulation. That is, the strength of the upward STE in the tropics and downward STE in the extratropics is controlled by the hemispheric Brewer-Dobson circulation rather than by smaller-scale, local transport processes at the tropopause boundary. For material descending into the troposphere, once it crosses from the overworld into the lowermost stratosphere, the timescale for it to cross the tropopause is on the order of a season. The actual transport from the lowermost stratosphere into the troposphere is governed by smaller-scale extratropical processes such as blocking anticyclones, cutoff lows and tropopause folds.

8.3 Brewer-Dobson Circulation and Stratosphere–Troposphere Exchange

Associated with the large-scale Brewer-Dobson circulation, the net mass exchange between the troposphere and stratosphere takes place, with a net upward flux in the tropics balanced by a net downward flux in the extratropics. However, near the tropopause, the exchange process is more complex, with two-way mixing across the extratropical tropopause at and below synoptic scales, and vertical mixing in the tropical tropopause layer (TTL) resulting from convective processes. Above the lowermost extratropical stratosphere and at the top of the TTL, the exchange is more one-way, with air slowly rising into the stratosphere above the TTL.

Model studies indicate that climate change will impact the mass exchange across the tropopause. An increase of 30% mass flux is estimated due to a doubling of atmospheric CO_2 amounts (Rind et al. 2001), and the net upward mass flux above the TTL would increase by about 3% decade^{-1} due to climate change (Butchart and Scaife 2001). For a doubled CO_2 concentration, all 14 climate change model simulations resulted in an increase in the annual mean troposphere-to-stratosphere mass exchange rate, with a mean trend of about 2% decade^{-1} (Butchart et al. 2006). The predicted increase occurred throughout the year but was, on average, larger during the boreal winter than during the austral winter.

8.3.1 Tropical Upwelling

Tropical upwelling is inversely related to model age of air (Austin and Li 2006), so that the age of air changes as the stratospheric climate changes. As noted for the tropical upwelling, the age of air does not change steadily. In Fig. 8.4 it is evident that the age of air decreased significantly from about 1975 to 2000, consistent with an increase in tropical upwelling, which is related with the ozone depletion.

The age of air remains constant for conditions of fixed greenhouse gases concentrations and sea surface temperatures (SSTs), indicating that changes in greenhouse gas concentrations and SSTs are major influences for age of air changes in the future. The overall decrease of age of air and increase in tropical upwelling on climate timescales imply a more rapid removal of the long-lived CFCs from the entire atmosphere, as well as source species such as CH_4 and N_2O. Once enhanced CH_4 concentrations reach the stratosphere, enhanced CH_4 oxidation would occur, leading to a faster increase in water vapor amounts than would be anticipated on the basis of the tropospheric CH_4 concentrations alone.

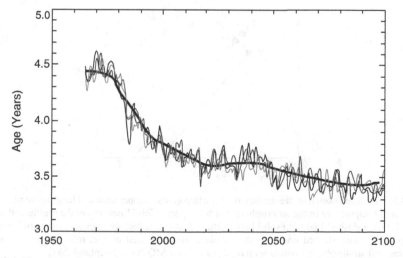

Fig. 8.4 Mean age of air in the tropical upper stratosphere for the period 1960–2100 computed from the CCM AMTRAC (Courtesy: WMO 2007; Austin and Li 2006)

8.3.2 Changes in Tropical Tropopause Layer

The tropical tropopause layer (TTL) is defined as the body of air extending from the level of the temperature lapse rate minimum at 11–13 km to the level of highest convective overshoot, slightly above the cold point tropopause (CPT) at 16–17 km. This includes the level of zero net radiative heating (z_0) that marks the transition from radiative cooling to radiative heating, divides the TTL into the lower and upper TTL and depends on the presence of clouds (Ramaswamy et al. 2006). Below z_0, the cool air sinks back into the troposphere, whereas above z_0 the warm air rises and eventually enters the stratosphere.

The TTL is sandwiched between a warm troposphere and a cool stratosphere, which makes it difficult to produce a theoretical estimate of the response of the CPT and stratospheric humidity. Figure 8.5 shows a simple conceptual picture of the tropopause. If one assumes a convectively controlled troposphere with a constant lapse rate, then tropospheric warming raises and warms the tropopause (cold point temperature increasing from T_1 to T_2). A cooling of the stratosphere further raises the tropopause but leads to a cooling from T_2 to T_3 (Shepherd 2002).

It seems likely that the tropospheric warming should be the dominant effect, because the enhanced infrared cooling by enhanced greenhouse gas concentrations is weak due to the very low temperatures close to the tropopause. The observed temperature trends just above the tropical tropopause correspond to a cooling of less than 0.4 K decade^{-1} but are not statistically significant (WMO 2003).

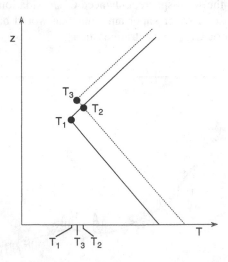

Fig. 8.5 Simplified sketch of the sensitivity of the tropopause temperature and height with regard to changes of temperature in the stratosphere and troposphere Solid lines: reference profile with cold point T_1; long-dashed line with solid line: perturbed profile reflecting tropospheric warming with cold point T_2; long-dashed line with short-dashed line: perturbed profile, reflecting tropospheric warming and stratospheric cooling with cold point T_3 (WMO 2007; Shepherd 2002)

The evolution of TTL temperatures is further complicated by the fact that there is a tropospheric amplification of surface warming. Conversely, a strengthening of the Brewer-Dobson circulation would imply a lowering of TTL temperatures. These temperatures are therefore at odds with tropospheric warming dominating the response of the CPT (Gettelman et al. 2004). Rather, they suggest that the CPT is being largely controlled by increases in the Brewer-Dobson circulation and by increased convection.

Changes in the TTL may also affect the abundance of many other species in the stratosphere. This may concern short-lived chemical species, such as biogenic bromine compounds that may be carried to the stratosphere via deep convection followed by transport through the TTL. Changes in deep convection may further affect the transport of longer-lived species produced by biomass burning, such as methyl bromide. Later, species may be transported in particulate form across the tropical tropopause. Little is known about these processes, and even less is known about climate-induced changes.

The TTL has undergone changes within the last few decades that are not well understood, and therefore our predictive capabilities remain extremely limited. There is no merged reference dataset of long-term global temperature observations for this height region.

8.3.3 Exchange Through the Tropical Transitional Layer

The existence of a secondary tropical tropopause well below the main tropopause gives the concept of tropical transitional layer (TTL), or tropical tropopause layer. The tropical tropopause acts as a transition layer between about 140 and 60 hPa instead of a discrete boundary. The TTL concept is revitalized by Highwood and Hoskins (1998), suggesting that the TTL is a key to understanding the water balance of the stratosphere.

Tropical tropopause is considered a principal source region of air entering through the stratosphere. Most of the tropical deep convection does not reach the cold point tropopause but ceases few kilometers lower, sandwiching the TTL. This relatively undisturbed body of air, subject to prolonged chemical processing of air parcels may slowly ascend into the stratosphere. Schematic representation of the tropical tropopause layer (TTL) above the active and nonactive convection regions is illustrated in Fig. 8.6.

Observational evidences show that cirrus anvil tops over oceans are rarely observed above 14 km, implying that the convective mass flux decreases rapidly above 13 km. The probability for convective clouds to penetrate above 14 km in the tropics is roughly 1%. Increase in ozone begins well below the cold point tropopause, where the temperature is the lowest. Below 14 km the lapse rate departs strongly from moist adiabat.

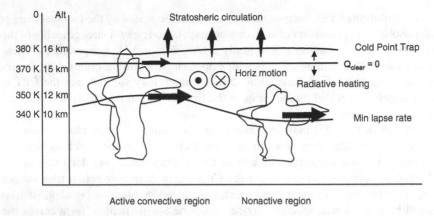

Active convective region   Nonactive region

Fig. 8.6 Schematic figure of the tropical tropopause layer (TTL). The horizontal motion is indicated as vectors in and out of the page (it also occurs along the zonal dimension). Thin black arrows represent radiative heating, large black arrows are the stratospheric (Brewer-Dobson) circulation, and horizontal thick arrows are the detrainment from convection (Gettelman and Forster 2002)

8.3.4 Changes in Extratropical Tropopause Layer

Extratropical tropopause layer (ExTL) is a layer of air adjacent to the local extratropical (thermal) tropopause, which has been interpreted as the result of irreversible mixing of tropospheric air into the lowermost stratosphere or as the result of two-way stratosphere–troposphere exchanges.

The ozone in the ExTL changes markedly with season, with photochemical production dominating in summer and transport from the stratosphere dominating in winter and spring. A general upward trend of the extratropical tropopause height has been identified and has been related to ozone column changes. This long-term change provides a sensitive indicator of human effects on climate and may be related to changes. However, it is presently not clear how these changes have affected, or will affect, the dominant transport pathways of air between the troposphere and the stratosphere on the synoptic scale or mesoscale.

8.4 Exchange Processes Near the Midlatitudes and Subtropics Tropopause

There is significant variability in the existence and strength of the subtropical transport barriers, which depend on season and are significantly modulated by various stratospheric processes. In the midlatitudes, exchange appears to be dominated by processes associated with tropopause folds and cut-off lows. They also include turbulent mixing, quasi-isentropic mixing, convectively breaking gravity waves, deep convection, and radiative heating. The development of a tropopause fold is a reversible process and thus irreversible processes must occur for the permanent transfer of material across the tropopause boundary. Specifically, exchange across a dynamically defined tropopause may occur through processes that modify potential vorticity. The mechanisms and the rate of transfer of mass and chemical species between the stratosphere and troposphere are currently uncertain.

It is well established that planetary wave-breaking and formation of a *surf zone* disturb the midlatitude stratosphere. The nature of this breaking is consistent with the predictions of Rossby wave critical layer theory (McIntyre and Palmer 1983). From this perspective, the subtropical transport barrier represents the tropical limit of the midlatitude surf zone. Similary, the polar vortex transport barrier represents the polar limit of the surf zone.

8.4.1 Blocking Anticyclones (Highs)

Large anticyclones or high pressure areas in the troposphere may persist for days or weeks. These are known as blocking anticyclones or blocking highs. The effect of such features is to lower column ozone in that region through two processes. First, the anticyclonic flow around the high brings lower latitude air poleward. This air will retain characteristics of its tropical or subtropical source region, including concentrations of trace gases, such as ozone.

Tropical and subtropical air has a lower ozone mixing ratio, reflected in a lower total column ozone, than polar air. This lower ozone air is transported into a region where ozone amounts are usually higher. Second, the warmer temperatures associated with a blocking high cause isentropic surfaces to bend upward, which have the effect of bending the tropopause upwards. Ozone density in the troposphere is lower than in the stratosphere, of course, so an increase in the vertical scale of the troposphere results in a lower column ozone density.

By pushing transporting tropospheric air poleward and bending the tropopause upwards, blocking highs can increase the time it takes STE processes to occur.

8.4.2 Cutoff Low Pressure Systems

Cutoff lows are upper level cyclones which are cut off or separated from the main flow of the upper tropospheric jet stream. They are usually associated with blocking patterns. The majority of cutoff lows form during summer months and can last for several days. In general, they form as the jet stream becomes distorted as an upper tropospheric trough elongates meridionally. As the system is cut off, it isolates air with characteristics of its polar source region. That is, it will contain cold air, high in potential vorticity, and trace gases with concentrations characteristic of higher latitudes.

Cutoff lows are obvious mechanisms for horizontal transport and are potentially important for STE. Cutoff lows are capable of large-scale cumulus convection. Within convective updrafts, tropospheric air can be transported across the
tropopause. This action can eventually erode the tropopause itself, creating a vertically mixed region of stratospheric and tropospheric air. The tropopause may then
reestablish itself at a higher altitude above the mixed layer, thereby capturing the
stratospheric ozone below.

8.4.3 Tropopause Folds

Another mechanism for STE is tropopause folding events. A stratospheric intrusion
of air that sinks into the baroclinic zone beneath the upper tropospheric jet stream
is known as a tropopause fold (see section 1.6.5). They form by a steepening of
the tropopause (and isentropes) at a jet core (see Fig. 8.7). Tropopause folds are the

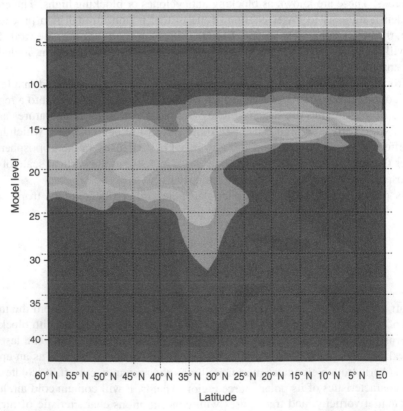

Fig. 8.7 ECMWF model simulated tropopause folding for a cross section at 81E on 1 February
1992, which shows both the tropopause and the ozone transport into the troposphere (Courtesy:
SPARC 2004)

dominant and most efficient form of STE in the middle latitudes. Folds usually occur off the western flank of cutoff low systems. Clean, dry stratospheric air, rich in ozone and potential vorticity, is transported downward to tropospheric levels. Observations of the circulation near folding events reveal that tropospheric air is being advected upward as well. This tropospheric air contains large amounts of water vapor, carbon monoxide, aerosols, and low values of potential vorticity.

Tropospheric folds are tongues of stratospheric air that get drawn into the troposphere around jet streams. They generally occur in winter and spring. They are observed near upper-tropospheric jets, usually downstream of a ridge. As the fold reaches downward into the troposphere, it can become irreversibly stretched to smaller and smaller scales, eventually becoming irreversibly mixed with the tropospheric air. Of course, such air brings with it higher values of ozone. This stratospheric ozone is thought to contribute about 10% of the global tropospheric ozone.

8.5 Exchange Processes Near the Tropical Tropopause

Exchange processes near the tropical tropopause have been a subject of much debate for a few decades, and it is still ongoing. Possible mechanisms include: (i) overshooting convection (as shown in Fig. 8.8) in *Micronesia*, where the air may overshoot, but getting it to stay at higher potential temperature may be difficult, and (ii) a *stratospheric fountain*, where the air is lifted into the stratosphere by heating. Observations show that the air enters the stratosphere throughout the year. This can be seen by the seasonal cycle of CO_2 and H_2O.

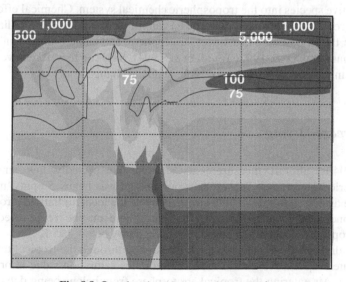

Fig. 8.8 Overshooting of air into the stratosphere

Table 8.1 Mass flux exchange between stratosphere and troposphere

Mass flux across the 100 hPa surface in 10^8 kg s^{-1}

	DJF	JJA	Annual mean
NH extratropics	−81	−26	−53
Tropics	114	56	85
SH extratropics	−33	−30	−32

The relative importance of these different mechanisms is not yet known. Many of them manifest themselves during the life cycles of extratropical cyclones. The quantification of the transport of stratospheric ozone into the troposphere is essential to our understanding of the tropospheric ozone budget. The understanding of the processes that transport water vapor from the tropical troposphere into the stratosphere is essential for the correct modeling of stratospheric water vapor mixing ratios. Also, the residence time of aircraft emissions depends on whether they are emitted in the troposphere or in the stratosphere, and on the intensity of mixing across the tropopause. The transport of air across the tropopause is not well quantified.

It can be seen from Table 8.1 that the northern hemispheric downward mass flux, especially during the winter season is much greater than the southern hemispheric downward mass flux.

Anthropogenic species transported from the troposphere into the stratosphere initiate much of the chemistry responsible for stratospheric ozone depletion. On the other hand, downward transport from the stratosphere not only constitutes the main removal mechanism for many stratospheric species, including those involved in ozone depletion, but also represents a significant input of ozone and other reactive species into the tropospheric chemical system. Chemical effects from stratosphere–troposphere exchange (STE) can, in turn, influence the radiative flux balance in the troposphere and lower stratosphere, and can do more than one way (e.g., Ramaswamy et al. 1992; Toumi et al. 1994). STE can therefore have a significant role in the radiative forcing of global climate change.

8.5.1 Freeze-Drying

In the 1940s, aircraft measurements showed unexpectedly low amounts of stratospheric water vapor. The water vapor content was below 5 ppmv, much less than if the air were saturated at the temperature of the local midlatitude tropopause. Brewer concluded that all air entering the stratosphere must be freeze-dried near the tropical tropopause, where temperatures are regularly below 195 K .

Freeze drying is a process in which air passing through the tropical tropopause has its water vapor mixing ratio reduced to the ice saturation value at or near the tropopause. Air entering the tropical stratosphere from below, being dehydrated as it enters, and then being gradually moistened, by methane oxidation and perhaps by weak mixing from the extratropics, is drawn upward in the tropics before being pushed poleward and eventually downward in middle and high latitudes.

The satellite data show the lowest zonal mean water vapor mixing ratios in the tropical lower stratosphere during northern hemisphere winter (about 23 ppmv), with values increasing with altitude and latitude to 46 ppmv. The sharpness of tropical temperature and water vapor minima is difficult to determine with satellite measurements that resolve only vertical scales greater than 23 km.

8.5.2 Atmospheric Tape Recorder

The water vapor mixing ratios of air entering the tropical stratosphere vary seasonally in phase with the annual cycle of tropical tropopause temperature. Just as a signal is recorded on a magnetic tape as it passes the head of a tape recorder, the minimum saturation mixing ratio near the tropical tropopause should be recorded on each layer of air moving upward in the large-scale tropical stratospheric circulation. Since, horizontal eddy transport by isentropic mixing between the tropics and extratropics is observed to be relatively weak, layers of air passing upward through the tropical tropopause should retain their water vapor signal for some time, perhaps many months, as they are advected upward by the large-scale circulation. With sufficient vertical resolution, the indication of the seasonally varying saturation mixing ratio of the tropical tropopause can then be observable in a time–height section as alternating layers of low and high water vapor mixing ratio anomalies that move upward at the speed of the large-scale circulation.

This tape recorder effect has confirmed by analysis of water vapor mixing ratio data obtained from upper atmosphere research satellite (UARS) observations. The annually varying mixing ratio signal produced by the freeze-drying process remains noticeable over an altitude range from 15 km to the middle stratosphere (see Fig. 8.9). The ascent rates just above the tropical tropopause are approximately consistent with the mass fluxes varying from about 0.2 mm s^{-1} in northern summer to about 0.4 mm s^{-1} in northern winter. For chemical reasons connected with methane oxidation chemistry, total hydrogen mixing ratios are uniform in the extratropical overworld with values close to, or weakly fluctuating about, the time-averaged tape signal. It follows that the total hydrogen signal can be due only to the recording head near the tropopause, and that, in the overworld, isentropic entrainment or mixing-in of extratropical air can hardly do other than attenuate the signal. Thus the total-hydrogen results confirm the weakness of isentropic entrainment from the extratropics.

About the amplitude of the observed tape signal in water vapor mixing ratios, it is suggested that the subtropical barrier is relatively permeable near the base of the overworld and is less permeable a few kilometres further up, where mixing-in is minimal. In the middle stratosphere, it again becomes more permeable, especially

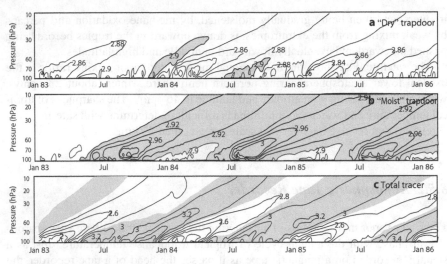

Fig. 8.9 Synthesized tape recorder (Mote et al. 1995, Courtesy: American Geophysical Union)

during strong westerly QBO episodes. Such a pattern is consistent with the horizontal eddy motion extending into the base of the overworld from the subtropical troposphere.

It is believed that the upward flow across the tropical tropopause is associated with the mass flow from tropical deep convection that penetrates into the lower stratosphere. The cumulonimbus penetrations could hyperventilate the lower tropical stratosphere (SPARC 2000). Such hyperventilation is most likely in the northern summer monsoon season, when surface conditions tend toward strengthening and deepening the cumulonimbus convection but the global-scale upwelling is slowest (Mote et al. 1995).

The deepest penetrations act like recording heads that are located furthest along the direction of tape travel. The dehydration signals from such recording heads have the greatest chance of surviving, i.e., of not being erased, or displaced back downward or sideways, by subsequent, higher injections, as the notional tape moves upward. The hyperventilation could be a contributing factor in determining the amplitude of the observed tape signal in total hydrogen. It would reinforce the effects of isentropic entrainment or mixing-in during northern summer, and work against them during northern winter.

8.5.3 Stratospheric Fountain

A stratospheric fountain is an area where air enters the stratosphere from the troposphere. The fountain occurs over the western tropical Pacific, northern Australia, Indonesia, and Malaysia in the November–March period and over the Bay of Bengal and India during the monsoon. It is suggested that the major portion of the

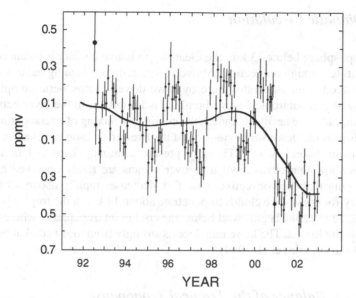

Fig. 8.10 Water vapor anomalies observed in the lower stratosphere (WMO 2007)

stratospheric air supply enters through these areas with most of the exchange occurring in the November–March period (Geller et al. 2002). The water vapor anomalies observed at 82 hPa level by satellite measurements is shown in Fig. 8.10.

The stratosphere contains less water than it would if the air entered at the average minimum tropical temperature. To explain this observation, it is hypothesized that air only enters the stratosphere at the coldest locations and times of the year. The stratospheric fountain exists over the Indonesian maritime continent during northern hemispheric winter.

The stratospheric fountain hypothesis is further reviewed by comparing estimates of the annually and zonally averaged volume mixing ratio (VMR) of water vapor entering the stratosphere to annually and zonally averaged estimates of the saturation VMR of the tropical tropopause region (Newell and Gould-Steuwart 1981). It is found that the VMR of water vapor entering the stratosphere (3.8 ± 0.3 ppmv) agrees well with the saturation VMR of the tropical tropopause region (4.0 ± 0.8 ppmv). However, the stratospheric fountain hypothesis fails in regions where the tropical tropopause is colder.

However, there are conceptual problems with the fountain mechanism. Satellite data show that air enters the stratosphere throughout the year. The stratospheric fountain idea is also contradicted by observational evidence that the average motion in the lower stratosphere is downward over the Indonesian subcontinent. Furthermore, long-term increases in midlatitude lower stratospheric water vapor have been observed while the tropical tropopause and lower stratosphere have cooled (Hartmann et al. 2001). If the fountain mechanism were the only important process acting, one would expect a decrease in water vapor coupled with a temperature decrease. Hence, other processes must also be important.

8.5.4 Diabatic Circulation

In the troposphere below 13 km, the clear sky radiative cooling is balanced by the latent heating through convection. Above 16 km, radiative heating balances the upwelling forced from the stratosphere by wave-driven stratospheric pumping. It is also realized that subtropical wave-breaking is important. In the layer between 13 and 16 km, radiative heating balances the cooling by mixing of overshooting cumulus clouds plus the downward extension of the Brewer-Dobson circulation.

The lower boundary of TTL may present a mixing barrier. Observational evidences show that cirrus anvil tops over oceans are rarely observed above 14 km, implying that the convective mass flux decreases rapidly above 13 km. The probability for convective clouds to penetrate above 14 km in the tropics is roughly 1%. Increase in ozone begins well below the cold point tropopause, where the temperature is the lowest. The lapse rate departs strongly from moist adiabat beginning below 14 km (Folkins et al. 1999).

8.5.5 Heat Balance of the Tropical Tropopause

Cirrus clouds near the tropopause are heated by radiation unless they lie above the cold deep convective anvils. In that case, radiative cooling can balance significant subsidence. Holton and Gettleman (2001) hypothesized that horizontal advection of air entering the stratosphere at other longitudes leads to passage through the cold trap. An air parcel is likely to visit the cold trap before it ascends into the stratosphere by radiative heating. Therefore, air is freeze-dried to cold trap temperatures even though stratospheric entry was elsewhere.

The relation between temperatures near the tropical tropopause and stratospheric entry conditions for water is thus not as simple as postulated by Brewer. Researchers are therefore looking for a dehydration mechanism that explains both the mean values and temporal changes observed in stratospheric water vapor. Two mechanisms are currently considered to be the leading candidates for tropical tropopause layer dehydration. One involves convective processes that loft ice into the upper troposphere/lower stratosphere (UT/LS). The second assumes that air parcels gradually ascend, eventually passing though the region of coldest temperatures in the tropical tropopause layer. Two-step processes are also possible, with convection lofting parcels and ice to the upper troposphere, followed by gradual dehydration as the parcels ascend into the stratosphere. If only one mechanism is active, it should leave a distinct isotopic signature in the water vapor that enters the stratosphere.

8.6 Stratospheric Drain

The idea of stratospheric drain first proposed by Sherwood (2000) based on the observational evidence is that there is subsidence, not rising motion, in the West Pacific. Sherwood and Dessler (2001) addressed this issue with processing by

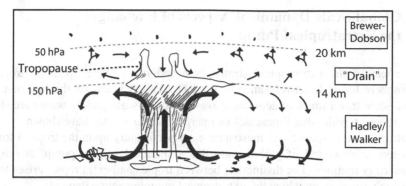

Fig. 8.11 Illustration of the diabatic mean flow (arrows) and its relationship to the locations of deep convection and convective outflows (Sherwood 2000, Courtesy: American Geophysical Union)

overshooting convection in the TTL. Hartmann et al. (2001) pointed at the radiative cooling from subvisible cirrus to account for the thermodynamic balance.

While one may argue whether the data provide convincing evidence for a stratospheric drain, there is certainly no evidence of a West Pacific fountain. Thus, temperature at entry to stratosphere is likely higher than ice saturation for observed stratospheric mixing ratios. The question still remains how to account for the observed dehydration. We have a puzzle that observed stratospheric aridity apparently requires freeze-drying by upward passage through West Pacific cold trap, while the observations suggest mean subsidence at cold trap instead of rising motion.

Figure 8.11 illustrates the circulations in the troposphere and stratosphere. This circulation is qualitatively similar to that deduced by Gage et al. (1991), though substantially weaker. The implications of the circulation for the energy budget of the layer will now be considered.

The calculated vertical velocity profiles indicate that the maritime continent region, which has been dubbed a stratospheric fountain, seems to contain what may be described as a stratospheric drain, where more mass sinks around convective towers than rises in them. This descent was calculated from sounding data and its uncertainty carefully estimated. Compensating ascent through the tropopause must occur elsewhere in the tropics; it was not observed directly, but may be inferred from the net radiative heating above 150 hPa calculated by radiation models.

Cooling by overshooting convective drafts has been offered as an explanation for the unusual circulation pattern, but this hypothesis is only supported qualitatively as yet. Models and more detailed observations will be needed to test it further. If the hypothesis is true, then nonhydrostatic convective effects are driving a small but possibly significant component of the general circulation. This would have important implications for climate and global forecast models, since existing parameterizations of convection are generally hydrostatic. Unlike the subtropical trade wind inversion, where motion is downward, the net upward motion at the tropopause makes detailed understanding of the entrainment mechanisms essential to quantifying the transport into the quiescent layer above.

8.7 Global-Scale Dynamical Aspects of Exchange: the Extratropical Pump

The effect of the extratropical stratosphere and mesosphere upon the tropical stratosphere has a direct relevance to the STE. This depends on the fact that the global-scale travel times of acoustic waves and large-scale gravity waves are short in comparison with other timescales of interest. Earlier studies have shown that the extratropical stratosphere and mesosphere act persistently upon the tropical lower stratosphere as a kind of global-scale fluid-dynamical suction pump, driven by certain eddy motions. The distinction between tropics and extratropics arises from the Earth's rotation, on which the extratropical pumping action depends.

The most important of the eddy motions is the one known as the breaking Rossby waves and related potential vorticity transporting motions. They have one-signed or ratchet-like character related to the sense of the Earth's rotation. These eddy effects have a strong tendency to add up and to give a persistently one-way pumping action, whose strength varies seasonally and interannually, in which air is gradually withdrawn from the tropical stratosphere and pushed poleward and ultimately downward. Where air parcels are being pulled upward (as in the tropical stratosphere), adiabatic cooling pulls temperatures below their radiative values; where air parcels are being pushed downward (as happens most strongly in high latitudes in winter and spring), adiabatic warming pushes temperatures above their radiative values.

This mechanically pumped global-scale circulation is often referred to as the wave-driven circulation. Figure 8.12 illustrates the schematic cross section of the wave-driven circulation in the stratopshere (Holton 2004). Its existence explains the observed temperature difference from the radiative values and it is essential in controlling the tropospheric lifetimes of long-lived trace chemicals, such as CFCs, nitrous oxide, and methane that have photochemical sinks in the stratosphere. The physical reality of the extratropical pumping action is well known, but its precise strength and its seasonal and interannual variability, and certain nonlinear aspects of the resulting circulation dynamics that bear on the precise way the pumping action reaches into the tropics are uncertain.

8.8 Equation for Wave-Driven Circulation

Based on the theoretical formulation (McIntyre 1992; Holton et al. 1995; Haynes 2005) the underlying physical principles for the global-scale dynamical aspects of exchange processes can be discussed. For a zonally symmetric atmosphere, the wave-induced force per unit mass is $\overline{G}$.

(i) The zonal momentum equation is

$$\frac{\partial \overline{u}}{\partial t} - 2\Omega \sin \phi \overline{v}^* = \overline{G} \tag{8.1}$$

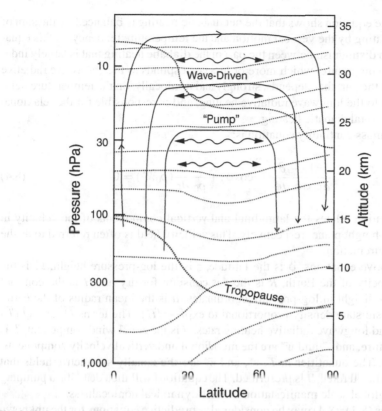

Fig. 8.12 Schematic diagram illustrating the wave-driven circulation in the middle atmosphere. Thin dashed lines represent isentropic surfaces. Solid lines are meridional-circulation-driven by the wave-induced forcing. Wavy arrows indicate meridional circulation and mixing by eddy motions (Holton 2004; Courtesy: Elsevier)

It states that the wave-induced force is balanced by the zonal flow acceleration plus the Coriolis force due to latitudinal motion.

(ii) The thermal wind equation is:

$$2\Omega \sin\phi \frac{\partial \bar{u}}{\partial z} + \frac{R}{aH}\frac{\partial \bar{T}}{\partial \phi} = 0 \qquad (8.2)$$

It suggests that the vertical shear of the zonal wind is proportional to the latitudinal gradient of the temperature. The coupling between the wind and temperature fields follows from the assumption that the zonal flow is in hydrostatic and geostrophic balance, thereby including all important nonlocal effects.

(iii) The thermodynamic energy equation is:

$$\frac{\partial \bar{T}}{\partial t} + \bar{w}^* \left[\frac{HN^2}{R}\right] = \bar{Q}_s + \bar{Q}_l(\bar{T}) \qquad (8.3)$$

The above equation shows that the net diabatic heating is balanced by the sum of adiabatic heating by the vertical motion and the temperature tendency. In this equation, one can distinguish between the part of the diabatic heating that is largely independent of temperature, which more or less corresponds to the shortwave radiative heating, and the part that depends, strongly and nonlocally, on the temperature field, corresponds to the longwave radiative heating, and is responsible for the relaxional nature of the total diabatic heating.

(iv) The mass continuity equation is:

$$\frac{1}{a\cos\phi}\frac{\partial}{\partial\phi}(\bar{v}^*\cos\phi)+\frac{1}{\rho_0}\frac{\partial}{\partial z}(\rho_0\bar{w}^*)=0 \qquad (8.4)$$

This equation relates the latitudinal and vertical components of the velocity in the latitude–height plane, respectively. This velocity field is often referred to as the meridional circulation.

In the above equations, ϕ is the latitude, z is the log-pressure height, Ω is the angular velocity of the Earth, R is the gas constant for dry air. H is the constant density scale height in log-pressure coordinates, a is the mean radius of the earth, ρ_0 is the basic state density proportional to $\exp(-z/H)$. The terms $\bar{Q}_s$ and $\bar{Q}_l(\bar{T})$ are short- and longwave radiative heating rates, $\bar{u}$ is the zonal wind component, $\bar{T}$ is the temperature, and $\bar{v}^*$ and $\bar{w}^*$ are the meridional and vertical velocity components, respectively. The quantities $\bar{u}$, $\bar{T}$, $\bar{v}^*$, and $\bar{w}^*$ are the zonally symmetric fields that the wave-induced force $\bar{G}$ is prescribed. The equations will then describe a pumping action of a global scale manifestation of fluid-dynamical nonlocalness.

Equations (8.1)–(8.4) may be considered as predictive equations for the unknown quantities, $\partial\bar{u}/\partial t$, $\partial\bar{T}/\partial t$, $\bar{v}^*$, and $\bar{w}^*$. In order to simplify calculation and to focus on timescales of importance, let us assume that the time dependence is harmonic, with constant frequency σ. So we can represent

$$\bar{G}=Re(\hat{G}e^{i\sigma t}); \qquad (8.5)$$

and

$$\bar{Q}_s=Re(\hat{Q}e^{i\sigma t}) \qquad (8.6)$$

where $\hat{G}$ and $\hat{\theta}$ are complex amplitudes.

Now characterize the response by the single variable $\hat{w}$, such that the vertical velocity

$$\bar{w}^*=Re\left[\hat{w}(\phi,z)e^{i\sigma t}\right] \qquad (8.7)$$

To parameterize the relaxational dependence of the diabatic field on temperature, let us assume $\bar{Q}_l=\alpha\bar{T}$, where α^{-1} is the Newtonian cooling with constant timescale.

With theses assumptions Eqs. (8.1)–(8.4) can be combined into a single equation as

$$\frac{\partial}{\partial z}\left(\frac{1}{\rho}\frac{\partial(\rho_0\hat{w})}{\partial z}\right) + \left(\frac{i\sigma}{i\sigma + \alpha}\right)\frac{N^2}{4\Omega^2 a^2 \cos\phi}\frac{\partial}{\partial\phi}\left(\frac{\cos\phi}{\sin^2\phi}\frac{\partial\hat{w}}{\partial\phi}\right)$$
$$= \left(\frac{i\sigma}{i\sigma + \alpha}\right)\frac{(R/H)}{4\Omega^2 a^2 \cos\phi}\left[\frac{\partial}{\partial\phi}\left(\frac{\cos\phi}{\sin^2\phi}\frac{\partial\hat{Q}}{\partial\phi}\right)\right] + \frac{1}{2\Omega a \cos\phi}\frac{\partial}{\partial\phi}\left(\frac{\cos\phi}{\sin\phi}\frac{\partial\hat{G}}{\partial z}\right)$$

$$(8.8)$$

This equation provides a suitable model for studying nonlocal control of the meridional circulation by the wave-induced force and by the shortwave heating $\overline{Q}_s$.

In Eq. (8.8) the first term on the left-hand side is elliptic, so that the vertical velocity response to a forcing localized to a region spreads away from that region. In the case of potential vorticity inversion, the flow is assumed balanced and the nonlocal response to the applied forcing may be deduced by applying the appropriate inversion operator to the potential vorticity field which changes because of the applied force.

In the high-frequency range, when thermal relaxation is negligible, the factor $[i\sigma/(i\sigma + \alpha)]$, which appears both on the left-hand side and right-hand side of Eq. (8.8), becomes approximately 1. The shortwave heating term on the right-hand side can be significant because both wave force and shortwave heating may be effective in driving a mean meridional circulation.

Questions and Exercises

8.1. What is Stratosphere-Troposphere Exchange (STE)? How is the Brewer Dobson circulation involved in stratosphere troposphere exchange?

8.2. Discuss the part played by the upper troposphere and lower stratosphere in the exchange processes between troposphere and stratosphere.

8.3. Discuss the exchange processes through the midlatitude and extratropical tropopause layers.

8.4. What is Freeze drying mechanism? Explain the exchange processes taking place nearer to the tropical tropopause region.

8.5. What conditions are required for the cross-tropopause flow? What are the difference in characteristics of the exchange through the tropopause, between tropics and extratropics?

8.6. How do blocking highs affect STE? In what ways do blocking anticyclones (highs) lower the total ozone over a region?

8.7. Discuss the conditions for the formation of tropopause folds and its role in mixing ozone-rich stratospheric air with the dense and moist tropospheric air.

8.8. What is meant by 'Atmospheric tape recorder'? Why it is called so? What information will we get about the lower stratosphere?

8.9. Explain the terms, 'stratospheric fountain' and 'stratospheric drain'. How they contribute to the stratosphere troposphere exchange processes in tropics.

8.10. What is called as 'extratropical pump'? Discuss its relevance to the STE in terms of the effect of extratropical stratosphere on tropical stratosphere.

References

Austin J, Li F (2006) On the relationship between the strength of the Brewer-Dobson circulation and the age of stratospheric air, Geophys Res Lett, 33, L17807, doi 10.1029/2006GL026867

Butchart N, Scaife AA (2001) Removal of chlorofluorocarbons by increased mass exchange between the stratosphere and the troposphere in a changing climate, Nature, 410 (6830): 799–802

Butchart N, Scaife AA, Bourqui M, Grandpre MJ de, Hare SHE, Kettleborough J, Langematz U, Manzini E, Sassi F, Shibata K, Shindell D, Sigmond M (2006) Simulations of anthropogenic change in the strength of the Brewer-Dobson circulation, Clim Dyn, 27 (7–8): 727–741, doi 10.1007/s00382-006-0162-4

Folkins I, Loewenstein M, Podolske J, Oltmans SJ, Proffitt M (1999), A barrier to vertical mixing at 14 km in the tropics: Evidence from ozonesondes and aircraft measurements, J Geophys Res, 104(D18): 22095–22102

Gage KS, Balsley BB, Ecklund WL, Carter DA, McAfee JR (1991) Wind-profiler related research in the tropical Pacific, J Geophys Res, 96: 3209–3220

Geller MA, Zhou XL, Zhang MH (2002) Simulations of the interannual variability of stratospheric water vapour, J Atmos Sci, 59 (6): 1076–1085

Gettelman A, Forster PM. de F, (2002) A climatology of the tropical tropopause layer, J Meteorol Soc Japan, 80 (4B): 911–924

Hartmann DL, Wallace JM, Limpasuvan V, Thompson DWJ, Holton JR (2000) Can ozone depletion and global warming interact to produce rapid climate change?, Proc Natl Acad Sci, 97 (4): 1412–1417

Hartmann DL, Moy LA, Fu Q (2001) Tropical convection and energy balance at the top of the atmosphere, J Climate, 14: 4495–4511

Haynes PH (2005) Stratospheric dynamics, Annu Rev Fluid Mech, 37: 263–293

Highwood EJ, Hoskins BJ (1998) The tropical tropopause, Quart J Roy Met Soc, 124: 1579–1608

Holton JR (2004) An Introduction to Dynamic Meteorology, fourth edition, Elsevier

Holton JR, Gettleman A (2001) Horizontal transport and dehydration of the stratosphere, Geophys Res Lett, 28: 2799–2802

Holton JR, Haynes PH, McIntyre ME, Douglass AR, Rood RB, Pfister L (1995) Stratosphere-troposphere exchange, Rev Geophys, 33 (4): 403–440

Hoskins BJ (1987) Towards a PV- view of the general circulation, Tellus, 43B: 27–35

McIntyre ME (1992) Atmospheric dynamics, some fundamentals, with observational implications. In Proc. Internatnl School Phys Enrico Fermi, CXV Course ed. J.C. Gille, G. Visconti, Amsterdam

McIntyre ME, Palmer TN (1983) Breaking planetary waves into the stratosphere, Nature, 305: 593–600

Mote PW, Rosenlof KH, McIntyre ME, Carr ES, Kinnersley KH, Pumphrey HC, Harwood RS, Holton JR, Russel III JM, Waters JW, Gille JC (1995) An atmospheric tape recorder: the imprint of tropical tropopause temperatures on stratospheric water vapour, J Geophys Res, 100: 8873–8892

Newell RE, Gould-Steuwart S (1981) A stratospheric fountain?, J Atmos Sci, 38: 2789–2796

Ramaswamy V, Schwarzkopf MD Shine KP (1992), Radiative forcing of climate from halo-carbon induced global stratospheric ozone loss, Nature, 355: 810–812

Ramaswamy V, Schwarzkopf MD, Randel WJ, Santer BD, Soden BJ, Stenchikov GL (2006), Anthropogenic and natural influences in the evolution of lower stratospheric cooling, Science, 311 (5764): 1138–1141, doi 10.1126/science.1122587

Rind D, Lerner J, McLinden C (2001) Changes of tracer distributions in the doubled CO_2 climate, J Geophys Res, 106 (D22): 28061–28080, doi:10.1029/2001JD000439, 2001

Shepherd TG (2002) Issues in stratosphere-troposphere coupling, J. Meteorol. Soc. Japan, 80 (4B): 769–792

Sherwood SC (2000) A "stratospheric drain" over the maritime continent, Geophys Res. Lett., 27: 677–680

Sherwood SC, Dessler AE (2001) A model for transport across the tropical tropopause, J Atmos Sci, 58(7): 765–779

SPARC (Stratospheric Processes And their Role in Climate) (2000), SPARC Assessment of Upper Tropospheric and Stratospheric Water Vapour, D. Kley, J.M. Russell III, and C. Phillips (eds.), World Climate Research Progam Report 113, SPARC Report No. 2, 312 pp., Verrires le Buisson, France

Toumi R, Bekki S, Law KS (1994) Indirect influence of ozone depletion on climate forcing by clouds, Nature, 372: 348–351

UGAMP (The UK Universities Global Atmospheric Modelling Program), A new understanding of stratosphere troposphere exchange, NERC. http://www.ugamp.nerc.ac.uk/research/brochure/pictures/wave_pump.gif

WMO (World Meteorological Organization), Scientific Assessment of Ozone Depletion: 2002, Global Ozone Research and Monitoring Project Report No. 47, Geneva, Switzerland, 2003

WMO (World Meteorological Organization), ScientificAssessment of Ozone Depletion: 2006, Global Ozone Research and Monitoring Project Report No. 50, Geneva, Switzerland, 2007

Chapter 9
Stratospheric Influence on Tropospheric Weather and Climate

9.1 Introduction

Changes in the tropospheric weather and climate will affect the ozone-rich stratosphere through modification in radiation, dynamics, transport, and chemical composition. In turn, changes to the stratosphere will affect climate through radiative processes, and consequential variation in temperature gradients will influence atmospheric dynamics. Therefore, tropospheric weather and climate system is coupled with the stratospheric thermal structure, dynamics, and the evolution of the ozone layer. Understanding all the processes involved is made more complex by the fact that many of the interactions are nonlinear.

There is increasing evidence that stratospheric processes influence the tropospheric circulation across a wide range of timescales. The tropospheric circulation may also be linked to the stratosphere on longer timescales. For example, it has been found that stratospheric forcings in relationship with ozone depletion, volcanic aerosols, or the quasi-biennial oscillation exhibit a signature in surface climate. Such a coupling may be important for more realistic simulations of anthropogenic climate change in relation with secular changes of greenhouse gases. Various factors involved in stratosphere–troposphere interactions are schematically shown in Fig. 9.1 as discussed by Holton et al. (1995).

The troposphere influences the stratosphere mainly through the vertical propagation of atmospheric waves. The stratosphere coordinates this chaotically varying wave forcing from below to create long-lived changes in the stratospheric circulation. These changes in the stratosphere can feedback to waves returning to the troposphere and create long-lasting effects on tropospheric weather and climate. A key to understanding stratospheric influence on weather and climate is the awareness that the stratosphere usually changes very slowly, and that conditions in the lower stratosphere affect surface weather and climate.

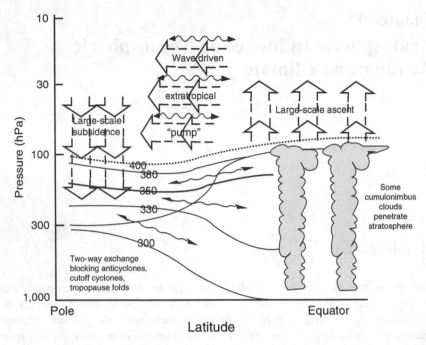

Fig. 9.1 Basic mechanism for the dynamical interaction and exchange between the stratosphere and troposphere (Holton et al. 1995, Courtesy: American Geophysical Union)

9.2 Radiative Forcing of Stratosphere–Troposphere Interactions

Troposphere is heated by the absorption of outgoing terrestrial radiation by greenhouse gases, mainly CO_2 and water vapor. Troposphere is in radiative equilibrium between latent heating and radiative cooling by the greenhouse gases. In the stratosphere, however, increased greenhouse gases lead to a net cooling as they emit more infrared radiation out to space than they absorb. Infrared emission increases with local temperature, so the cooling effect increases with altitude, maximizing near the stratopause, where stratospheric temperatures are highest.

The effect of stratospheric cooling varies with latitude, as it depends on the balance between absorption of terrestrial radiation from below and local emission. The net cooling effect of greenhouse gases extends to lower levels at high latitudes, roughly following the tropopause.

Any change in radiatively active gas concentrations will change the balance between incoming solar and outgoing terrestrial radiation in the atmosphere. Radiative forcing is conventionally given as the net change in radiative fluxes at the tropopause, which can be a reasonable indicator of the surface temperature response.

To determine radiative forcing from stratospheric ozone changes, it is important to distinguish between instantaneous effects and the effects after the stratospheric temperature has adapted. Depletion of ozone in the lower stratosphere causes an instantaneous increase in the shortwave solar flux at the tropopause and a slight reduction of longwave radiation. The net instantaneous effect is a positive radiative forcing.

However, the decrease in ozone causes less absorption of solar and terrestrial radiation, leading to a local cooling. After the stratosphere has adjusted, the net effect of ozone depletion in the lower stratosphere is a negative radiative forcing. In contrast, ozone depletion in the middle and upper stratosphere causes a slight positive radiative forcing. The maximum sensitivity of radiative forcing for ozone changes is found in the tropopause region. The maximum sensitivity of surface temperatures to ozone changes also peaks near the tropopause (Forster and Shine 1997). Quantifying the impact of stratospheric ozone changes on surface temperatures is more complicated than estimating radiative forcing.

Figure 9.2 shows the global and annual mean temperature trends for the period 1980–2000, from an average of model results using observed changes in ozone and greenhouse gases, and idealized water vapor trends. The stratospheric cooling observed during the past two decades has slowed in recent years (WMO 2007). Satellite and radiosonde measurements reveal an overall cooling trend in the global-mean lower stratosphere of approximately 0.5 K decade^{-1} over the 1979–2005 period, with a slowdown in the temperature decline since the late 1990s. The overall temperature decrease is disrupted by transient warmings of the stratosphere associated

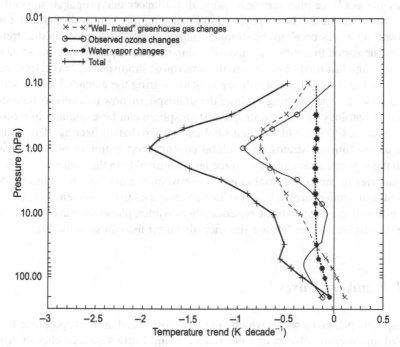

Fig. 9.2 Global and annual mean temperature trends from 1980 to 2000 (Shine et al. 2003, Courtesy: W J Randell)

with the major volcanic eruptions in 1982 and 1991. Model calculations suggest that the observed ozone loss is the predominant cause of the cooling observed over this period. The lower stratospheric cooling is evident at all latitudes, in particular in both Arctic and Antarctic winter/spring lower stratosphere but with considerable interannual variability.

9.3 Wave-Induced Interactions

Although the stratosphere and troposphere are in many ways distinct, the atmosphere is continuous, allowing vertical wave propagation and a variety of other dynamical interactions between these regions. The dynamical coupling of the stratosphere and troposphere is primarily mediated by wave dynamics. A variety of waves originate in the troposphere, propagate upward into the stratosphere and above, and then dissipate, shaping the spatial and temporal structure of the stratospheric flow. This traditional view of a passive stratosphere has more recently given way to a greater appreciation of the stratosphere's ability to shape not only its own evolution but that of the troposphere as well. The thermal structure of the stratosphere and its seasonal variability basically depend on the dynamics of waves that are generated in the troposphere.

During northern winter, air flowing over mountain ranges and continental landmasses induces large planetary scale wave disturbances that propagate upward, refract, and reflect in the stratosphere, transferring momentum as they interact with the flow. The waves break in the stratosphere and create fluctuations in the strength of the polar vortex that move downward within the stratosphere on a timescale of a few weeks and last for 1–2 months in the lowermost stratosphere. This effect occurs only when the stratospheric winds are westerly during the extended winter season from October to April. During summer the stratospheric flow is relatively quiescent.

The climatology of the extratropical stratosphere can be explained in terms of wave dynamics along with the seasonal cycle of radiative heating. The easterly winds of the summer stratosphere inhibit upward propagation of planetary waves and so the summer stratosphere is much less disturbed than the winter stratosphere. Asymmetries in the continental landmass between the northern hemisphere (NH) and southern hemisphere (SH) necessitate asymmetries in the efficiency of planetary wave generation mechanisms. Consequently in winter, planetary wave disturbances in the stratosphere of the NH are significantly larger than those in the SH.

9.3.1 Planetary Waves

Atmospheric planetary waves which are mainly produced in the troposphere by dynamical and thermal effects of solar radiation and Earth's surface characteristics, in certain conditions will propagate to the middle atmosphere and play important

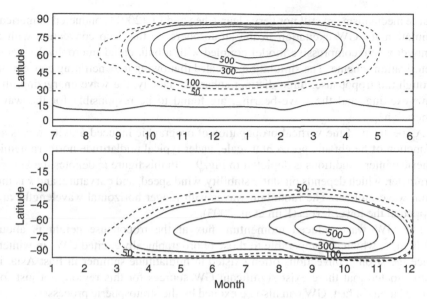

Fig. 9.3 Monthly amplitude of deviations in the 10 hPa geopotential height for northern hemisphere (top panel) and southern hemisphere (bottom panel) (Courtesy: NASA)

roles in regulating stratospheric circulation and ozone and distribution of other constituents. Figure 9.3 shows the average monthly amplitude of deviations in the 10 hPa geopotential height field for a full year for northern hemisphere (top panel) and southern hemisphere (bottom panel) – time axes shifted for comparison of northern and southern hemisphere winters.

Planetary waves are often neither steady nor conservative. There exists planetary wave–mean flow interaction and coupling. A noticeable difference of planetary wave propagation emerges clearly at middle and high latitudes with obvious larger upward and equatorward EP fluxes in the easterly phase than in the westerly phase in winter. By modulating the propagation of planetary waves, the tropical QBO introduces significant variability of the planetary wave amplitudes and the residual circulation in the northern hemisphere. The planetary wave amplitudes are shown to be greater during the easterly phase than during westerly winter. Planetary wave activity that induces stratosphere–troposphere coupling through QBO, stratospheric warming, etc. are discussed in the following sections.

9.3.2 Gravity Waves

Gravity waves (GW) play an important role in the stratospheric circulation. It is commonly recognized that sources of the middle atmosphere GW are mainly from the troposphere but the detailed information and quantitative description of the source mechanism are still not very clear. Chen and Lu (2001) conducted numerical simulation for GW excitation in the stratosphere by the deep convection with a nonhydrostatic compressive model coupled with a bulk cloud microphysics parameterization scheme. The simulation revealed that owing to penetrating convection through the tropopause, three distinct subareas, namely, the wave-energizing, the wave-exciting, and the wave-bearing, are found to be responsible for the wave generation.

A spectrum of the vertical propagation of orographic induced gravity waves as a function of height and horizontal scale, under typical midlatitude northern hemispheric winter conditions is depicted in Fig. 9.4. In this figure l^2 denotes the Scorer parameter, which depends on static stability, wind speed, and curvature, and k is the zonal wave number. The mountain waves with longer horizontal wavelengths are trapped in the lower levels (Kim et al. 2003).

The GW wave-induced momentum flux at the tropopause height is about 0.3 N m^{-2}, which is comparable to those of orographically excited GWs in winter. Since the strong convections often existed in midlatitude summer in East Asia, it is anticipated that there exist significant GW sources for this region, not just for tropical area. In fact, GW can also be excited by the stratospheric processes.

Based on observation of stratospheric sudden warming (SSW), it is noted that owing to upward propagation of quasi-stationary planetary waves, strong ageostrophic motion existed in the stratosphere, which caused strong divergence field and strong long period GWs. Huang and Chen (2002) simulated such a mechanism of GW excitation with a linear barotropic spherical-spectral model. This

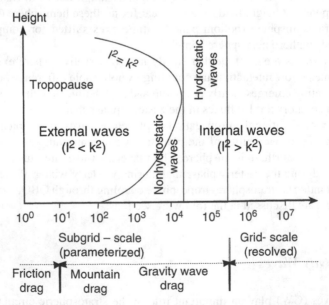

Fig. 9.4 Gravity wave-forcing in the troposphere and stratosphere (Kim et al. 2003, Courtesy: Canadian Meteorological and Oceanographic Society)

simulation study reveals that the stratospheric geostrophic adaptation process is also a GW excitation mechanism within the middle atmosphere.

Results of numerical simulation by Yue and Yi (2001) show that after several periods, the gravity wave packets propagate steadily upward and keep their shape well. The interior points and boundaries of the simulated regions are stable. During the propagation the amplitude of wave-associated perturbation velocities increases with the increasing height. It is also found that under nonlinear condition, a gravity wave packet is still able to keep some characterizations under linear propagation.

9.4 Role of Quasi-Biennial Oscillation in Coupling Process

In the tropical stratosphere, the dominant form of variability is a quasi-periodic (2–3-year) wave-driven descending zonal mean wind reversal, called the quasi-biennial oscillation (QBO). The period of the QBO at any level varies from 2 to 3 years, and it could probably be predicted for about 1 year. The QBO is observed to affect the global stratospheric circulation. It modulates a variety of tropical and extratropical phenomena including the strength and stability of the wintertime polar vortex, and the distribution of ozone and other gases (Baldwin et al. 2001).

Figure 9.5 illustrates the schematic picture of the dynamical overview of the QBO during northern winter (Baldwin et al. 2001). In the tropics, the stratospheric QBO is driven by the upward propagating gravity, inertia gravity, and Kelvin and Rossby gravity waves. In the middle and high latitudes, it is maintained by the

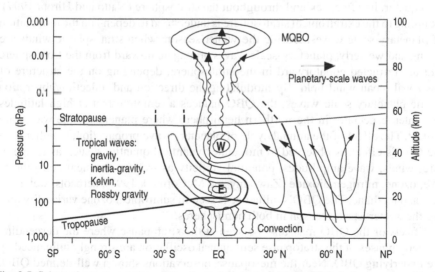

Fig. 9.5 Schematic representation of the dynamical control of the QBO during northern hemispheric winter (Baldwin et al. 2001, Courtesy: American Geophysical Union)

planetary-scale waves. The contours in the tropics are similar to the observed wind values when the QBO is easterly. It can be seen that the QBO extends to the mesospheric region and even above 80 km.

The QBO is driven by the dissipation of a variety of equatorial waves and gravity waves that are primarily forced by deep cumulus convection in the tropics. The stratospheric QBO effects extend to Earth's surface during northern midwinter. There is also observational evidence that the QBO modulates the depth of the troposphere in the tropics and subtropics, affecting convection, monsoon circulations, and hurricanes.

Although the amplitude of the QBO decreases rapidly away from the equator, observations and theory show that the QBO affects a much larger region of the atmosphere. Through wave coupling, the QBO affects the extratropical stratosphere during the winter season, especially in the northern hemisphere where planetary wave amplitudes are large. These effects also appear in constituents such as ozone. In the high-latitude northern winter, the QBOs modulation of the polar vortex may affect the troposphere through downward penetration. Tropical tropospheric observations show intriguing quasi-biennial signals which may be related to the stratospheric QBO. The QBO has been linked to variability in the upper stratosphere, mesosphere, and even ionospheric F layer (see Fig. 9.5).

9.4.1 Effects of the QBO in the Stratosphere and Mesosphere

Although the QBO is observed to be confined to within about 25° north and south of the equator, its effects extend throughout the stratosphere (Naito and Hirota 1997). Coupling to the extratropical stratosphere is understood to depend on the modulation of planetary-scale waves. During the winter season, when stratospheric winds are strong and westerly, planetary-scale waves propagate upward from the troposphere and are refracted equatorward in the stratosphere, depending on the structure of the zonal mean wind field. By modulating the direction and reflection/absorption of the planetary scale waves, the QBO induces a remote effect at high latitudes in winter, especially in the northern hemisphere where planetary-scale waves are largest. The effect of the annual cycle of planetary wave propagation, together with the QBO's effect, is observed to modulate dynamical quantities, such as temperature, winds, wave amplitudes, potential vorticity as well as chemical constituents like, ozone, nitrogen dioxide (Zawodny and McCormick 1991), aerosols, water vapor, and methane. The QBO's influence is seen in subtropical ozone variability during the winter–spring season in both hemispheres.

Effects of the QBO may also be seen at the stratopause where the descending westerly phases of the stratopause semiannual oscillation are strongly influenced by the underlying QBO. Near the mesopause, observations show a well-defined QBO, which may be driven by the selective filtering of small-scale gravity waves by the underlying winds they traverse. Chen and Robinson (1992) documented a statistical link between the phase of the QBO and the ionospheric equatorial ionization anomaly. A mechanism was proposed in which the planetary waves modulate the tidal wind, and by means of the dynamo effect, change electric fields in the E region, which can be transferred to the F region along geomagnetic field lines to cause variations in the equatorial ionization anomaly.

9.4.2 QBO Effects on the Troposphere

The QBO's effects on the troposphere are suggestive, but are not well understood. Gray (1984) has demonstrated an intriguing and significant link between the phase of the QBO and hurricane formation. The equatorial troposphere shows variability on the timescale of the QBO, but a direct link to the stratospheric QBO has not been established. It is possible that the QBO may influence the high-latitude northern troposphere through its effect on the stratospheric polar vortex. Coupling between stratospheric zonal mean wind and the mid-tropospheric North Atlantic oscillation is strong, but the cause and effect are not clear. It is possible that QBO-induced high-latitude wind anomalies penetrate downward into the troposphere (Gray 2003; Gray et al. 2004).

9.4.3 Stratospheric QBO and Tropospheric Biennial Oscillation

Signals of biennial oscillations with periods ranging from 20 to 32 months are noted in the tropospheric temperature over the tropics (Sathiyamurthy and Mohanakumar 2002). The phase of the tropospheric biennial oscillation (TBO) in temperature does not vary with height from surface to the level of tropopause and is found to be associated with the intensity of the monsoon rainfall. Temperature over the tropical region shows Quasi-Biennial Oscillation (QBO) in lower stratosphere. Phases of the QBO and TBO in temperature meet at tropopause level (see Fig. 9.6). Where they meet, phases of the QBO and TBO are unsynchronized during the decade 1971–1981 and synchronized during next decade, 1982–1992. The QBO in zonal wind has neither interdecadal variability nor disturbances.

9.5 Sudden Stratospheric Warming and Its Association in STI

The sudden stratospheric warming (SSW) is caused by a rapid amplification of planetary waves propagating upward from the troposphere. Planetary waves deposit westward momentum and create a strong meridional circulation which produces a large warming in the polar stratosphere due to adiabatic heating.

During winter, tropospheric planetary waves propagate into the stratosphere along the westerly jet. On the contrary, the zonal-mean zonal wind anomalies slowly

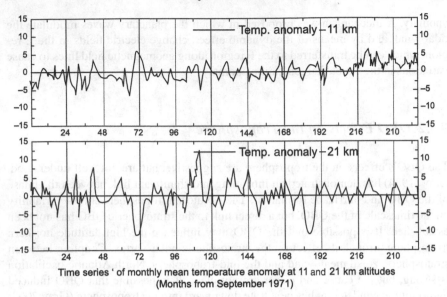

Time series ' of monthly mean temperature anomaly at 11 and 21 km altitudes
(Months from September 1971)

Fig. 9.6 Time series of monthly mean temperature anomaly at 11 km (troposphere) and 21 km (stratosphere) (Sathiyamurthy and Mohanakumar 2000)

propagate from the subtropical upper stratosphere to the polar region of the lower stratosphere and the troposphere during the boreal winter. It has been shown that SSWs occur in association with slowly propagating zonal-mean zonal wind anomalies, and the related changes in the troposphere exhibit the annular mode-like structure (Kodera et al. 2000). Baldwin and Dunkerton (1999) also showed that the downward propagation of the AO from the stratosphere to the troposphere occurs in association with SSWs.

9.5.1 Stratospheric Warmings and Downward Control

Most of the variability in the stratosphere is associated with variations in the large-scale wave driving, which is strongly associated with the occurrence of stratospheric warmings. In the troposphere wave-driving is also necessary to move the westerly jets in latitude, but this driving is provided by transient baroclinic waves as well as quasi-stationary planetary waves. Although stratospheric annular variability and tropospheric annular variability are coupled at times, tropospheric annular variations also occur independent of stratospheric annular variations (Kodera and Kuroda 2000). It appears that low-frequency quasi-barotropic waves are important for producing changes in the polarity of annular modes of variation in the troposphere, but high-frequency baroclinic waves are vital for maintaining the persistence of these anomalies (Lorenz and Hartmann 2001, 2003).

Limpasuvan et al. (2005) have taken the 44-year NH data from the NCEP/NCAR reanalysis and composited the flow relative to stratospheric warming events. The stratospheric warming events are selected as anomalies of the first mode of 50 hPa zonal flow, the times when the zonal vortex is weak. These dates correspond approximately to major or minor stratospheric warmings. These composites show statistically significant forerunner and follower structures to the NH wintertime warming events.

The downward propagation of the signal from the stratosphere is closely associated with the concept of downward control in which wave-driving effects are projected downward to the flow below the level of wave-driving. In major warmings, the wave-driving propagates downward to the lower stratosphere and forces a response in the troposphere. Once a zonal wind anomaly is projected into the troposphere, the resulting changes in wave propagation and baroclinic instability can result in the positive feedback that reinforces the initial signal through the intermediary of the synoptic scale waves.

The possibility of rather weak forcing from stratospheric changes produces much larger than expected changes in tropospheric climate and has been successfully simulated in modeling studies (Hartmann et al. 2000; Shindell et al. 2001). Because of the nonlinearity of the stratospheric warming dynamics, small changes in wind in the stratosphere or troposphere can lead to changes in the probability of major stratospheric warmings, which can have large effects on stratospheric and tropospheric climate.

The stratospheric warming events in both hemispheres seem particularly important in enforcing a stratosphere–troposphere connection through the annular modes. Compositing of the NH warmings in a 44-year dataset suggests that the synoptic-scale waves are especially important in producing the shift in tropospheric wind patterns, which in the stratosphere are driven primarily by wave-forcing from planetary-scale waves. This further suggests that the stratospheric wave drag and zonal wind responses are able to induce a transition in the naturally occurring tropospheric mode of variation.

9.6 Stratospheric Polar Vortex and Tropospheric Weather

Observational evidences suggested that the strength of the stratospheric polar vortex influences circulation in the troposphere. There is a weak downward forcing in the stratosphere that is pinging the lower atmosphere and stimulating modes of existing variability in the troposphere.

The polar vortex is a wintertime feature of the stratosphere. Consisting of winds spinning counterclockwise above the pole, the vortex varies in strength on long timescales because of interactions with planetary waves generated in the troposphere. The polar vortex acts like a big flywheel. When it weakens, it tends to stay weakened for a while. A strong connection has been noted between periods when the polar vortex is weak and outbreaks of severe cold in the northern hemisphere.

When the vortex is strong, the westerlies descend all the way to Earth's surface. This carries more warm air from the ocean onto the land. When the vortex is weak, then the really deep cold occurs. Such correlation could prove useful for weather forecasting.

To explain this behavior of the atmosphere, a dynamical mechanism in which stratospheric forcing, through the mechanism of downward control, weakly forces the Arctic oscillation, has been proposed (Song and Robinson 2004). This forcing is then reinforced in the troposphere by interactions with transient eddies in the lower atmosphere, creating a substantial amplification of the signal.

The polar vortex does not create new modes of variability in the troposphere. It stimulates preexisting modes that are fundamental to the dynamics of the lower atmosphere.

The physical mechanisms responsible for the exchange of mass and trace constituents between the stratosphere and troposphere include diabatic processes, temporal movement of the tropopause, and transport along isentropic surfaces which intersect the tropopause. The mechanism of the isentropic transport operates in several meteorological settings of different spatial scales, such as tropopause folding near extratropical and subtropical jet streams, cutoff lows, as well as transverse circulations around quasi-steady jet streams.

9.7 Stratosphere–Troposphere Coupling and Downward Propagation

The dynamical mechanisms by which stratospheric circulation anomalies can influence the troposphere are poorly understood, and most current numerical models for weather or climate do not include a well-resolved stratosphere. Observations indicate that the eddy heat flux at the tropopause correlates well with the zonal mean winds in the middle stratosphere, indicating that stratospheric anomalies are controlled by anomalous wave activity fluxes from the troposphere (Baldwin et al. 2003a, b). It is not yet clear whether the stratosphere simply responds to upward wave activity fluxes from the troposphere or actually modulates such wave activity fluxes.

An important manifestation of this downward influence is the propagation of low-frequency zonal wind variations from the upper stratosphere to the lower troposphere. Sigmond et al. (2004) have given evidence for the presence in circulation data of a dynamical mechanism through which the stratosphere influences the troposphere. Analysis of ECMWF data for the NH extratropical winters has indicated a downward influence through variations in a meridional circulation cell. For all winters the meridional mass flux variations in the stratosphere precede those in the troposphere, by about 1 day. Through mass conservation, this lag is associated with variations in the meridional surface pressure gradient and the associated surface zonal wind in the midlatitudes.

9.7.1 Arctic Oscillation's Impact on Climate

A growing body of evidence indicates that the Arctic oscillation has wide-ranging effects in the northern hemisphere and operates differently from other known climate cycles. The evidence indicates that the acceleration of a counterclockwise spinning ring of air around the polar region could be responsible for warmer winters in Scandinavia and Siberia, thinning of the stratospheric ozone layer, and significant changes in surface winds that might have contributed to Arctic ice thinning.

The Arctic oscillation is a seesaw pattern in which atmospheric pressure at polar and middle latitudes fluctuates between positive and negative phases. The negative phase brings higher-than-normal pressure over the polar region and lower-than-normal pressure at about 45°N latitude. The positive phase brings the opposite conditions, steering ocean storms farther north and bringing wetter weather to Alaska, Scotland, and Scandinavia and drier conditions to areas such as California, Spain, and the Middle East. A time series standardized 3-month running mean AO index from 1950 to 2007 is shown in Fig. 9.7.

The existence of a virtually identical phenomenon in the southern hemisphere is known as the Antarctic oscillation . The Antarctic oscillation affects the even faster

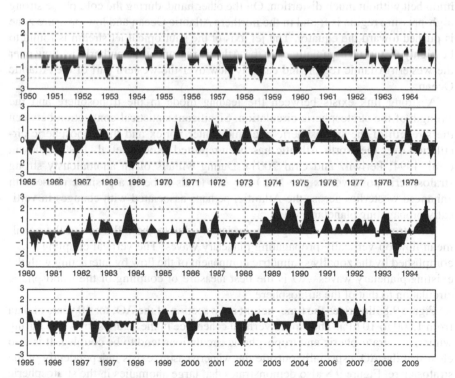

Fig. 9.7 Standardized 3-month running mean AO index from 1950 to 2007 (National Weather Service, NOAA)

spinning ring of air that encircles the south pole. It is believed that the presence of large landmasses in the north prevents the ring of air flowing around the Arctic from becoming as strong as that in the Antarctic.

During the winter, the Arctic oscillation extends up through the stratosphere. The stratosphere's effect on the Arctic oscillation's behavior appears particularly fascinating because it is the opposite of what happens in other major climate systems. When the oscillation changes its phases, the strengthening or weakening of the circulation around the pole tends to begin in the stratosphere and work its way down through lower levels of the atmosphere. In the case of El Niño in the equatorial Pacific Ocean, the changes begin in the ocean and work their way up through the atmosphere (Baldwin 2000).

North Atlantic oscillation is a part of the Arctic oscillation, which involves atmospheric circulation in the entire hemisphere. The trend toward a stronger, tighter circulation around the north pole could be triggered by the processes in the stratosphere as by those in the ocean. The trend in the NAO has been reproduced in climate models with increasing concentrations of greenhouse gases.

During the cold phase low-pressure areas are located over North America and the Atlantic region, and the high-pressure area is seen over the equatorial region. The subtropical jet stream is found to be oriented west to east almost in the same latitude belt without much distortion. On the other hand, during the cold phase strong high-pressure region is found in the northern Atlantic Ocean and low-pressure zone is pushed toward the equator. The jet axis of the subtropical jet stream is found to be highly distorted, especially over the Atlantic region, with southerly jet flow over the western Atlantic region and easterly flow over the western part of the Atlantic Ocean.

A strong link exists between the leading mode in the troposphere and the strength of the stratospheric circulation. Significant statistical connection between the strength of the stratospheric cyclonic winter vortex and the tropospheric circulation over the North Atlantic has been reported (Kodera et al. 1990; Baldwin et al. 1994; Perlwitz and Graf 2001). During winters of an anomalously strong stratospheric polar vortex, the NAO (or AO) tends to be in a positive phase with enhanced westerlies across the Atlantic, perhaps associated with changes in vertically propagating planetary waves.

The strong vertical coupling may be largely due to the influence of the zonal-mean flow on vertically propagating planetary waves. The vertical coupling is accomplished by the zonally symmetric component of the flow by interacting with the existing planetary waves. AO is the best measure of coupling of the tropospheric circulation to that of the stratosphere.

Figure 9.8 illustrates the time–height cross section of the AO, for 40 years of data from 1,000 to 10 hPa. The strong vertical coherence in the figure shows that during winter, the stratospheric and tropospheric circulations tend to be coupled. It is also clear that the AO, especially outside the winter season, exists independently of the stratosphere. Figure 9.8 also demonstrates that large anomalies in the stratospheric circulation tend to propagate downward to the Earth's surface, with an average propagation time of approximately 3 weeks. The downward propagation is variable, and

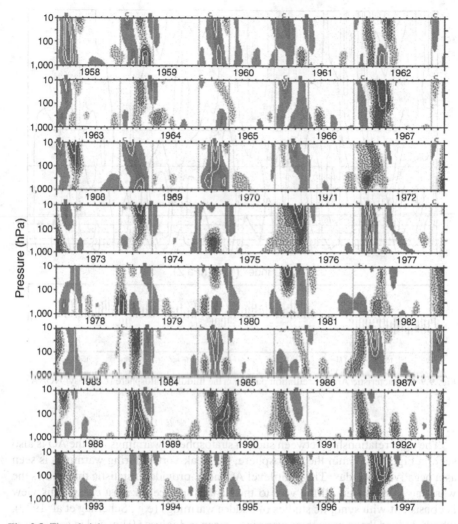

Fig. 9.8 Time–height cross section of the Arctic oscillation, 1958–1997 (Baldwin et al. 2003a, Reprinted with permission from AAAS)

is not apparent during some winters. The mechanism for downward propagation appears to be a result of wave mean–flow interaction, but the details are not well understood.

Figure 9.9 shows a nondimensional measure of the magnitude of the patterns as a function of height and time, in a 75-day low pass filtered data. The diagram illustrates that the characteristic Arctic oscillation pattern in the middle stratosphere tends to occur before the corresponding pattern in the troposphere, with a variable downward propagation time averaging about 3 weeks. The downward propagation occurs for either positive or negative anomalies, but does not always occur.

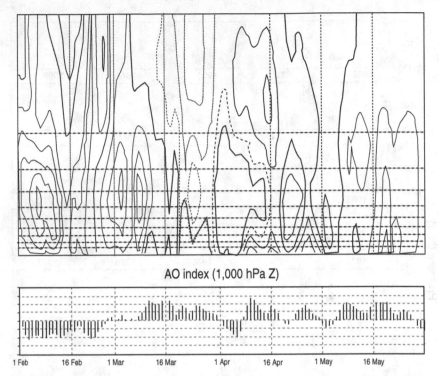

AO index (1,000 hPa Z)

1 Feb   16 Feb   1 Mar   16 Mar   1 Apr   16 Apr   1 May   16 May

Fig. 9.9 Characteristic Arctic oscillation pattern in the middle stratosphere (Courtesy: NOAA)

The tight relationship between sudden stratospheric warmings and the AO is also seen in Fig. 9.9. Within the stratosphere, the weak vortex during warmings is seen as a negative AO index. The multilevel AO index provides a statistic that allows the warming to be traced all the way to the Earth's surface, in most cases. This view is consistent with synoptic studies of sudden warmings (e.g., Scherhag et al. 1970), which showed that some major warmings began as high as 60 km and in some cases progressed downward to the Earth's surface.

The quasi-biennial oscillation (QBO) in the equatorial stratosphere acts to force the AO. It appears that the phase and strength of the AO, especially in the stratosphere, are influenced by the QBO in a manner consistent with the QBO's effect on the strength of the polar vortex. The AO tends to be in its negative phase (weak vortex) when the QBO is easterly.

Within the stratosphere, anomalies in zonal mean-zonal wind tend to propagate northward and downward during the winter season (Kodera et al. 2000). The results have been confirmed by numerical experiments in which circulation anomalies begin as high as the lower mesosphere, and move slowly northward and downward, over a period of 2–3 months. The propagation is described well in the phase space of the first two EOFs of zonal mean-zonal wind, which tend to trace out a circular path, especially during sudden warmings. A synoptic interpretation of this process is the development of a sudden warming with a reduction in strength of the polar vortex and downward progression of the warming, sometimes to the Earth's surface. The resulting surface signature of this process, denoted as the *Polar Jet Oscillation (PJO)* (Kodera et al. 2000), is that of the Arctic oscillation. The relationship between the AO and the PJO is not clear; in the lower stratosphere one could call the leading mode of variability the AO or the PJO.

It appears that an amplifier is needed in order for the stratosphere to have a significant effect on surface pressure. Synoptic-scale waves associated with frontal systems are stronger in the upper troposphere, but extend several kilometers into the stratosphere. There is a region where these waves and the stratospheric wind anomalies overlap. In this region, synoptic-scale waves can be affected by stratospheric wind anomalies. The altered waves could affect the tropospheric circulation, with surface pressure changes corresponding to the AO. The connection between the stratospheric polar vortex and the AO has important implications for climate trends and the prediction of the response of tropospheric climate. A detailed understanding is lacking in this regard, and a complete explanation is needed.

9.7.2 Annular Modes

Annular modes are hemispheric spatial patterns of climate variability characterized by north–south shifts in mass between polar and lower latitudes. Tropospheric signatures of stratospheric variability are often well described by annular mode patterns. In both the stratosphere and troposphere, the annular modes explain a larger fraction of variance than any other pattern of climate variability in their respective hemisphere. On month-to-month timescales, annular variability at tropospheric levels is strongly coupled with annular variability at stratospheric levels. Time series of the annular modes provide a convenient way to describe some aspects of stratosphere–troposphere coupling. The northern hemisphere annular mode (NAM) near the Earth's surface is alternatively known as the arctic oscillation (AO) and the North Atlantic oscillation (NAO). The southern hemisphere annular mode (SAM) is also referred to as the Antarctic oscillation and high latitude mode.

The annular mode in the troposphere is believed to arise from the two-way interaction between baroclinic eddies and the tropospheric midlatitude jet (Haynes 2005a; Robinson 1991). At the same time the annular variability in the stratosphere is a manifestation of the variation in the strength of the polar night jet, driven by the wave force. The variability of the tropospheric wave-forcing also plays an important role for the two-way interaction between waves and mean flow. Figure 9.10 shows the composites for the periods of high and low NAM index in longitudinal wind and Eliassen-Palm flux. The EP flux is calculated only for longitudinal wave numbers 1, 2, and 3. In the high phase, wave fluxes tend to be directed equatorward within the troposphere and to converge in the subtropical troposphere. In the low

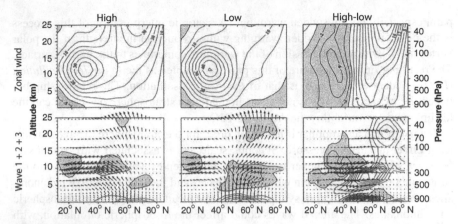

Fig. 9.10 Composites for periods of high and low NAM index and their difference in longitudinal wind (top) and Eliassen-Palm flux, which indicates wave propagation and transport of westward momentum, and its divergence, which indicates eastward wave force (Haynes 2005a, Courtesy: SPARC)

phase, wave fluxes tend to be directed upward from troposphere to stratosphere and to converge, indicating an anomalous westward wave force in the midlatitude and polar stratosphere (Hartmann et al. 2000).

On intraseasonal timescales, observations show that anomalies of the northern hemisphere wintertime stratospheric polar vortex frequently precede persistent changes to the tropospheric circulation that resemble the Arctic oscillation (Fig. 9.11). This may be useful in improving the predictability of the extratropical troposphere on timescales on the order of several weeks. The tropospheric circulation may also be linked to the stratosphere on longer timescales. It has been found that stratospheric forcings in relation with ozone depletion, stratospheric warming, volcanic aerosols, or the quasi-biennial oscillation exhibit a signature in surface climate. Such a coupling may be important for more realistic simulations of anthropogenic climate change in relation with secular changes of greenhouse gases.

Thermally forced temperature changes in the stratosphere associated with human-induced ozone depletion and greenhouse gas increases are larger than temperature changes nearer the surface of Earth. The dynamical response to these changes can be important and can be translated into changes in surface climate that are larger, or structured differently than those expected from direct forcing in the troposphere. The response to ozone depletion seems to consistently give a stronger and more persistent stratospheric vortex, which expresses itself as a positive anomaly of the annular mode variability.

Later studies suggest the annular modes reflect feedbacks between eddies and zonal flow at middle latitudes (Lorenz and Hartmann 2003). Other studies necessitate that the annular modes are expected in any rotating planetary fluid system that conserves momentum and mass, and that has some smoothness property (Gerber and Vallis 2005). The key dynamics that underlie the annular modes are still under investigation.

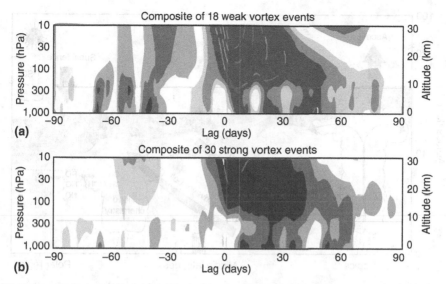

Fig. 9.11 Observed composites of time–height development of the northern annular mode for (a) 18-week vortex events and (b) 30 strong vortex events (Baldwin et al. 2003b, Reprinted with permission from AAAS)

9.8 Anthropogenic Effect

It is now well recognized that anthropogenic pollutants such as chlorofluorocarbons (CFCs) released in the troposphere cause depletion of stratospheric ozone on a global scale. Stratospheric model results predict that increasing levels of atmospheric carbon dioxide, associated with global warming in the troposphere as a result of an enhanced greenhouse effect, should lead to global cooling of the stratosphere (Brasseur and Hitchmann 1998).

At stratospheric altitudes and above where infrared emission from CO_2 can escape to space, dramatic atmospheric cooling is expected. Tropospheric climate models generally predict a 1–4°C increase in the tropospheric temperature in a doubled-CO_2 scenario. Corresponding middle atmospheric models for the same scenario predict a 10–20°C decrease in stratospheric temperatures.

Such a large stratospheric signal of global temperature change suggests that the atmospheric effects of increasing levels of CO_2 may become apparent at higher altitudes before an unambiguous trend is observed in the troposphere. A large change in the stratospheric thermal structure would also drive significant changes in the stratospheric circulation and dynamics. This would affect tropospheric climate because stratospheric dynamics provide the upper boundary mechanical forcing of tropospheric weather patterns.

To understand the effects of anthropogenic pollutants in the stratosphere, the dynamic processes by which tropospheric trace gases are transported into the

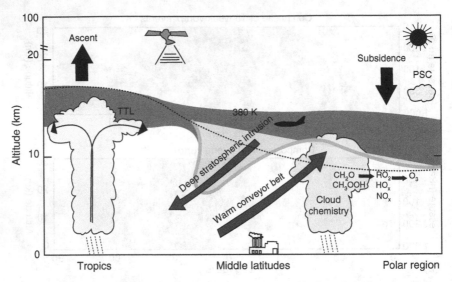

Fig. 9.12 Anthropogenic activity in the troposphere and stratospheric region (Courtesy: Barbara Summey, NASA)

stratosphere should be determined. Both observations and models have shown that most tropospheric trace gases enter the stratosphere through the tropical tropopause and are then transported through the stratosphere by winds and various mixing processes (see Fig. 9.12). For greenhouse gases like CO_2 and CH_4, which show increasing abundance and have a long atmospheric lifetime, the time difference between the appearance of a given abundance in the troposphere and the appearance of that same abundance in the stratosphere is known as the age of air.

The effect of greenhouse gases on stratospheric warming probabilities is less clear. The polar stratosphere will not obviously cool more significantly than the tropical stratosphere (Shindell et al. 2001). Moreover, a tendency for polar amplification of warming near the surface would seem to work against enhanced meridional temperature gradients in the stratosphere. One might expect that warming of the tropical upper troposphere and cooling of the polar stratosphere by greenhouse gas increases would lead to increased temperature gradients on constant pressure surfaces in the upper troposphere–lower stratosphere region, and it would increase the vertical shear and refractive index in midlatitudes. It is not clear that these changes would be far enough poleward to produce a positive feedback on the stratospheric polar night jet. Moreover, many other changes would occur in the troposphere that would produce effects. Gillett et al. (2003a, b) found consistent positive annular mode responses in reply to CO_2 increases, but the magnitudes of these changes were not large.

Chemical reaction rates in the atmosphere are dependent on temperature, and thus the concentration of ozone is sensitive to temperature changes. Decreases in upper stratospheric temperature slow the rate of photochemical ozone destruction in this region. Hence the concentration of upper stratospheric ozone increases in response to cooling. Cooling of the polar lower stratosphere would lead to more efficient chlorine activation on aerosol and polar stratospheric clouds and enhanced ozone destruction. Therefore, the concentration of ozone in the springtime polar lower stratosphere would decrease in response to cooling.

9.9 Impact of Ozone Changes on Surface Climate

The largest stratospheric ozone depletion is observed in the austral spring in the Antarctic stratosphere, and therefore it is here that one might expect any effect of stratospheric ozone depletion on tropospheric climate to be largest. While the largest stratospheric temperature and geopotential height trends over the Antarctic have been observed in November, coincident in space with the maximum ozone depletion, significant decreases in geopotential height also extend to the troposphere 1–2 months later (Thompson and Solomon 2002), as illustrated in Fig. 9.13.

Several modeling studies have examined the tropospheric response to prescribed changes in stratospheric ozone. The simulated and observed geopotential height and temperature changes agreed well, both in magnitude and in seasonality, supporting the hypothesis that the observed trends were largely induced by stratospheric ozone

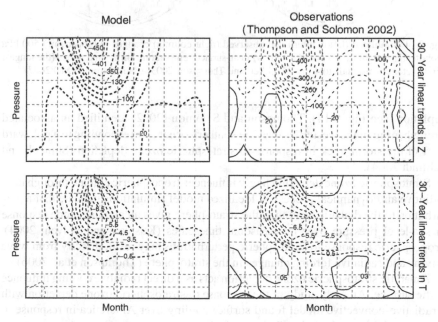

Fig. 9.13 Model simulation of the response to stratospheric ozone depletion (left) on Antarctic geopotential height (top) and temperature trends (bottom). Shading indicates regions of significant change (Courtesy: Gillett and Thompson 2003; Adapted from WMO 2007)

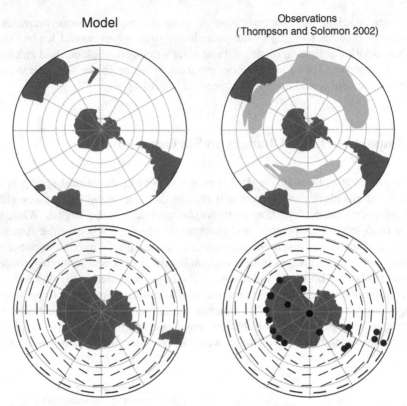

Fig. 9.14 Simulated (left column) and observed (right column) changes in (upper row) 500 hPa geopotential height (in m) and (lower row) near-surface temperature (in K) and winds (the longest wind vector corresponds to 4 ms⁻¹) (Gillett and Thompson 2003 Adapted from WMO 2007)

depletion (see Fig. 9.14). Thompson and Solomon (2002) identified an associated strengthening of the westerlies over the southern ocean, corresponding to a poleward shift of the storm track in response to stratospheric ozone depletion (Shindell and Schmidt 2004).

Stratospheric ozone depletion likely influences tropospheric climate through both radiative and dynamical processes. Idealized model simulations indicate that a perturbation to the diabatic heating in the stratosphere gives an annular-mode response in sea-level pressure over intraseasonal timescales (Kushner and Polvani 2004). Tropospheric response to stratospheric perturbations may result directly from wave driving and radiative forcing changes in the stratosphere (Thompson et al. 2006).

Antarctic stratospheric ozone depletion acts as a direct radiative cooling influence at the surface. An early study of the response to stratospheric ozone depletion with a radiative-convective model found surface cooling over Antarctica in response to stratospheric ozone depletion (Lal et al. 1987).

Maximum stratospheric ozone depletion close to the tropopause occurs in December to January, more than a month after the maximum ozone depletion at 70 hPa, due to the downward transport of ozone-depleted air (Solomon et al. 2005). Since surface temperature is particularly sensitive to changes in ozone concentration close to the tropopause, it suggests that surface cooling is radiatively induced. The apparent lag between stratospheric and tropospheric responses is due to the downward transport of ozone-depleted air toward the tropopause, rather than any dynamical effect.

Most studies suggest that Antarctic ozone depletion is likely to peak sometime in the current decade, and that a recovery is likely to follow over the next 50 years. Therefore, over the coming decades, increases in stratospheric ozone should drive a decrease in the Southern Hemisphere annular mode (SAM) index toward values seen before ozone depletion (WMO 2007). However, increasing greenhouse gases (GHGs) will likely have the opposite effect, contributing to an increase in the SAM index (Shindell and Schmidt 2004; Arblaster and Meehl 2006). The magnitudes of the future ozone and GHG effects on the SAM are therefore uncertain.

A summary of the stratospheric ozone and climate interaction is presented in Fig. 9.15. Stratospheric ozone depletion causes an increase in UV irradiance which impacts tropospheric chemistry, the biosphere, and primary organic production. Such UV increases lead to an enhanced OH production, which reduces the lifetime of methane and influences of ozone, both of which are important greenhouse gases.

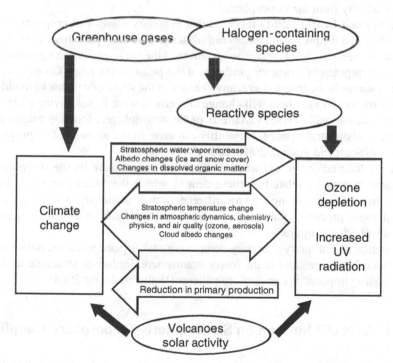

Fig. 9.15 Schematic illustration of stratospheric ozone variability and climate change interaction (EUR 2003)

Measurements and model calculations show long-term increases in erythermal irradiance of about 6–14% over the last 20 years. It is evident that the long-term UV changes are not driven by stratospheric ozone alone, but also changes in cloudiness, aerosol particles, and surface albedo (EUR 2003).

Climate change will have an important influence on future surface UV irradiance. Any changes in stratospheric ozone will modulate the amount of UV penetrating to the troposphere. Further, the transmission through the troposphere will be affected by any changes in other variables, such as clouds, aerosols, and local albedo (e.g., snow cover), which will be strongly influenced by climate change. The level of understanding of UV changes and their prediction is high for ozone changes, moderate for aerosols and albedo and low for cloudiness. The level of understanding of the combined effect of all factors is not well understood.

9.10 Stratosphere Affects Its Own Variability

It was believed that the stratospheric variability was being caused directly by variability in tropospheric wave sources. Now it is widely accepted that the configuration of the stratosphere itself plays an important role in determining the vertical flux of wave activity from the troposphere.

Charney and Drazin (1961) reported that planetary waves can propagate vertically only when the prevailing winds are westerly. This theory has been extended to account for the strongly inhomogeneous nature of the stratospheric background state and the steep potential vorticity gradients at the polar vortex edge. Given a steady source of waves in the troposphere, any changes in the stratospheric background potential vorticity (PV) gradient will change the vertical wave fluxes, giving rise to the possibility of internally driven variability of the stratosphere. Modeling studies suggest that realistic stratospheric variability can arise in the absence of tropospheric variability (Scott and Polvani 2004, 2006).

The modulation of vertical wave flux into the stratosphere by the stratospheric configuration may be related to the extent to which the stratosphere can act as a resonant cavity, involving downward reflection of stationary planetary waves. Through these processes, the tropospheric circulation itself is also influenced by the stratospheric configuration.

Reflection of stationary planetary wave energy takes place when the polar vortex exceeds a critical threshold in the lower stratosphere, leading to structural changes of the leading tropospheric variability patterns (Walter and Graf 2005).

9.11 Effects of Monsoon on Stratosphere–Troposphere Coupling

Monsoon is an important global atmospheric general circulation system. The large areas involved in monsoons and the grand scale of the weather within monsoons suggest that they play a significant role in modulating global climate.

The Asian summer monsoon is the most energetic system of the global circulation in northern summer. It mainly affects the Indian subcontinent and Southeast Asia, and is probably the most noted of the monsoons. Now as monsoons have become better understood, the definition now indicates climatic systems anywhere in which the moisture increases dramatically in the warm season.

The monsoon occurs as part of a larger phenomenon characterized by the Intertropical convergence zone (ITCZ). The ITCZ separates the wind circulations of the northern and southern hemispheres. This zone moves north and south with the annual changes of the Sun's declination, and is defined where the northeast and southeast trade winds flow together. This region is characterized by strong upward motion and heavy rainfall, as a result of intense heating by the direct rays of the Sun. The heating also results in a three-dimensional atmospheric circulation referred to as a Hadley cell. The Hadley cell consists of rising air at the equator and descending air at 30° north and south. This results in strong equatorial surface winds which blow from the east due to the Earth's Coriolis force. The Indian monsoon is unique in that the winds blow from the southwest.

The strong Indian summer monsoon years are generally associated with positive tropospheric temperature anomalies over Eurasia and negative temperature anomalies over the Indian Ocean and the eastern Pacific but positive sea surface anomalies in the western Pacific. Several connections between Eurasian snow cover, the Indian monsoon, and the El Niño/southern oscillation (ENSO) have been established. Based on 100 years' data, major droughts have been associated with warmer than normal sea surface temperatures (SST) in the equatorial Eastern Pacific for time periods spanning a monsoon season. Floods, on the other hand, have been associated with cooler SST events in the tropical Eastern Pacific. Also, anomalously high winter Eurasian snow has been linked to weak rainfall in the following summer monsoon.

A three-dimensional representation of the Asian summer monsoon circulation which extends from the troposphere to the lower stratosphere is presented in Fig. 9.16. The Indian summer monsoon rainfall averaged over the whole country is found to be stable over the last century, without any significant long-term trend, but is dominated by high interannual variability. This variability is generally attributed to the slowly varying surface boundary forcings of sea surface temperature, soil moisture, snow cover, and the circulation features of the upper troposphere and the mid-troposphere. During the last two decades links between stratospheric features and Indian monsoon rainfall have been suggested (Mukherjee et al. 1985; Kripalani and Kulkarni 1997). However, due to the sparsity of data at stratospheric levels, studies have been limited.

9.11.1 Stratospheric QBO and Monsoon

The discovery of the QBO in tropical stratospheric winds stimulated the search for stratospheric–tropospheric links. In the late 1970s some evidence for a link between

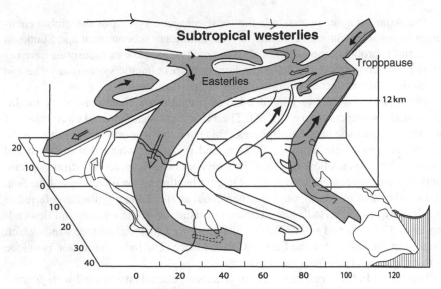

Fig. 9.16 A three-dimensional schematic representation of the Asian monsoon circulation (Subbramayya and Ramanatham 1981 Reprinted with permission from Cambridge University Press)

the Indian summer monsoon rainfall (ISMR) and stratospheric zonal winds was produced. Thapliyal (1984) has shown that in winters of the easterly QBO years, a subtropical ridge is situated over the northern hemisphere around 20°N. Such winters are followed by normal monsoon activity. Thus, the January circulation features at 50 hPa level can indicate the deficient (normal) monsoon rainfall in westerly (easterly) QBO years.

An important feature of the QBO is the downward phase propagation. The wind reversal first appears above 30 km and propagates downward at a speed of about 1 km month^{-1}. Using this fact, Bhalme et al. (1987) related the January 10 hPa zonal wind anomalies at Balboa with ISMR and found a correlation of 0.52 during the period 1958–1985. They found that ISMR tends to be less (more) than normal during an easterly (westerly) anomaly. The India Meteorological Department uses 16 predictors in an operational long-range forecasting model (Gowariker et al. 1991). Two of these predictors are related to the stratosphere, namely the 50 hPa wind pattern in winter and the 10 hPa zonal wind pattern in January.

Mohanakumar (1996) has attempted to find the combined effect of the solar activity and phase and the speed of the equatorial QBO on annual as well as monsoon rainfall over the Indian subcontinent. The effect of solar activity and phases of equatorial QBO at 15–20 hPa level during the previous winter (January–February) on Indian monsoon rainfall is schematically represented in Fig. 9.17. During the period of solar minimum, excess rainfall is associated with the westerly phases of QBO and deficient rainfall is connected with the easterly phase. Normal rainfall is evenly distributed in both the easterly and westerly phases of QBO during solar minimum.

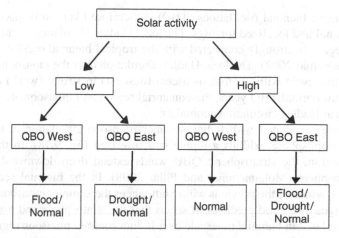

Fig. 9.17 Modulation of QBO phases on solar activity induced effect on Indian summer monsoon (Mohanakumar 1995)

On the other hand, during the period of high solar activity and when QBO is in its westerly phase, only normal rainfall occurs without much deviation from its long-term mean value. Extreme rainfall events are absent during this period. But in the easterly phase of the QBO during solar maximum, all the three possibilities can occur. Easterly phases of the QBO normally lead to extreme monsoon conditions.

Stratospheric QBO is driven by the distinct mechanism associated with interactions between the zonal mean flow and equatorial waves like Kelvin waves, mixed Rossby gravity waves, and internal inertial gravity waves. The 10–20-day mode originates over the equatorial central Pacific and propagates west/north-westward, and intensifies over the northwest Pacific modulating both the oceanic and continental convection simultaneously. This mode is more regional in structure and the associated latent heating influences only the cross-equatorial flow (Kodera 2004). QBO signal is also detected in the temporal variations of SST in the tropical western Pacific Ocean.

On the interannual timescales, Asian summer monsoon exhibits a clear biennial tendency of years of heavy rainfall trend to be followed by years of reduced rainfall. This biennial component is known as tropospheric biennial oscillation (TBO) and appears in a wide range of parameters such as precipitation, sea surface temperature, sea-level pressure, and wind. It is believed that dynamic coupling between the ocean and the atmosphere, in conjunction with the seasonal cycle of convection over the tropical Indian and Pacific oceans and associated large-scale east–west circulation in the atmosphere is responsible for TBO (Meehl 1993; Meehl and Arblaster 2001, 2002). Chang and Li (2000) and Li et al. (2001) demonstrated that the TBO is an inherent monsoon mode, resulting from multiregion interaction between Asian and Australian monsoons and adjacent tropical oceans.

Tropospheric biennial oscillations (TBO) is therefore likely to be pace-marked by the seasonal and localized forcings. The local Hadley circulation over the Asian monsoon region is strongly connected with the tropical biennial oscillation (Pillai and Mohanakumar 2007). The local Hadley circulation over the monsoon area follows the TBO cycle with anomalous ascent (descent) in strong (weak) monsoon years. During normal TBO years, the equatorial region and monsoon areas exhibit opposite local Hadley circulation anomalies.

TBO cycle associated with Indian summer monsoon circulation in the lower and upper troposphere exhibits a dipole structure (see Fig. 9.18). In the Indian monsoon region, the stratospheric QBO winds extend deep downward into the lower troposphere (Mohanakumar and Pillai 2008). In the biennial scale lower stratospheric winds of the previous winter influences the summer monsoon rainfall, which emerges as a good predictor. It seems that the stratosphere and troposphere are closely related in biennial timescale over Indian summer monsoon region.

Meehl et al. (2003) suggested that the coupled interaction between ocean and atmosphere contributes to a mechanism that produces a biennial component of interannual variability in the tropical Indian and Pacific regions. The basic idea is that a wet (dry) monsoon year will be followed by a dry (wet) monsoon year. The ocean

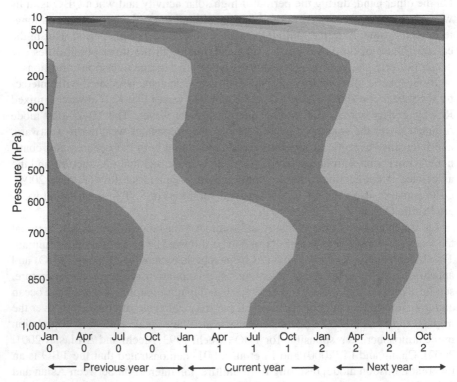

Fig. 9.18 Interaction of stratospheric QBO and tropospheric biennial oscillation over monsoon region (Mohanakumar and Pillai 2008, Courtesy: Elsevier)

heat content seems to provide the 1-year timescale memory necessary for the biennial mechanism. Whether this mechanism has any relationship with stratospheric
QBO needs a separate investigation.

9.12 Tropical Convection and Water Vapor Forcing

The stratosphere–troposphere interactions in the tropics start with large regions of
cloud-scale convection. This convection locally mixes air and humidity across potential temperature (isentropic) surfaces. The mixing may be well off the equator.
Large-scale motions subsequently redistribute the moist air across the tropics, and
the air may rise to higher potential temperature surfaces through heating in the
course of these circulations, spiraling upward. Some of these isentropic surfaces
may lie above the cold point tropopause. The large-scale circulation itself is strongly
affected by the distribution of convective heating.

Figure 9.19 gives a mechanism in which tropospheric water vapor entering into
the stratosphere is demonstrated. Water vapor in the upper troposphere and lower
stratosphere has significant influence on the climate system and plays a crucial role
in stratospheric chemistry (Austin and Li 2006). Therefore it is important to understand the processes which determine its transport and distribution in these regions

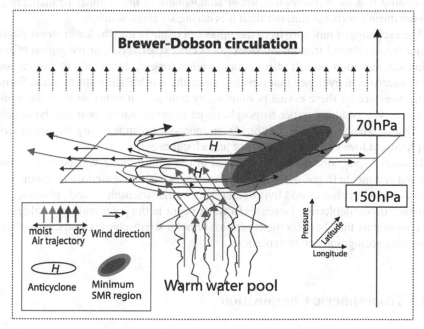

Fig. 9.19 Mechanism through which water vapor enters into the stratosphere (Courtesy: SPARC)

of the atmosphere, including the processes by which moisture can be exchanged between the troposphere and the stratosphere (Gettelman et al. 2002).

The Asian monsoon region is one of the regions with the highest frequency of convective clouds at and above the tropopause. The results of the simulation indicate that monsoon air can bypass the tropical tropopause, and enter the stratosphere in the subtropics (10–30°N in boreal summer). Some of this air enters the tropics above the tropopause, and can be lofted into the stratosphere. It would indicate a pathway for water vapor to enter the stratosphere without passing through the region of cold tropical tropopause temperatures. This might have a significant impact on water vapor trends. Model simulations do not show any trends in the monsoon flux of water vapor in a 22-year simulation, but the interannual variability is high.

The Indian region (including the equatorward branch of the Asian monsoon anticyclone over the western Pacific) outside of the deep tropics is responsible for nearly three fourth of the net tropical upward flux of water vapor during July–September at 95 hPa, and over half at 66 hPa in the simulation. The meridional fluxes also show significant transport of this air into the deep tropics, both below and above the tropopause. This transport into the tropics in the simulation appears to bypass the cold point tropopause, consistent with observed climatological flow.

In the southern hemisphere, the seasonal moisture input is strongest during southern summer. Because the southern hemisphere does not have such an intense monsoon system in summer, the moistening is weaker than in the northern hemisphere. Therefore, if this mechanism proves to be quantitatively important, it may explain the finding that the extratropical lower stratosphere in the summer hemisphere is moister during northern summer than it is during southern summer.

The exchange of moisture from the upper troposphere into the lower stratosphere across the dynamical tropopause does not occur exclusively near the region of the Asian summer monsoon. It also occurs at other longitudes, mainly in association with extratropical cyclones that develop along the north Pacific storm track. Some of the moisture for these events is supplied by transport of moist air from the monsoon anticyclone along upper tropospheric jet streams that lie near the dynamical tropopause. Hence, the role of the Asian monsoon in moistening the upper troposphere and lower stratosphere is not just a local one.

If water vapor concentration increases in the future, there will be both radiative and chemical effects. Modeling studies suggest that increased water vapor concentrations will enhance odd hydrogen (HO_x) in the stratosphere and subsequently influence ozone depletion. Increases in water vapor in the polar regions would raise the temperature threshold for the formation of polar stratospheric clouds, potentially increasing springtime ozone depletion.

9.13 Tropospheric Composition

Many of the chemical constituents present in the stratosphere have sources that originate in the troposphere. Any changes in the chemical composition of the troposphere can affect the composition of the stratosphere. The chemical constituents are either directly emitted in the troposphere, mostly at or near the surface, or they are oxidation products of emitted species.

The predominant source gases for stratospheric hydrogen, halogen, and nonvolcanic sulfur are long-lived species, such as water vapor, methane (CH_4), nitrous oxide (N_2O), organic halogen gases such as chlorofluorocarbons (CFCs), halons, and carbonyl sulfide (COS). Surface emissions of short-lived species, like sulfur dioxide (SO_2), dimethyl sulfide, are also important sources of sulfur to the stratosphere, as are occasional large volcanic eruptions. Emissions of short-lived species such as biogenic hydrocarbons also increase with temperature.

Climate changes can also alter other key processes in the exchange of chemical constituents between the troposphere and stratosphere. The vertical transport of surface emissions to the tropopause is largely dependent on the intensity of convective activity, and the flux of tropospheric air into the stratosphere is mostly determined by the strength of the upwelling from the troposphere that is linked to the strength of the Brewer-Dobson circulation.

However, it is difficult to quantify climate-driven changes in natural sources because the temperature is not the only driving factor. Other factors such as water table level, soil moisture, vegetation cover, photosynthetically active radiation, biogenic productivity, or exposure to atmospheric pollutants can play a role, depending on the emitting substrate and the emitted species. Ignoring changes in land use, most natural emissions are expected to increase as the Earth's surface warms.

9.14 Widening of the Tropical Belt

Tropical climate is quite distinct compared to that of the extratropical regimes. However, the boundaries of the tropics are not well defined. Tropical temperatures are warm, and, except for the major monsoon regions, both seasonal and diurnal changes are small compared with extra tropical climates.

In the tropics, the Hadley circulation, manifests itself in several ways throughout the atmosphere, as illustrated in Fig. 9.20. In the equatorial region, the ascending part of the Hadley circulation carries moisture into the air and produces rainfall, whereas the descending branches are drier at the outer boundary of the tropics. Within the Hadley circulation cell, atmospheric mass in the lower atmosphere moves toward the equator, whereas outside the cell it moves toward the poles. The latitude at which the net north–south flow is zero can be considered the poleward extent of the Hadley cell and therefore can be used to estimate the width of this tropical circulation (Seidel and Randel 2007).

The evidence of the widening of the tropical belt is reported by Fu et al. (2006) based on troposphere and stratospheric temperature observations from the microwave sounding unit for 1979–2005. The troposphere warming and stratospheric cooling trends in the 15–45° latitudes in both the hemispheres involves poleward shift of the latitude of maximum meridional tropospheric temperature gradient and

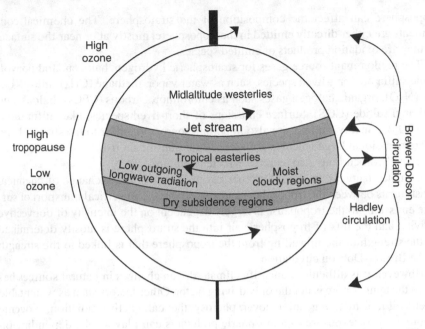

Fig. 9.20 Diagram illustrating widening of the tropics (Siedel and Randel 2008; Courtesy: Nature Geoscience)

associated poleward movement of the jet streams. Based on the outgoing longwave radiation Qu and Fu (2007) demonstrated that the tropical Hadley circulation expanded from 2° to 4.5° of latitude since 1979.

A possible latitudinal expansion of the tropical belt as reported by Fu et al. (2006) is found to be consistent with the tropospheric warming in the subtropics, rise of subtropical tropopause, and a poleward shift of the subtropical westerly jet stream (Seidel and Randel 2007). Further study by Seidel and Randel (2007) shows that the tropopause heights in the subtropics exhibit a bimodal distribution with maxima in frequency occurrence above 15 km, which is the characteristic height of the tropical tropopause, and below 13 km, which is the normal height of the subtropical tropopause. The frequency of occurrence of high tropopause days in the subtropics with the characteristics of the tropical tropopause has increased during the last three decades in both the hemispheres. The increase in the number of days of high tropical tropopause in the subtropical region in both the hemispheres suggests that the tropical belt is widening.

Figure 9.21 shows the time series of NCEP/NCAR reanalysis data of the mean latitudes averaged over all longitudes at which the days are indicated when tropopause height goes beyond 15 km in a year and it can be seen that the long-term poleward trends of the 300 days year^{-1} isolines are 1.6° latitude per decade in the southern hemisphere and 1.5° decade^{-1} in the northern hemisphere (Seidel and Randel 2007). The expansion to SH is more zonally uniform than in NH, where it appears to be restricted to the western hemisphere. Tropical belt expansion

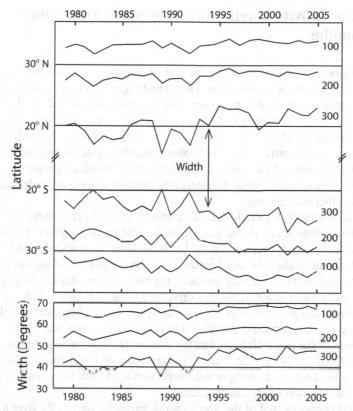

Fig. 9.21 Mean latitudes averaged over all longitudes at which the days are indicated when tropopause height goes beyond 15 km in a year based on NCEP/NCAR reanalysis data (Courtesy: Bill Randel)

(contraction) is associated with low (high) tropical lower stratospheric temperatures and high (low) tropical tropospheric temperatures. This tropopause-based analysis of the width of the tropical belt indicates a 5–8° latitude since 1979.

Widening of the tropical belt would probably be associated with poleward movement of major extratropical climate zones due to changes in the position of jet streams, storm tracks, mean position of high and low pressure systems, and associated precipitation regimes. An increase in the width of the tropics could bring an increase in the area affected by tropical storms, or could change climatological tropical cyclone development regions and tracks.

Tropical belt expansion may also affect the stratosphere by changing the distribution of climatically important trace gases. Since the Brewer-Dobson circulation pumps air upward from the troposphere to the stratosphere in the tropics, the area over which this upwelling occurs increases, thereby enhancing the transport of water vapor into the stratosphere. This could lead to an enhanced greenhouse effect, including tropospheric warming and stratospheric cooling, and reduced ozone (Seidel et al. 2008).

9.15 Solar Activity Forcing on Stratosphere Troposphere Coupling

The stratosphere is postulated to be the seat of many phenomena that are directly related to the rest of the atmosphere. There appears to be a link between the upper part of the atmosphere that is more sensitive to solar changes and the dense lower atmosphere, where the weather phenomenon occur. Since the Sun is the primary source of energy for driving the global energy circulation, one may safely deduce that the solar activity affects the tropospheric weather systems only after the variation in the intensity of the solar radiation has been modified during the passage through the atmosphere. The stratosphere thus holds the key to an understanding of the Sun–weather relationship.

Solar ultraviolet (UV) irradiance variations in 11-year cycles have a direct impact on the radiation and ozone budget of the middle atmosphere. During years with maximum solar activity, the solar UV irradiance is enhanced, which leads to additional ozone production and heating in the stratosphere and above. By modifying the meridional temperature gradient, the heating can alter the propagation of planetary-and smaller-scale waves that drive the global circulation. Although the direct radiative forcing of the solar cycle in the upper stratosphere is relatively weak, it could lead to a large indirect dynamical response in the lower atmosphere through a modulation of the polar night jet and the Brewer-Dobson circulation (Kodera and Kuroda 2002). Such dynamical changes can feedback on the chemical budget of the atmosphere because of the temperature dependence of both the chemical reaction rates and the transport of chemical species.

The total energy output of the Sun's energy varies by only ~1% over a 11-year solar cycle. But the extreme UV and X-ray part of the solar spectrum shows more than 2% variations in a solar cycle. Since UV radiation is absorbed by ozone in the stratosphere, the concentration of ozone varies with the intensity of UV radiation (Haigh 1994). This radiative-photochemical mechanism effectively amplifies the solar cycle through a positive feedback with the ozone concentration. Ozone variations thus have a radiative impact on the stratosphere and troposphere, and observations and modeling studies (Matthes et al. 2003) are broadly consistent with the expected radiative forcing.

Observations (Dunkerton 2001; Kodera and Kuroda 2002) and modeling (Gray 2003) studies show that circulation anomalies in the upper and middle stratosphere move poleward and downward during the winter season and are linked to anomalies in wave-induced momentum transport. The anomalous wave forcing by interacting with the mean flow can be regarded as a dynamical mechanism to maintain or even to enhance the amplitude of anomalies as they migrate downward into regions of higher density. Due to the solar cycle, perturbations originating in the tropical upper stratosphere may be transmitted to higher latitudes and lower altitudes by the dynamical mechanism.

It has already been seen that ozone changes have a direct radiative impact on the stratosphere and troposphere. Tropospheric Hadley cell is maintained by processes internal to the troposphere by way of moist deep convection and fluxes of momentum and heat due to synoptic-scale baroclinic waves, and other factors, such as tropical ozone. Owing to the size of the ozone heating anomaly, it is plausible that changes in tropical ozone have a significant effect on the tropospheric jet streams. This direct influence is transmitted to midlatitudes by anomalous fluxes associated with synoptic scales (Haigh et al. 2005). On astronomical timescales, much larger direct solar influence is possible to alter the weather systems in the equatorial region.

Potential response of the solar activity in the 11-year cycle on lower and middle atmosphere is schematically shown in Fig. 9.22. The solar UV irradiance can penetrate into the mesosphere and upper stratosphere and photodissociate the molecular oxygen. The photodissociation rate of molecular oxygen can vary by about 15–20% during a 11-year solar cycle (Brasseur and Solomon 2006). A feedback exists between changes in ozone and their effect on the penetration of UV radiation into the

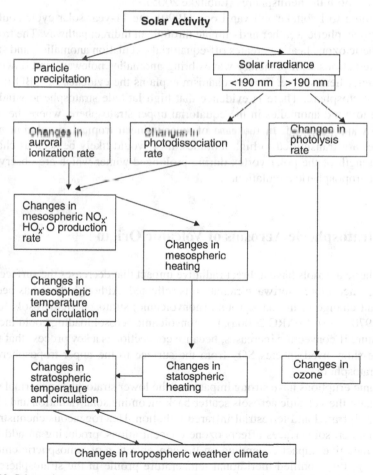

Fig. 9.22 Schematic representation of the potential response of solar activity in troposphere and middle atmosphere

stratosphere (Mohanakumar 1995). High solar activity levels are also accompanied by energetic particles which can increase atmospheric ionization and produce heating and chemical perturbations. The effect of these processes can transport downward and influence the dynamics and chemistry of the stratosphere. Changes in stratospheric structure and dynamics in turn affect the tropospheric circulation and its dynamics (Mohanakumar 1985,1988).

The Arctic lower and middle stratosphere tend to be cold and undisturbed during west-wind phases of the equatorial quasi-biennial oscillation, while they are warm and disturbed during QBO east-wind phases (Holton and Tan 1980). Further analysis (Labitzke 1987; van Loon and Labitzke 1987; Labitzke and van Loon 1988) showed that this relationship is strong during solar minimum conditions, while during solar maximum years the relationship does not hold. Followed by the discovery of Labitzke (1987) on this solar–QBO connection, several studies reported strong correlation between the solar cycle and stratospheric geopotential heights and temperatures in both the hemispheres (Labitzke 2005).

According to Labitzke and van Loon (1999), the 11-year solar cycle could influence tropospheric weather and climate through an indirect pathway. The tropical stratospheric ozone heating creates off-equatorial circulation anomalies, and subsequent interactions with planetary waves bring anomalies poleward and downward in the winter hemisphere. This mechanism explains the evidence of QBO in high latitude stratosphere. There is evidence that high latitude stratospheric winds are sensitive to wind anomalies in the equatorial upper stratosphere, where the ozone anomalies are observed. In the case of solar-induced tropical circulation where anomalies are transmitted to high latitudes, they would likely be seen as changes to the strength of the polar vortex during prolonged winter which are observed to affect the tropospheric circulation.

9.16 Stratospheric Aerosols of Volcanic Origin

Stratospheric aerosols have a direct radiative impact that decreases the surface temperature, since more shortwave radiation is reflected. Although there has been no significant change in the background (nonvolcanic) stratospheric aerosols for the period 1970–2004 (SPARC 2006a), the nonvolcanic aerosol loading could increase in the future if convection increases, because convection is a key process that transports the short-lived species SO_2 from the surface to the upper troposphere and lower stratosphere.

Volcanic eruptions have strong impacts on the lower stratospheric thermal structure because the volcanic aerosols scatter back incoming solar radiation and absorb solar near-infrared and terrestrial infrared radiation. Heterogeneous chemistry occurring on aerosol surfaces affects ozone concentrations, producing an additional indirect radiative impact depending on the concentration of atmospheric chlorine. In addition, the modified meridional temperature profile in the stratosphere may result in colder polar vortices in winter.

The ozone impact of a given volcano depends on the amount of material, in particular sulfur, injected by the volcanic eruption and on whether the material reaches the stratosphere, as well as the phase of the QBO and the latitude of the eruption. The height reached by ejected materials depends on the intensity of the eruption, not on its location. Ejected substances in the tropics will be carried upward and poleward by the Brewer-Dobson circulation and, therefore, they will spread throughout much of the stratosphere with a long residence time, whereas the materials from mid- to high-latitude eruptions will more quickly be returned to the troposphere by the descending branch of the Brewer-Dobson circulation.

The effect of a future major volcanic eruption will depend on chlorine levels. For low-chlorine conditions, heterogeneous chemistry can lead to ozone increases in the stratosphere, whereas for high chlorine conditions, as observed in recent years, volcanic aerosols lead to additional ozone depletion. Moreover, volcanic aerosol affects photolysis rates and therefore ozone concentrations (Timmreck et al. 2003).

If a major volcanic eruption occurs while stratospheric halogen loading is elevated, ozone will temporarily be depleted. Strong volcanic eruptions enhance stratospheric aerosol loading for 2–3 years. A global average increase of lower stratospheric temperature (about 1 K at 50 hPa) was observed following the eruptions of El Chichon and Mt. Pinatubo, and globally averaged total ozone significantly decreased by about 2% before recovering after about 2–3 years (WMO 2007). Ozone destruction via heterogeneous reactions depends on halogen loading, so the effects on ozone of a major eruption are expected to decrease in the coming decades. For sufficiently low halogen loading, a large volcanic eruption would temporarily increase ozone. Long-term ozone recovery would not be significantly affected.

9.17 Future Scenario

The stratospheric effects on tropospheric weather and climate are thus found to have significant implications for many aspects of intraseasonal and interannual variability and systematic change in the tropospheric circulation. It suggests possible mechanisms for explaining apparent signals in the tropospheric circulation of the solar activity, volcanic dust, etc. to the stratosphere. It also strengthens the link between possible climate change in the stratosphere, due to ozone depletion or increasing greenhouse gases. At the same time as care is needed in interpreting observations as implying real downward propagation or influence, there is convincing evidence from numerical model simulations that changes in the stratosphere can sometimes lead to significant effects in the troposphere (Haynes 2005a,b).

Sudo et al. (2003) have reported that ozone transport from the stratosphere to the troposphere will be enhanced due to the global warming based on their studies by using chemical/climate model experiment. It is suggested that the enhanced intrusion of ozone to the troposphere will further accelerate global warming.

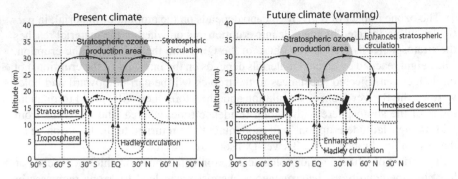

Fig. 9.23 Global warming and associated changes in the stratospheric circulation and tropospheric ozone variability (Sudo et al. 2003, Courtesy: American Geophysical Union)

In their study an increase in tropospheric ozone has been calculated, which will result in a big difference in vertical distribution of ozone between the cases with and without considering global warming. Since tropospheric ozone is recognized as the third most important greenhouse gas in the IPCC Third Assessment Report, its relevance to global warming/climate change attracts much attention. Results of experiment considering warming shows great enhancement of ozone in the upper troposphere at midlatitudes and tropics (see Fig. 9.23).

The reason was found to be due to the enhancement of atmospheric circulation both in the stratosphere and troposphere, resulting in the increase of stratospheric intrusion of ozone to the troposphere. It is known that ozone in the upper troposphere has a strong greenhouse effect and its increase has marked effect on the increase of earth surface temperature (Sudo et al. 2003). This result implies that global warming enhances ozone in the upper troposphere, and the ozone enhancement possibly further accelerates warming.

The troposphere, is no longer considered to be almost solely responsible for weather and climate on Earth. Emerging evidence indicates that the stratosphere shows an important influence on modulating the tropospheric circulation and dynamics, thereby controlling the weather and climate.

Thus, we have seen that stratosphere and troposphere interact in many ways. Stratosphere is now no more considered as a passive region, and has got the ability to shape not only its own evolution but that of tropospheric circulation and dynamics as well. Since the stratospheric effects vary slowly, a better understanding of the link between its physical mechanism and tropospheric changes will be an important tool for long-range forecasting of tropospheric weather systems and climate.

Questions and Exercises

9.1. Explain how the radiative effects in stratospheric region can modulate the tropospheric circulation and dynamics, thereby influencing the weather systems.

9.2. How do wave dynamics involve in the coupling between stratosphere and troposphere? Explain the role of planetary waves and gravity waves in stratosphere troposphere interactions.

9.3. Equatorial QBO is considered as a tropical stratospheric phenomenon. But its signal can be seen from troposphere to mesosphere and also from lower to higher latitudes. Explain the influence of QBO as a global phenomenon which act as an agent for the lower and middle atmosphere coupling.

9.4. What are the tropospheric changes in the polar region associated with sudden stratospheric event? Whether these changes can influence the tropical region? Explain.

9.5. How does the stratospheric polar vortex affect the tropospheric weather systems? Discuss the Arctic oscillation's impact on climate.

9.6. Discuss the man-made pollutants and industrialization in the troposphere cause depletion of stratospheric ozone on a global scale.

9.7. What are the impact of the ozone depletion in the stratosphere and its effect on the short-term and long-term changes in surface climate?

9.8. Discuss the role of tropical monsoon circulation in stratosphere troposphere interactions. Explain the modulation of stratospheric QBO on tropical monsoon rainfall.

9.9. How does water vapor reach the stratosphere? What happens, if more water vapor is added to the stratosphere?

9.10. Why is the temperature in the stratosphere decreasing and that in the lower troposphere increasing? If the present trends continue, what would happen to the circulation pattern of the troposphere and stratosphere?

9.11. Discuss the influence of the natural forcing, such as solar activity and volcanic eruptions on stratosphere troposphere coupling.

9.12. Can stratosphere be considered as a potential region for weather forecasting? Discuss the possibilities.

References

Arblaster JM, Meehl GA (2006) Contributions of external forcings to Southern Annular Mode trends, J Clim,19 (12): 2896–2905, doi 10.1175/JCL13774.1

Austin J, Li F (2006) On the relationship between the strength of the Brewer Dobsun circulation and the age of stratospheric air, Geophys Res Lett, 33, doi:10.1029/2006GL026867

Baldwin MP, Dunkerton TJ (1999) Propagation of the Arctic Oscillation from the stratosphere to the troposphere, J Geophys Res, 104 (D24): 30937–30946, doi 10.1029/1999JD900445

Baldwin MP (2000) The Arctic oscillation and its role in stratosphere-troposphere coupling, SPARC Newsletter 14

Baldwin MP, Cheng X, Dunkerton TJ (1994) Observed correlations between winter-mean tropospheric and stratospheric circulation anomalies, Geophys Res Lett, 21: 1141–1144

Baldwin MP, Dunkerton TJ (2001) Stratospheric harbingers of anomalous weather regimes, Science, 244(5542): 581–584

Baldwin MP, Dunkerton TJ (2005) The solar cycle and stratospheric tropospheric dynamical coupling, J Atmos Solar Terr Phys, 67: 71–82

Baldwin MP, Gray LJ, Dunkerton TJ, Hamilton K, Haynes PH, Randel WJ, Holton JR, Alexander MJ, Hirota I, Horinouchi T, Jones DBA, Kinnersley JS, Marquardt C, Sato K, Takahashi M (2001) The Quasi-Biennial Oscillation, Rev Geophys, 39(2): 179–229

Baldwin MP, Stephenson DB, Thompson DWJ, Dunkerton TJ, Charlton AJ, O'Neill A (2003a) Stratospheric memory and skill of extended-range weather forecasts. Science, 301: 636–640

Baldwin MP, Thompson DWJ, Shuckburgh EF, Norton WA, Gillett NP (2003b) Weather from the stratosphere, Science, 301: 317–319

Bhalme HN, Rahalkar SS, Sikdar AB (1987) Wind and Indian monsoon rainfall: Implications, J Climatol, 7: 345–353

Brasseur G, Hitchmann MH (1998) Stratosphere response to trace gas perturbations: Changes in ozone and temperature distribution, Science 240: 634-637

Brasseur G, Solomon S (2006) Aeronomy of the Middle Atmosphere- second edition, Springer Dordrecht, The Netherlands

Chang CP, Li T (2000) A theory for the tropical tropospheric biennial oscillation, J Atmos Sci, 57: 2209–2224

Charney JG, Drazin PG (1961) Propagation of planetary scale disturbances from the lower into the upper atmosphere, J Geophys Res, 66(1): 83–109

Chen P, Robinson WA (1992) Propagation of planetary waves between the troposphere and stratosphere, J Atmos Sci, 49(24): 2533–2545

Chen Z, Lu D (2001) Numerical simulation on stratospheric gravity waves above mid-latitude deep convection, Adv Space Res, 27(10): 1659–1666

Dunkerton TJ (2001) Quasi-biennial and sub-biennial variations of stratospheric trace constituents derived from HALOE observations, J Atmos Sci, 58(1): 7–25

EUR (European Commission) (2003) Ozone-climate interactions, Air pollution research report No. 81, EUR 20623, Belgium

Forster PM de F, Shine KP (1997) Radiative forcing and temperature trends from stratospheric ozone changes, J Geophys Res, 102(D9): 10841–10855

Fu Q, Johanson CM, Wallace JM, Reichler T (2006) Enhanced midlatitude tropospheric warming in satellite measurements, Science, 312, 1179

Gerber EP, Vallis GK (2005) A stochastic model for the spatial structure of annular patterns of variability and the Northern Atlantic Oscillations, J Clim, 18: 2102–2118

Gettelman A, Salby ML, Sassi F (2002) Distribution and influence of convection in the tropical tropopause region, J Geophys Res, 107(D10): 4080, doi 10.1029/2001JD001048

Gillett NP, Thompson DWJ (2003a) Simulation of recent Southern Hemisphere climate change. Science, 302: 273–275

Gillett NP, Allen MR, Williams KD (2003b) Modelling the atmospheric response to doubled CO_2 and depleted stratospheric ozone using a stratosphere-resolving coupled GCM. Quart J Roy Meteor Soc, 129: 947–966

Gowariker V, Thapliyal V, Kulshrestha SM, Mandal GS, Sen Roy N, Sikka DR (1991) A power regression model for long range forecast of southwest monsoon rainfall over India, Mausam, 42: 125–130

Gray WM (1984) Atlantic seasonal hurricane frequency. Part I - El Nino and 30 mb quasi-biennial oscillation influences, Mon Wea Rev, 112: 1649–1668

Gray LJ (2003) The influence of the equatorial upper stratosphere on stratospheric sudden warmings, Geophys Res Lett, 30 (4): 1166, doi 10.1029/2002- GL016430, 2003

Gray LJ, Crooks S, Pascoe C, Sparrow S, Palmer M (2004) Solar and QBO influences on the timing of stratospheric sudden warmings, J Atmos Sci, 61(23): 2777–2796

Haigh JD (1994) The role of stratospheric ozone in modulating the solar radiative forcing of climate, Nature, 370(6490): 544–546, 1994

Haigh JD, Blackburn M, Day R (2005) The response of tropospheric circulation to perturbations in lower stratospheric temperature, J Clim, 18: 3672–3685

Hartmann DL, Wallace JM, Limpasuvan V, Thompson DWJ, Holton JR (2000) Can ozone depletion and global warming interact to produce rapid climate change? Proc Nat Acad Sci USA, 97: 1412–1417

Haynes PH (2005a) Stratospheric dynamics, Annu Rev Fluid Mech, 37: 263–293

Haynes PH (2005b) Stratosphere-troposphere coupling, SPARC Newsletter 25, July 2005

Holton JR, Tan H-C (1980) The influence of the equatorial quasi-biennial oscillation on the global circulation at 50 mb, J Atmos Sci, 37: 2200–2208

Holton JR, Haynes PH, McIntyre ME, Douglass AR, Rood RB, Pfister L (1995) Stratosphere troposphere exchange, Rev Geophys, 33: 403–439

Huang R, Chen J (2002) Geotropic adaptation processes and excitement of inertia-gravity waves in the stratospheric spherical atmosphere, Chinese J Atmos Sci, 26(3): 289–303

Hu Y, Fu Q (2007) Observed poleward expansion of the Hadley circulation since 1979, Atmos Chem Phys, 7: 5229–5236

Kim YJ, Eckermann SD, Chun HY (2003) An overview of the past, present and future gravity wave drag parameterization for numerical climate and weather prediction models, Atmosphere Ocean 41: 65–98

Kodera K (2004) Solar influence on the Indian Ocean monsoon through dynamical processes, Geophys Res Lett, 31, L24209 doi:10.1029/2004GL020928

Kodera K, Kuroda Y (2000) Tropospheric and stratospheric aspects of the Arctic Oscillation. Geophy Res Lett, 27: 3349–3352

Kodera K, Yamazaki K, Chiba M, Shibata K (1990) Downward propagation of upper stratospheric mean zonal wind perturbation to the troposphere, Geophys. Res. Lett., 17: 1263–1266

Kodera K, Kuroda Y, Pawson S (2000) Stratospheric sudden warming and slowly propagating zonal-mean zonal wind anomalies, J Geophys Res, 105: 12351–12359

Kodera K, Kuroda Y (2002b) Dynamical response to the solar cycle, J Geophy Res, 107. 4749 doi.10.1029/2001PA000724

Kripalani RH, Kulkarni A (1997) Possible link between the stratosphere and Indian monsoon, SPARC Newsletter, No. 9, July

Kushner PJ, Polvani LM (2004) Stratosphere-troposphere coupling in a relatively simple AGCM: The role of eddies, J Clim, 17: 629–639

Labitzke K (1987) Sunspots, the QBO and the stratospheric temperatures in the north polar region, Geophys Res Lett, 14: 535–537

Labitzke K (2005) On the solar cycle-QBO relationship: A summary, J Atmos Solar Terr Phys, 67: 45–54

Labitzke K, van Loon H. (1988) Associations between the 11-year solar cycle, the QBO, and the atmosphere, Part I: Troposphere and stratosphere in the Northern Hemisphere in winter, J Atmos Terr Phys, 50: 197–206

Labitzke K, van Loon H (1999) The Stratosphere; Phenomena, History and Relevance, Springer, Berlin

Lal M, Jain AK, Sinha MC (1987) Possible climatic implications of depletion of Antarctic ozone, Tellus 39B: 326–328

Li T, Tham CW, Chang CP (2001) A coupled air-sea-monsoon oscillator for the tropospheric biennial oscillation, J Climate 14: 752–764

Limpasuvan V, Hartmann DL, Thompson DWJ, Jeev K, Yung YL (2005) Stratosphere-troposphere evolution during polar vortex intensification, J Geophys Res, 110 D24101, doi 10.1029/2005-JD006302

Limpasuvan V, Hartmann DL (2000) Wave maintained annular modes of climate variability, J Clim, 13: 4414–4429

Lorenz DJ, Hartmann DL (2001) Eddy-zonal flow feedback in the Southern Hemisphere, J Atmos Sci, 58: 3312–3327

Lorenz DJ, Hartmann DL (2003) Eddy-zonal flow feedback in the Northern Hemisphere winter, J Climate 16: 1212–1227

Matthes K, Kuroda Y, Kodera K, Langematz U (2006) Transfer of solar signal from the stratosphere to the troposphere: Northern winter, J Geophys Res, 111, doi 10.1029/2005JD006283

Meehl GA (1993) South Asian summer monsoon variability in a model with doubled carbon dioxide, J Climate, 6: 31–41

Meehl GA, Arblaster JM (2001) The tropospheric biennial oscillation and Indian monsoon rainfall, Geophys Res Lett, 28: 1731–1734

Meehl GA, Arblaster JM (2002) The tropospheric biennial oscillation and Asia-Australian monsoon rainfall, J Climate, 15(7):722–744

Meehl GA, Washington WM, Wigley TML, Arblaster JM, Dai A (2003) Solar and greenhouse gas forcing and climate response to the twentieth century, J Climate, 16(3): 426–444

Mohanakumar k (1985) An investigation on the influence of solar cycle on mesospheric temperature, Planet Space Sci, 33, 795–805

Mohanakumar K (1988) Response of an 11-year solar cycle on middle atmospheric temperature, Phys Sc, 37: 460–465

Mohanakumar K (1995) Solar activity forcing of the middle atmosphere, Ann Geophys, 13: 879–885

Mohanakumar K (1996) Effects of solar activity and stratospheric QBO on tropical monsoon rainfall, J Geomag Geoelectr, 48: 343–352

Mohanakumar K, Pillai PA (2008) Stratosphere troposphere interaction associated with biennial oscillation of Indian summer monsoon, J Atmos Terr Phys, 70(5): 764–773

Mukherjee BK, Indira K, Reddy SS, Ramana Murthy BhV (1985) Quasi-biennial oscillation in stratospheric zonal wind and Indian summer monsoon rainfall, Mon Wea Rev, 113: 1421–1424

Naito Y, Hirota I (1997) Interannual variability of the northern winter stratospheric circulation related to the QBO and solar cycle, J Meteorol Soc Japan, 75: 925–937

Perlwitz J, Graf H-F (2001) Troposphere-stratosphere dynamic coupling under strong and weak polar vortex conditions, Geophys Res Lett, 28(2): 271–274, doi 10.1029/2000GL012405

Pillai PA, Mohanakumar K (2007) Local Hadley circulation over the Asian monsoon region associated with the Tropospheric Biennial Oscillation, Theo Appl Climatol, DOI 10.1007/s00704-007-0305-5

Robinson WA (1991) The dynamics of the zonal index in a simple model of the atmosphere, Tellus, 43A: 295–305

Sathiyamurthy V, Mohanakumar K (2002) Characteristics of tropical biennial oscillation and its possible association with Stratospheric QBO, Geophys Res Lett, 7: 669–672

Scherhag R, Labitzke K, Finger FG (1970) Developments in stratospheric and mesospheric analyses which dictate the need for additional upper air data, Meteor Monogr, 11: 85–90

Scott RK, Polvani LM (2004) Stratospheric control of upward wave flux near the tropopause, Geophys Res Lett, 31: L02115, doi 10.1029/2003-GL017965

Scott RK, Polvani LM (2006) Internal variability of the winter stratosphere, Part I: Time independent forcing, J Atmos Sci, 63 (11): 2758–2776, doi 10.1175/JAS3797.1

Seidel DJ, Randel WJ (2007) Recent widening of the tropical belt: Evidence from tropopause observations, J Geophys Res, 112, doi 10.1029/2007JD008861

Seidel DJ, Fu Q, Randel WJ, Reichler TJ (2008) Widening of the tropical belt in a changing climate, Nature Geoscience, 1: 21–24, doi 10.1038/ngeo.2007.38)

Shindell DT, Schmidt GA (2004) Southern hemisphere climate response to ozone changes and greenhouse gas increases, Geophys Res Lett, 31, L18209 doi:10.1029/2004/GL020724

Shindell DT, Schmidt GA, Miller RL, Rind D (2001) Northern hemisphere winter climate response to greenhouse gases, ozone, solar and volcanic forcing, J Geophys Res, 106: 7193–7210

Sigmond M, Siegmund PC, Manzini E, Kelder H (2004) A simulation of the separate climate effects of middle atmospheric and tropospheric CO2 doubling, J Climate, 17: 2352–2367

Solomon S, Portmann RW, Sasaki T, Hofmann DJ, Thompson DWJ (2005) Four decades of ozonesonde measurements over Antarctica, J Geophys Res, 110: D21311, doi 10.1029/2005JD005917, 2005

Song Y, Robinson WA (2004) Dynamical mechanisms for stratospheric influences on the troposphere, J Atmos Sci, 61 (14): 1711–1725, 2004

SPARC (Stratospheric Processes And their Role in Climate) (2006) SPARC Assessment of Upper Tropospheric and Stratospheric Water Vapour, D Kley, JM Russell III, and C Phillips, World Climate Research Progam Report 113, SPARC Report No. 2, 312 pp., Verrires le Buisson, France

SPARC (Stratospheric Processes And their Role in Climate) (2006) SPARC Assessment of Stratospheric Aerosol Properties (ASAP), L Thomason and Th Peter (eds.), World Climate Research Progam Report 124, SPARC Report No. 4, 346 pp., Verrires le Buisson, France

Subbaramayya I, Ramanadham R (1981) On the onset of the Indian southwest monsoon and the monsoon general circulation, Monsoon Dynamics, J Lighthill and R. P. Pearce, Eds, Cambridge University Press, 213–220

Sudo K, Takahashi M, Akimoto H (2003) Future changes in stratosphere-troposphere exchange and their impacts on future tropospheric ozone simulations, Geophys Res Lett, 30: 2256 doi 10.1029/2003GL018526

Thompson DWJ, Solomon S (2002) Interpretation of recent Southern Hemisphere climate change. Science, 296: 895–899

Thompson DWJ, Solomon S (2005) Recent stratospheric climate trends as evidenced in radiosonde data: Global structure and and tropospheric linkages, J Clim, 18: 4785–4795

Thompson DWJ, Furtado JC Shepherd TG (2006) On the tropospheric response to anomalous stratospheric wave drag and radiative heating, J Atmos Sci, 63: 2616–2629

Timmreck C, Graf H-F, Steil B (2003) Aerosol chemistry interactions after the Mt. Pinatubo eruption, in Volcanism and the Earth's Atmosphere, A Robock and C Oppenheimer (eds.), Geophysical Monograph 139; 214–225, American Geophysical Union, Washington, D C

van Loon H, Labitzke K (1987) The Southern Oscillation, Part V: The anomalies in the lower stratosphere of the Northern Hemisphere in winter and a comparison with the Quasi-Biennial Oscillation, Mon Wea Rev, 115 (2): 357–369

Walter K, Graf HF (2005) The North Atlantic variability structure, storm tracks, and precipitation depending on the polar vortex strength, Atmos Chem Phys, 5;239-248

WMO (World Meteorological Organisation) (2007) Scientific assessment of ozone depletion:2006, Report No. 50, Geneva, Switzerland

Yue X, Yi F (2001) A study of non-linear propagation of 3rd gravity wave packets in a compressible atmosphere by using ADI scheme, Chinese J Space Science, 21(2): 148–158

Zawodny JM, McCormick MP (1991) Stratospheric aerosol and gas experiment II measurements of the quasi-biennial oscillations in ozone and nitrogen dioxide, J Geophys Res, 96: 9371–9377

Acronyms

APE	available potential energy
AO	annual oscillation, Arctic oscillation
CCN	cloud condensation nuclei
CFC	chlorofluorocarbon
CLAES	Cryogenic Limb Array Etalon Spectrometer
CPT	cold-point tropopause
CSRT	clear-sky radiative tropopause
DU	Dobson unit
ECMWF	European Center for Medium Range Weather Forecasting
EESC	Equivalent Effective Stratospheric Chlorine
ENSO	El Niño southern oscillation
EOF	empirical orthogonal function
EOS	Earth observing system
EP	Eliassen-Palm
ExTL	extratropical tropopause layer
GCM	general circulation model
GHG	greenhouse gas
GMT	Greenwich mean time
GW	gravity wave
GWP	Global Warming Potential
HALOE	halogen occultation experiment
HCFC	hydrochlorofluorocarbon
HIRS	High Resolution Infrared Sounder
ICAO	International Civil Aviation Organization
IPCC	Intergovernmental Panel on Climate Change
IR	infrared
ISMR	Indian summer monsoon rainfall
ITCZ	intertropical convergence zone
KE	kinetic energy
LIMS	Limb Infrared Monitor of the Stratosphere
LRT	lapse-rate tropopause

LTE	local thermodynamic equilibrium
MJO	Madden-Julian Oscillation
MLS	microwave limb sounder
MRG	mixed Rossby gravity
MSU	Microwave Sounding Unit
NAM	northern hemisphere annular mode
NAO	North Atlantic oscillation
NAT	nitric acid trihydrates
NASA	National Aeronautics and Space Administration
NCAR	National Center for Atmospheric Research
NCEP	National Center for Environmental Prediction
NH	northern hemisphere
NOAA	National Oceanic and Atmospheric Administration
ODC	ozone-depleting chemical
ODS	ozone-depleting substance
OMD	ozone mass deficit
PAN	peroxyacytyl nitrate
PFJ	polar front jet
PJO	polar jet oscillation
ppmv	parts per million by volume
PSC	polar stratospheric cloud
PV	potential vorticity
PVU	potential vorticity unit
QBO	quasi-biennial oscillation
SAGE	stratospheric aerosol and gas experiment
SAM	southern hemisphere annular mode
SAO	semiannual oscillation
SBUV	solar backscatter ultraviolet
SH	southern hemisphere
SPARC	Stratospheric Processes and its Influence in Climate
SST	sea surface temperature
SSU	Stratospheric Sounding Unit
SSW	sudden stratospheric warming
STE	stratosphere–troposphere exchange
STJ	subtropical jet stream
STT	secondary tropical tropopause
TBO	tropospheric biennial oscillation
TEJ	tropical easterly jet stream
TOMS	total ozone mapping spectrometer
TTL	tropical transitional layer/tropical tropopause layer
TTT	tropical thermal tropopause
UARS	upper atmosphere research satellite
UNEP	United Nations Environment Programme
UT/LS	upper troposphere and lower stratosphere
UV	ultraviolet

VMR	volume mixing ratio
VOC	volatile organic compound
WCRP	World Climate Research Programme
WMO	World Meteorological Organization

List of Symbols

A	Amplitude
A_n	Fourier coefficients
A_z	Mean available potential energy
B	Eddy heat flux
B_n	Fourier coefficient
B_λ	Blackbody radiance
C	Circulation
C_p	Specific heat at constant pressure
C_v	Specific heat at constant volume
D	Diffusion coefficient,
	Horizontal divergence
E	Irradiance
E_λ	Monochromatic irradiance
	Emitted spectral radiance
F_0	Shallow-water Froude number
F	Vertical molecular flux
	Eliassen-Palm (EP) flux
	Net flux
F_r	Frictional force
F_x	Zonal component of drag force
F_y	Meridional component of drag force
F_z	Vertical component of drag forcing, Restoring force
G	Universal gas constant
H	Scale height
HA	Hour angle
I	Flux of radiation
	Incident radiation
	Intensity or radiance
	Spectral intensity
I_∞	Solar intensity above Earth's atmosphere
I_λ	Actual emitted radiance
	Incident spectral radiance

I_v	Monochromatic Intensity
J	Photolysis rate coefficient
K_z	Zonal mean kinetic energy
KE	Kinetic energy
L	Wavelength
L_x, L_y, L_z	Wavelengths in zonal (x), meridional (y), and vertical (z) directions
M	Eddy momentum flux
M_x	Rate of accumulation of mass along the x direction
M_y	Rate of accumulation of mass along the y direction
M_z	Rate of accumulation of mass along the z direction
N	Molecular number density,
	Brunt-Vaisala frequency,
	Buoyancy frequency
$\dot{Q}$	Diabatic heating rate
R	Universal gas constant
R_e	The real part of the wave function φ
R_s	Radius of curvature
R_T	Radius of curvature of the trajectory
S_n	Component of diabatic heating
T	Brightness/Radiance/Absolute/Equivalent blackbody temperature
$\overline{T}$	Mean temperature of the layer
T_s	Global average temperature
T_λ	Transmissivity
U	Zonal mean flow
U_c	Critical mean wind speed
V	Relative velocity, Volume
$\vec{V}$	Three dimensional velocity vector
V_a	Absolute velocity
V_{ag}	Ageostrophic wind
V_G	Gradient wind
V_g	Geostrophic wind
$\vec{V}_H$	Horizontal velocity vector
V_n	Amplitude of perturbation component velocity in the y direction
V_T	Thermal wind
W	Energy associated with a photon
X	The mean zonal Eddy drag
$[X]$	Abundance of X in number density or volume mixing ratio
Z	Height
a	Mean radius of the Earth
a_λ	Absorptivity of the layer,
	Absorbance at a given wavelength,
	Monochromatic absorptivity
b	Wien's displacement constant
c	Mean wind speed, Speed of light
c_1, c_2	Empirical constants.

c_p	Specific heat, or specific enthalpy
$d\omega$	Represents an elemental arc of solid angle
f	Coriolis parameter,
	Planetary vorticity
f_{0v}	Coriolis forcing due to small eddies
g	Acceleration due to gravity
g^*	Gravitational force per unit mass
h	Planck's constant, height
h_n	Equivalent depth
$k(\lambda)$	Mass absorption or scattering coefficient
k	Boltzmann's constant
	Zonal wave number
	First-order rate coefficient,
k^*	Second-order rate coefficient
k_a	Absorption coefficient
$k_{a\lambda}$	Mass absorption coefficient
k_B	Boltzmann's constant
l	Meridional wave number
m	Mass, Vertical wave number
n	Molecular number density
	Number of molecules
	Refractive index
p	Pressure
p_o	Standard pressure level (1,000 hPa)
p_s	Standard reference pressure (1,000 hPa)
q	Quasi-geostrophic potential vorticity
q'	Eddy potential vorticity
r	Mass of the absorbing gas per unit mass of air,
	Distance from the center of Earth
	Rate of energy deposition
dl	Displacement vector
r_m	Rate of energy deposition at the maximum
r	Reflectivity
t	Time
t_λ	Transmissivity of the layer
z	Height above mean sea level
z_0	Number density at height $z = 0$
z_m	Height of maximum deposition rate, Altitude of maximum absorption
Ψ	Stream function
Γ	Environment lapse rate
Γ_d	Dry adiabatic lapse rate
Φ	Geopotential
Θ	Solar zenith angle
χ	Geopotential tendency
Ω	Angular speed of the rotation of the Earth

$\vec{\Omega}$	Angular velocity of Earth
α	Specific volume
α_r	Newtonian cooling rate
β	Planetary vorticity gradient (df/dy)
γ	Specific heat ratio
δ	Solar declination angle
ε_λ	Emissivity
η	Absolute vorticity
θ	Zenith angle
	Potential temperature
κ	Ratio of gas constant to specific heat at constant pressure (R/C_p)
λ	Wavelength
	Longitude, positive eastward
λ_{max}	Peak wavelength in meters
μ	Dynamic viscosity coefficient
ν	Frequency of light
ρ	Density of air
ζ	Relative vorticity
σ	Static stability parameter
σ_a	Absorption cross section
τ	Period of oscillation
τ_λ	Normal optical depth or optical thickness
τ_a	The optical thickness
φ	Latitude
	Radiant flux
	Wave function
ψ	Stream function
ω	Vertical velocity

Table of Physical Constants

Angular velocity of rotation of Earth	7.292×10^{-5} s^{-1}
Astronomical unit (AU)	1.496×10^{11} m
Boltzmann's constant (k)	1.38054×10^{-23} J K^{-1}
Earth's mean radius (a)	6371 km
Polar radius	6357 km
Equatorial radius	6378 km
Mass	5.973×10^{24} kg
Surface area	5.10×10^{14} m^2
Mean temperature	288 K
Gravitational constant (G)	6.6720×10^{-11} N m^2 kg^{-2}
Gravitational acceleration (g)	9.80665 m s^{-2}
Mass of atmosphere	5.136×10^{18} kg
Mass of ocean	1.4×10^{21} kg
Mean sea-level pressure	1013.25 hPa
Planck's constant (h)	6.6256×10^{-14} J s^{-1}
Solar constant	1367 W m^{-2}
Specific heat of dry air	1.005 J g^{-1} K^{-1}
Specific heat of water vapor	4.1855 J g^{-1} K^{-1}
Speed of light (vacuum) (c)	2.9979×10^{-8} m s^{-1}
Stefan-Boltzman constant (σ)	5.6697×10^{-8} W m^{-2} K^{-4}
Sun's radius	6.96×10^{-8} m
Mass	1.99×10^{30} kg
Surface area	6.087×10^{18} m^2
Luminosity	3.85×10^{26} W
Emission temperature	5783 K
Mean subtended full angle	0.532° (31.99 arc min)
Mean distance to Earth	1.496×10^{11} m
Universal gas constant (R)	8.31436 J mol^{-1} K^{-1}
Wien's displacement	2898×10^3 nm K^{-1}

Answers to Selected Problems

Chapter 1

1.1   985 hPa, 28 hPa
1.2   28.97 kg mol^{-1}
1.3   At a height of 8 km, atmospheric pressure drops below 400 hPa. If the cabin is not pressurized, the inflated balloon expands and bursts.
1.4   973.50 hPa
1.5   2.437 km, 39.834 km
1.6 $-27.5\,^{\circ}$C
1.7   29 days and 6 hours
1.8   2 hours 18 minutes
1.9   0.165 $^{\circ}$C/degree latitude
1.10 In January 2008, the phase of the zonal wind at 30 km is easterly and that at 18 km is westerly due to the QBO of zonal wind, which descends at a rate of 1 km per month.

Chapter 2

2.1   483 nm; 255 $^{\circ}$K
2.2 14.51×10^6 W m^{-2}
2.3   1372 W m^{-2}
2.4   288 K
2.7   69.28 kgm^{-2}; 0.867
2.8 6.7×10^{-3}
2.9   105 m
2.11 Flux density of ocean changes to 1.2 times that of ice
2.13 0.07

Chapter 3

3.1 -4.9×10^{-5} ms^{-2}
3.2   0.53 °C per hour
3.3   34.3 ms^{-1}
3.4   27.2 ms^{-1}
3.5   12 kg m^3 s^{-1}
3.6   1.64 ms^{-1}
3.7   10 ms^{-1}
3.8   11.0 ms^{-1} km^{-1}
3.9   28.0 ms^{-1} Northerly; Zonal thermal wind component is absent
3.10 10^{-1} s^{-1}
3.11 1.414×10^{-5} s^{-1}; 1.155×10^{-5} s^{-1}; 1.00×10^{-5} s^{-1}
3.12 6.2×10^{-7} m^{-1} s^{-1}
3.13   1.176 PVU

Chapter 4

4.2   306.9 m s^{-1}
4.4   0.14 m s^{-1}
4.5   221.4 m s^{-1}
4.8   Period = 2 minutes; Time taken = 8 hours
4.11   28 m s^{-1} Westerly

Chapter 5

5.1   Two additional chemical and transport processes are to be considered. (1) Ozone loss reactions with gases containing chlorine, bromine, nitrogen, and hydrogen that contribute to overall ozone loss, (2) Transport of ozone from the photochemical source region in the tropics to the middle and high latitudes by Brewer Dobson circulation.

5.2   Atomic oxygen decays rapidly after sunset, the amount of ozone stays close to its sunset value. X decays, XO tends to a constant amount.

5.5 1.81×10^{-14} cm^{3}molecule^{-1} s^{-1}; 6.09×10^{-34}cm^{6}molecule^{-2}s^{-1}

5.7   The lifetime for an O atom is about 0.002 second and that for an O$_3$ molecule is about 1000 seconds. Eventhough the rapid photolysis of ozone (less than 20 minutes), photolysis of molecular oxygen, O$_2$, also occurs in the upper atmosphere, which frees two O atoms. An O atom quickly reacts with another O$_2$ molecule to create an O$_3$ molecule. On an average, the local amount of stratospheric ozone does not vary much.

5.8 6.9×10^{-4}
5.9   38.1 minutes
5.10   10.2 years

Index

413